New Frontiers in Photochromism

Masahiro Irie • Yasushi Yokoyama • Takahiro Seki
Editors

New Frontiers in Photochromism

Springer

Editors
Masahiro Irie
Professor
Department of Chemistry
Rikkyo University
Tokyo, Japan

Yasushi Yokoyama
Professor
Department of Advanced Materials Chemistry
Yokohama National University
Kanagawa, Japan

Takahiro Seki
Professor
Department of Molecular Design & Engineering
Nagoya University
Aichi, Japan

ISBN 978-4-431-54290-2 ISBN 978-4-431-54291-9 (eBook)
DOI 10.1007/978-4-431-54291-9
Springer Tokyo Heidelberg New York Dordrecht London

Library of Congress Control Number: 2013939352

Printed on acid-free paper

Springer is part of Springer Science+Business Media (www.springer.com)

Preface

Photochromism is defined as a reversible transformation of a chemical species between two forms having different absorption spectra induced in one or both directions by photoirradiation. Chemical bond rearrangement during the phototransformation induces electronic as well as geometrical structure changes of the molecules. This process has captivated many chemists and physicists over several decades. The reversible molecular structure changes upon photoirradiation can be applied to various photonic devices, such as erasable memory media, photo-optical switch components, and display devices. To promote activities in this research field based on accumulated top-level work in Japan, we conducted a Grant-in-Aid for Scientific Research (*Kakenhi*) Project on the Priority Area "New Frontiers in Photochromism" (Masahiro Irie, project leader, and 58 researchers) from 2007 to 2011 supported by the Ministry of Education, Culture, Sports, Science and Technology, Japan.

The research project focused not only on the above-mentioned primary subjects but also on unprecedented application fields that had not yet been explored in photochromism research, such as light-driven mechanical motion of single crystals, photomodification of surface properties of solid materials, photomanipulation of microstructures of polymers, photocontrol of chiral properties, and multiphoton photochromic reactions, among others. More specifically, the following research targets were established and attained:

1. To improve the photochromic reactivity of molecules and ultimately achieve performances such as 100 % quantum yield reactivity and 100 % stereoselectivity
2. To search for new types of photochromic molecules. One of the target molecules is a chromophore exhibiting very fast photocoloration–thermal decoloration
3. To develop new, unprecedented photoresponsive properties such as light-driven mechanical motion of single crystals and photomodification of surface properties
4. To study detailed reaction mechanisms of photochromic reactions using femtosecond lasers

This book compiles the accomplishments of the research project including those of the research targets mentioned above. We believe that the achievements of the project will be a significant milestone and turning point in the stream of studies on photochromism, photoresponsive molecules, and photofunctional materials.

The editors wish to express their thanks to outstanding contributors and to the Ministry of Education, Culture, Sports, Science and Technology, Japan, for their support of the project.

Tokyo, Japan Masahiro Irie
Kanagawa, Japan Yasushi Yokoyama
Aichi, Japan Takahiro Seki

Contents

Part I
New Photochromic Materials

Chapter 1
Photomechanical Response of Diarylethene Single Crystals

Masahiro Irie

Abstract It is a dream for chemists to construct molecular systems, which can transform shape changes of individual molecules induced by chemical or physical stimuli to macroscopic motion of materials and perform mechanical work. During the course of a study of single-crystalline photochromism of diarylethenes, we found that the surface morphology of the single crystals, as well as the bulk crystal shape, reversibly changes upon photoisomerization of component diarylethene molecules. Single crystals of 1,2-bis(2-ethyl-5-phenyl-3-thienyl)perfluorocyclopentene (**3a**) and 1,2-bis(5-ethyl-2-phenyl-4-thiazolyl)perfluorocyclopentene (**4a**) with sizes ranging from 10 to 100 μm change their shape from a square to a lozenge, whereas a rectangular single crystal of 1,2-bis(5-methyl-2-phenyl-4-thiazolyl)perfluorocyclopentene (**5a**) contracts in length. X-ray crystallographic analysis revealed that the geometrical structure changes of individual molecules in densely packed crystals induce the crystal shape deformation. A rod-like crystal prepared from **5a** reversibly bends upon alternate irradiation with ultraviolet (UV) and visible light due to the gradient in the extent of the photoisomerization in the crystal. The fatigue resistance of the crystals is remarkably improved by mixing two diarylethene derivatives. Upon UV irradiation the mixed crystal can repeat the light-driven bending cycle more than 1,000 times and lift a metal load, which is several hundred times heavier than that of the crystal. A two-component mm-size cocrystal composed of 1,2-bis(2-methyl-5-(1-naphthyl)-3-thienyl)perfluorocyclopentene (**8a**) and perfluoronaphthalene (**FN**) also performs fatigue-resistant mechanical work. The robust light-driven actuators made of diarylethene molecules having substantial mechanical properties can be potentially applied in various micro- and nano-mechanical devices.

Keywords Atomic force microscope • Diarylethene • Light-driven actuator • Photomechanical effect • Single crystal

---

M. Irie (✉)
Department of Chemistry and Research Center for Smart Molecules,
Rikkyo University, Nishi-Ikebukuro 3-34-1, Toshimaku, Tokyo 171-8501, Japan
e-mail: iriem@rikkyo.ac.jp

M. Irie et al. (eds.), *New Frontiers in Photochromism*,
DOI 10.1007/978-4-431-54291-9_1, © Springer Japan 2013

1.1 Introduction

It is of particular interest from both scientific and technological points of view to have molecular systems make mechanical motion based on geometrical structure changes of individual molecules induced by chemical or physical stimuli, such as chemicals, photons, or electrons (or holes), and to link the motion to macroscopic mechanical work. Mechanical work means controlled, large amplitude, or directional motion of materials, which results in a net task being performed. In biological systems, molecular-scale movements of actin–myosin proteins are artfully linked to macroscopic motion of muscles [1]. Although elaborate molecular machines [2, 3], such as molecular shuttles [4], molecular muscles [5, 6], molecular elevators [7], and molecular rotors [8, 9], have been extensively studied, attempts to link the molecular-scale movements of these man-made devices to macroscopic motion of materials have failed. On the other hand, various types of polymers [10–16] and carbon nanotubes [17] have been reported to convert molecular phenomena into macroscopic motion of materials. The motion, however, relies not on individual molecular behavior but on the response of bulk materials. The photoinduced contraction of liquid-crystal elastomers, for example, is attributed to the order–disorder phase transition of the liquid-crystal materials [13–16]. It is a challenge to construct molecular systems that perform macroscopic mechanical work that stems from photostimulated geometrical structure changes of individual molecules.

1.2 Reversible Surface Morphology Changes of Diarylethene Single Crystals

In general, photochromic reactions rarely occur in crystals, because in crystals large geometrical structure changes are prohibited. For example, azobenzene and most of spiropyran derivatives cannot undergo photochromism in the single-crystalline phase, because they require relatively large geometrical structure changes during the photoisomerization reactions. Although the structure changes are limited in crystals, some compounds exist which undergo photoisomerization reactions even in crystals. If the reactions take place in crystals, the crystals exhibit efficient and fatigue-resistant color changes and provide characteristic optical properties. During the study of diarylethene photochromism, we accidentally discovered that some of diarylethene derivatives exhibit photochromism even in the single-crystalline phase [18]. The crystals undergo thermally irreversible and fatigue-resistant photochromic reactions with a very high cyclization quantum yield [19, 20]. The single-crystalline photochromism of diarylethene derivatives is ascribed to their rather small geometrical structure changes during the photoisomerization and thermal stability of the photogenerated isomers.

Figure 1.1 shows an example of the geometrical structure change of a diarylethene derivative in the crystalline phase determined by X-ray crystallographic

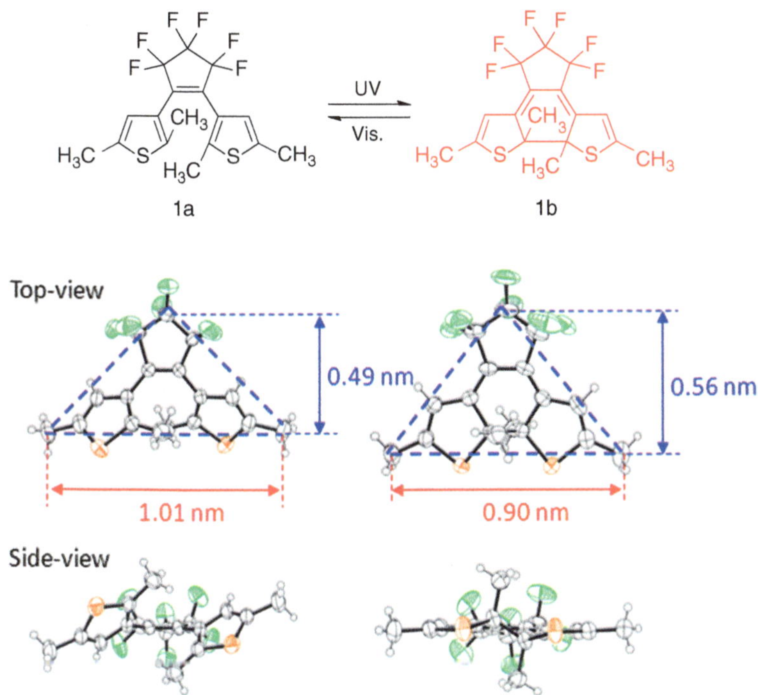

Fig. 1.1 Chemical structures of the open- (**1a**) and closed-ring (**1b**) isomers of 1,2-bis(2,5-dimethyl-3-thienyl)perfluorocyclopentene and *top* and *side views* of the geometrical structures of the two isomers

analysis [21, 22]. 1,2-Bis(2,5-dimethyl-3-thienyl)perfluorocyclopentene (**1a**) exhibits photochromism in the crystalline phase. As can be seen from the top view, the base width of the triangle shape (blue broken line) decreases from 1.01 to 0.90 nm and the height increases from 0.49 to 0.56 nm. The side view of the molecule indicates that the thickness of the molecule becomes thinner when the molecule converts from **1a** to **1b**. The molecule undergoes a small but distinct shape change. Molecular crystals make up a densely packed organized system. Therefore, if the photoisomerization takes place in the densely packed crystal, the crystal shape is anticipated to follow the shape changes of the component diarylethene molecules. In other words, the stress generated by the shape changes of individual molecules causes a macroscopic shape deformation of the crystal.

The most convenient way to elucidate the crystal shape deformation is to measure the surface morphology by using an atomic force microscope (AFM) [23]. We carried out an AFM measurement of a single crystal of 1,2-bis(2,4-dimethyl-5-phenyl-3-thienyl)perfluorocyclopentene (**2a**). The surface morphology change of the (1 0 0) face upon irradiation with UV (366 nm) and visible (>500 nm) light was monitored. Figure 1.2 shows the chemical structures of **2a** and the closed-ring isomer **2b**, AFM images, and the molecular packing of **2a** determined via X-ray

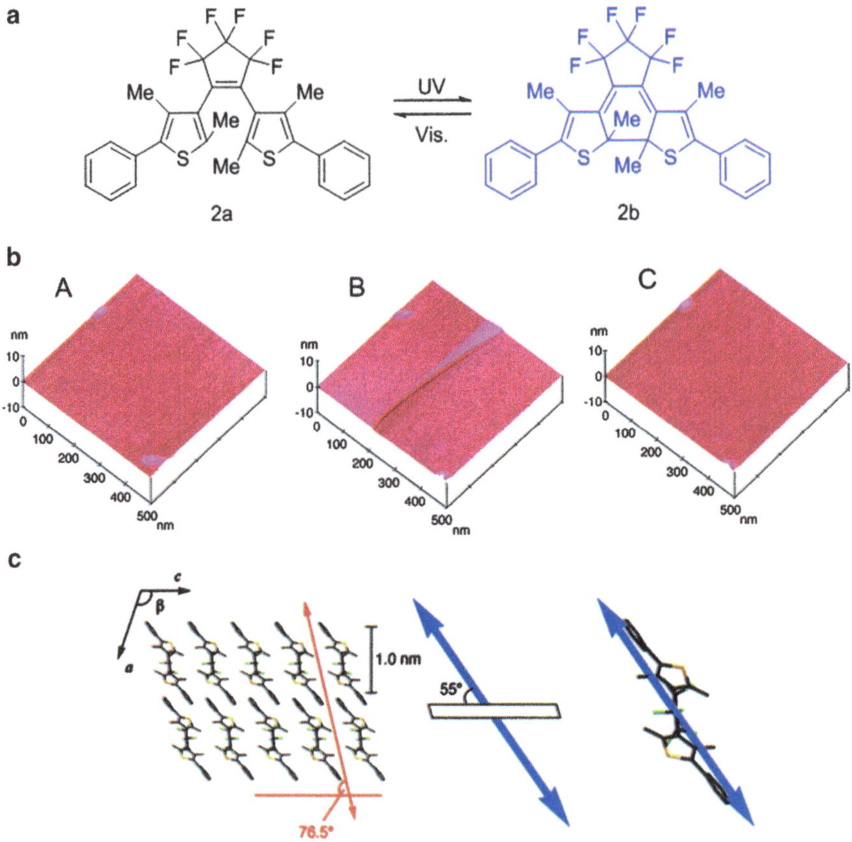

Fig. 1.2 (**a**) Chemical structures of 1,2-bis(2,4-dimethyl-5-phenyl-3-thienyl)-perfluorocyclopentene (**2a**) and the closed-ring isomer **2b**. (**b**) AFM images of the (100) crystal surface: before photoirradiation (*A*), after irradiation with 366-nm light for 10 s (*B*), and after irradiation with visible ($\lambda > 500$ nm) light (*C*). (**c**) Molecular packing of the molecules

crystallographic analysis. Upon irradiation for more than 10 s with UV light, new steps appeared on the surface. No step formation was discerned during the irradiation for the initial 10 s but appeared after the induction period. The step height was about 1.0 nm, which corresponds to one molecular layer. The step disappeared by bleaching upon irradiation with visible light. When the irradiation time was prolonged, the number of steps increased and steps with the height of 2.0 and 3.0 nm appeared. The step height was in multiples of the minimum step height of 1.0 nm. Any step with a height lower than the unit height (~1.0 nm) was not observed. The morphology change was reversible and correlated with the color change.

The molecular structures of **2a** and **2b** were determined via X-ray crystallographic analysis to establish the relation between the reversible formation of the steps and the photoisomerization of molecules. The end-to-end distance between the two phenyl rings decreases from 1.41 to 1.39 Å when **2a** converts to **2b**. The shortening in the distance between the phenyl ring edges upon the cyclization reaction results in the decrease in the thickness of the molecular layers. The shrinkage of the molecular layers produces vacancies in the crystal bulk. Accumulation of the vacancies deep within the crystal allows the surface molecular layer to sink as much as one layer (~1.0 nm). Prolonged irradiation increases the vacancies and causes subsidence of multilayers. Irradiation for 10 s induces the photocyclization reactions of 600 molecular layers. The random distribution of the reactions as deep as 5×10^5 layers (=500 μm) produces the one molecular layer vacancy. This explains the step formation on the prominent (1 0 0) surface.

The surface morphology change is attributed to the shape and size changes of component diarylethene molecules densely packed in the single crystal. This result indicates that the geometrical structure change of each molecule is linked to the shape deformation of the bulk crystal.

1.3 Reversible Shape Changes of Diarylethene Single Crystals

The striking results above encouraged us to search for photostimulated bulk shape deformation of crystals [24, 25]. So far, molecules in crystals are considered to be stationary and they never move in crystals. However, recently some crystals have been reported to readily change their bulk shape under external mechanical stress [26]. Molecules in crystals have a potential for movement when external or inner stress is applied. The photostimulated shape changes of component molecules in crystals generate the inner stress. In polymer films and gels, the strain energy generated by the shape changes of molecules is released by their free volume [24]. Therefore, the bulk shape of the polymer films does not change. On the other hand, in densely packed crystals the strain energy is anticipated to directly influence the bulk shape of crystals.

In order to deform the bulk shape of crystals upon photoisomerization of component molecules without breaking, the following conditions are required:

1. The whole crystal must undergo a homogeneous photoreaction. In other words, the whole crystal is homogeneously irradiated.
2. Phase separation must not take place between the photoreacted isomers and unreacted isomers.
3. The photoreactions randomly take place. In other words, local or cooperative reactions should be avoided.

The most convenient way to fulfill the above conditions is to use small-size crystals. When the crystal size is in the range of 10–100 μm and the thickness is less than 1 μm, UV irradiation can induce a homogeneous reaction. We prepared such small-size crystals by sublimation.

Figure 1.3 shows the color and shape changes of three small-size single crystals, 1,2-bis(2-ethyl-5-phenyl-3-thienyl)perfluorocyclopentene (**3a**) [27], 1,2-bis(5-ethyl-2-phenyl-4-thiazolyl)perfluorocyclopentene (**4a**) [28], and 1,2-bis(5-methyl-2-phenyl-4-thiazolyl)perfluorocyclopentene (**5a**) [27]. When crystals **3a** and **4a** were irradiated with UV (365 nm) light, their color and shape changed. Upon irradiation with UV light, the colorless single crystal of **3a** turned blue and its corner angles changed from 88° and

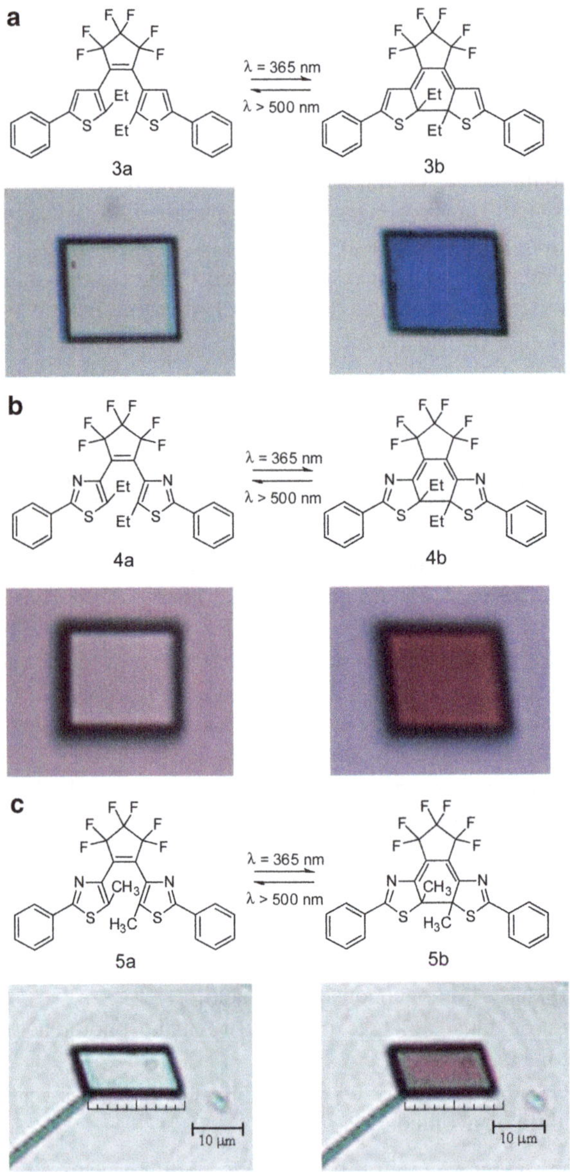

Fig. 1.3 Chemical structures and deformation of three diarylethene crystals upon UV (365 nm) and visible ($\lambda > 500$ nm) light irradiation. (**a**) 1,2-Bis(2-ethyl-5-phenyl-3-thienyl)perfluorocyclopentene (**3a**), (**b**) 1,2-bis(5-ethyl-2-phenyl-4-thiazolyl)perfluorocyclopentene (**4a**), and (**c**) 1,2-bis(5-methyl-2-phenyl-4-thiazolyl)perfluorocyclopentene (**5a**)

92° to 82° and 98°, while crystal **4a** changed its color from colorless to red and the corner angles changed from 90° and 90° to 86° and 94°. Both crystals changed in shape from a square to a lozenge. The blue and red colors are ascribed to the colors of the respective photogenerated closed-ring isomers. The deformed blue and red crystals returned to the initial colorless and square shape by bleaching upon irradiation with visible (>500 nm) light. Although the color changes were different, both crystals showed a similar bulk shape deformation. The result suggests that the deformation of crystal bulk shape is due to the geometrical structure changes of component molecules and that the electronic structure changes do not influence the shape deformation.

The time dependence of the color and the corner angle of crystals **3** and **4** upon alternate irradiation with UV and visible light was monitored. The corner angles of both crystals initially remained constant and then gradually changed as much as 3–6° upon irradiation with UV light and returned to the initial values upon irradiation with visible light. No hysteresis between the forward and reverse processes was observed. The induction period of the shape changes is ascribed to the accumulation time of the closed-ring isomers in the crystals. The interaction between the adjacent closed-ring isomers was confirmed by the absorption spectral shift of the closed-ring isomers. The absence of hysteresis indicates that no cooperative phase change effect takes place in the shape deformation process.

To evaluate the crystallinity of crystal **3** used in the above switching experiment, the melting point and the order parameters of the visible absorption were measured at the photostationary state. The melting point of crystal **3** before UV irradiation was 164 °C and decreased to 45–55 °C upon irradiation with UV light. Upon irradiation with visible light, the melting point returned to 164 °C. The melting point of a small-size crystal is the same as observed with a large crystal. The order parameter of the absorption at 600 nm, $(A/\!/-A\perp)/(A/\!/+2A\perp)$, at the photostationary state was 0.53, which is identical to the value measured for a large crystal. The recovery of the melting point on visible light irradiation indicates that crystal **3** remains highly crystalline even after a cycle of irradiation with UV and visible light. The decrease of the melting point upon UV irradiation is due to the coexistence of two isomers, open- and closed-ring isomers, in the same crystal. The reversibility of the photostimulated angle changes was also examined. No deterioration in performance of the angle changes even after 20 cycles of alternate irradiation with UV and visible light was observed.

To elucidate the relation between the crystal shape deformation and the shape changes of individual molecules, X-ray crystallographic study was carried out [24, 25]. Figure 1.4 shows the molecular packing of crystal **3a** before UV irradiation. The shape deformation from a square to a lozenge indicates that the crystal expands along the *b*-axis and contracts along the *c*-axis. *In situ* X-ray crystallographic analysis of micrometer-size crystal **3** (20 × 15 × 8 μm) confirmed that the unit cell length of *b*-axis expanded from 10.971 to 11.026 Å and that of *c*-axis contracted from 10.585 to 10.542 Å. Figure 1.5 shows the geometrical structures of **3a** and **3b**. These two isomers were isolated by HPLC and independently crystallized in hexane. When the molecule converts from **3a** to **3b**, the height of the triangle shape (blue broken line) of the molecule increases from 0.61 to 0.73 nm, which is in the direction of the *b*-axis. On the other hand, the thickness of the molecules is reduced. The cofacial packing of thin layers of the planar closed-ring isomers along the *c*-axis, as shown in Fig. 1.4c, allows the molecules to be stacked one by one,

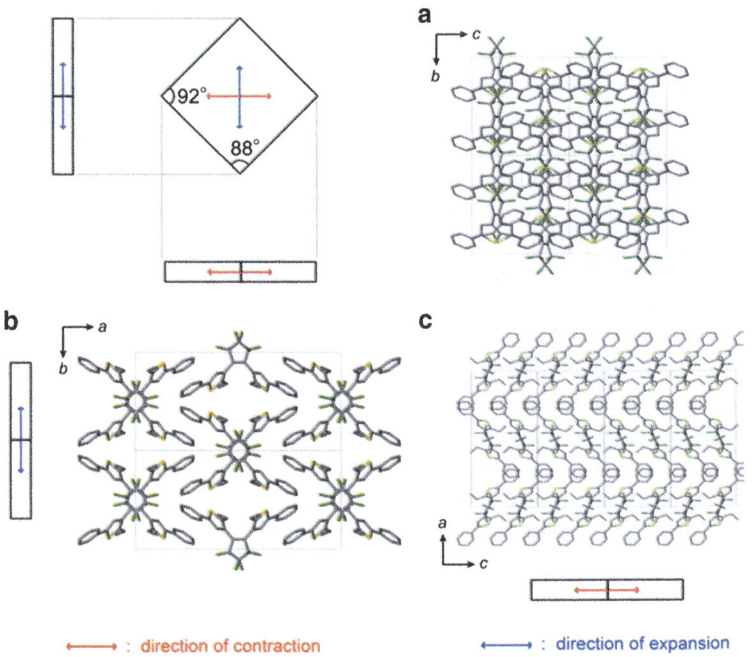

Fig. 1.4 Molecular packing of a single crystal of 1,2-bis(2-ethyl-5-phenyl-3-thienyl)perfluorocyclopentene (**3a**) before UV irradiation. The *red arrows* indicate the direction of contraction and the *blue arrows* the direction of expansion of the crystal upon UV irradiation

resulting in contraction along the *c*-axis. The structure changes of diarylethene molecules directly influence the bulk crystal shape.

Figure 1.3c shows another type of crystal shape deformation [27]. Irradiation of the rectangular single crystal of **5a** with UV light and visible light induced contraction and expansion, respectively, by as much as 7 %. Just as observed with crystals **3** and **4**, after an induction period during the initial stage of UV irradiation, the crystal length decreased linearly. No hysteresis between the forward and reverse processes was observed. The melting point of the crystal decreased from 158 °C to 95–105 °C upon UV irradiation and returned to 158 °C upon irradiation with visible light, whereas the order parameter of the absorption at 550 nm at the photostationary state was as high as 0.75. These observations indicate that the crystal remains highly crystalline after a cycle of irradiation with both UV and visible light.

A rod-like crystal prepared from **5a** bent toward the direction of the incident UV light, as shown in Fig. 1.6. The rod-like crystal and the thin-plate crystal above have the same crystal structure. This effect is due to the gradient in the extent of the photoisomerization caused by the high absorbance of the crystal. The shrinkage of the irradiated part of the crystal causes the bending as observed in a bimetal. The bent

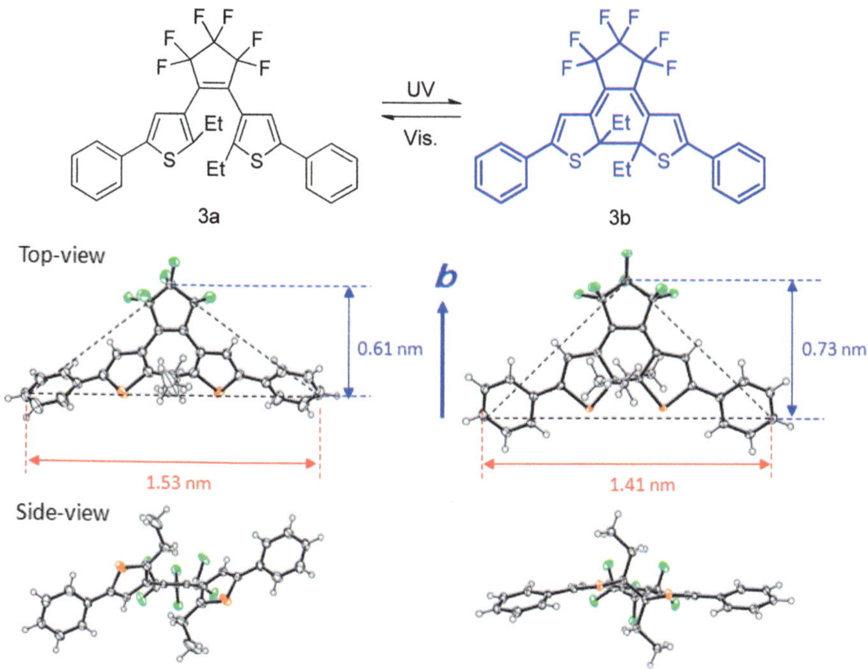

Fig. 1.5 Chemical structures of the open- (**3a**) and closed-ring (**3b**) isomers of 1,2-bis(2-ethyl-5-phenyl-3-thienyl)perfluorocyclopentene and *top* and *side views* of the geometrical structures of the two isomers. The two isomers were isolated by HPLC and independently recrystallized

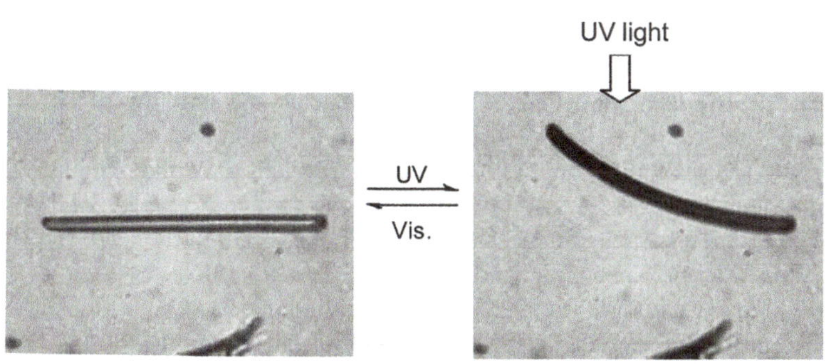

Fig. 1.6 Light-driven reversible bending of a rod-like crystal of 1,2-bis(5-methyl-2-phenyl-4-thiazolyl)perfluorocyclopentene (**5a**) upon irradiation with UV and visible light. The right-side end of the rod was fixed to the glass plate

rod-like crystal again became straight upon irradiation with visible light. The rod-like crystal can launch a tiny silica particle. The relatively large Young's modulus of the crystal (around 10 GPa) can hit the silica particle as if it were a tennis ball.

1.4 Light-Driven Bending of Rod-Like Mixed Crystals

For practical applications, the crystal actuators should have sufficient durability and substantial mechanical properties. The above single-component diarylethene crystals lack the durability and break in less than 100 deformation cycles. To improve the fatigue-resistant property, we prepared two-component mixed crystals composed of 1-(5-methyl-2-phenyl-4-thiazolyl)-2-(5-methyl-2-*p*-tolyl-4-thiazolyl)perfluorocyclopentene (**6a**) and 1,2-bis(5-methyl-2-*p*-tolyl-4-thiazolyl)perfluorocyclopentene (**7a**) (Fig. 1.7) [29].

Microcrystals were prepared by recrystallization from ethanol. Plate-like crystals were mainly formed from the ethanol solution containing only **6a**. When **7a** was added to the solution in amounts exceeding 30 mol%, formation of rod-like crystals became dominant. In the ethanol solution containing equimolar amounts of **6a** and **7a**, rod-like crystals composed of **6a** (63 mol%) and **7a** (37 mol%) grew. The two-component rod-like crystals bent toward the UV (365 nm) light source and became straight upon irradiation with visible (>500 nm) light. When the content of **7a** in the mixed crystals was increased over 60 mol%, the photostimulated bending was suppressed and the bending was not observed for the rod-like crystal of **7a**. The bending is due to the gradient in the extent of the photoisomerization caused by the high absorbance of the crystal. Figure 1.8 shows the bending cycles of a rod-like mixed crystal (1.3 mm × 25 μm × 13 μm) containing **6a** and **7a** (**6a/7a** = 63/37) upon alternate irradiation with UV and visible light. The crystal can repeat the bending cycles more than 1,000 times. The crystal surface remained clear even after 1,000 cycles and no damage was discerned. The melting point of the mixed crystals showed the minimum of 131 °C at the molar ratio of 1:1. The low melting point of the crystal containing equimolar amounts of the two-component molecules suggests that intermolecular interaction among the component molecules is weakened in the mixed crystals and the weakened intermolecular interaction is considered to improve the durability of the crystals.

The rod was irradiated from the right, lower, and left sides with UV and visible light, and the bending behavior was monitored. The crystal bent toward the UV light source irrespective of the irradiation direction, as shown in Fig. 1.9b. The bent rod-like crystal returned to a straight position upon irradiation with visible light. When the light intensity of UV and visible light was controlled, the edge of the crystal exhibited rotation movement, as shown in Fig. 1.9c. Figure 1.9d shows curling of the rod-like crystal. When the mixed crystal had a length as long as 3.0 mm, the rod exhibited curling into a hairpin shape upon irradiation with UV light. The crystal remained crystalline even after the curling and returned again to its straight shape upon irradiation with visible light. Characteristic features of the rod-like crystal include the rapid response time and performance at low temperature. The bending rate was measured using a high-speed camera, and the bending rate was determined to be less than 5 μs. The crystal exhibited light-driven bending even at 4.6 K. Polymer artificial muscles, such as polymer gels, conductive polymers, and liquid-crystal elastomers, are rather slow acting (typically in seconds and minutes), and the working temperature is limited above T_g. The present molecular crystals are superior to the existing polymer muscles in their short response time (<5 μs) and a wide range of working temperature (4.6 K < T < 370 K). Another feature of the

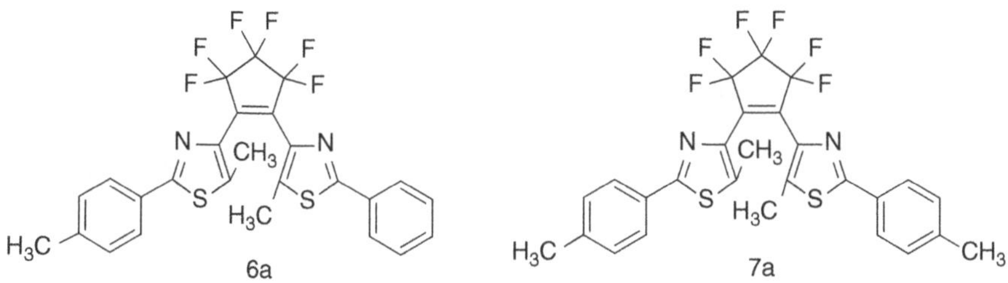

Fig. 1.7 Chemical structures of **6a** and **7a**

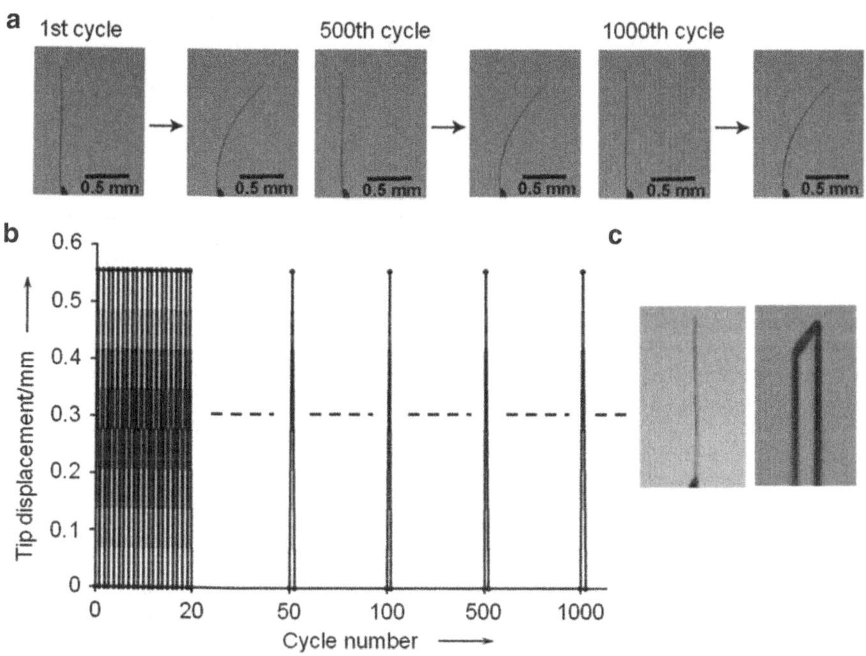

Fig. 1.8 Reversible bending of a rod-like two-component mixed crystal of 1-(5-methyl-2-phenyl-4-thiazolyl)-2-(5-methyl-2-*p*-tolyl-4-thiazolyl)perfluorocyclopentene (**6a**) and 1,2-bis(5-methyl-2-*p*-tolyl-4-thiazolyl)perfluorocyclopentene (**7a**) (**6a/7a**=63/37) upon alternate irradiation with UV (365 nm) and visible (>500 nm) light. (**a**) Pairs of photographs of the crystal (1.3 mm×25 μm×13 μm) showing 1st, 500th and 1,000th cycles. (**b**) Tip displacement of the crystal during reversible bending cycles. (**c**) An expanded photograph of the crystal after the 1,000th cycle. Reproduced from [29]

molecular-crystal actuator is their light-driven bending motion in water. The crystals exhibited the bending behavior in water similar to that observed in air.

We tried to use the rod-like crystal to lift a metal load upon UV irradiation. The rod-like crystal (2.5 μg) was fixed at the edge of a glass plate as a cantilever arm, and a metal weight (2.2710 mg) was loaded onto the rod. The weight is 908 times heavier than the crystal. Upon irradiation with UV light, the weight was lifted as high as 0.10 mm. The tiny rod-like crystal, which weighs only 2.5 μg, performed lifting

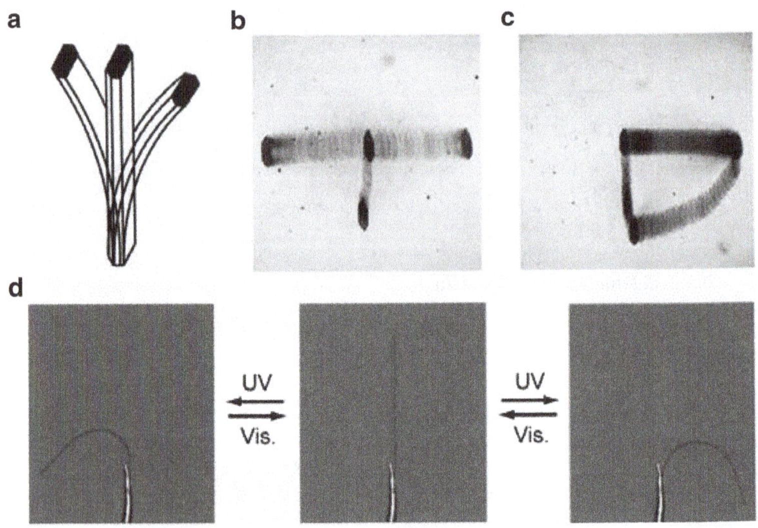

Fig. 1.9 Bending of a rod-like two-component mixed crystal containing **6a** and **7a** (**6a**/**7a** = 63/37) in all directions. (**a**) Schematic illustration of the bending rodlike crystal. (**b**) The crystal (1.3 mm × 40 μm × 15 μm) was irradiated from the *right*, *lower*, and *left sides* with UV and visible light, and the movement of the rod-like crystal edge was monitored. The edge moved toward the UV light source and then returned to the original position upon irradiation with visible light. (**c**) First, the crystal was irradiated from the *lower side* with UV light, and then from the *right side* with a controlled intensity of UV, and of visible light. The edge exhibited rotation movement. (**d**) Reversible curling to a hairpin shape upon irradiation with UV light (crystal length: 3.0 mm). The crystal kept the crystalline state even after the curling and returned to the original straight shape upon irradiation with visible light. Reproduced from [29]

work as large as 2.2 nJ. The large mechanical work is ascribed to a large Young's modulus of the crystal. The modulus of the two-component crystal (**6a**/**7a** = 63/37) was measured by means of a manual beam-bending test to be as large as 8.5 GPa. This value is much larger than those of typical polymer materials (~1 GPa). The large Young's modulus enables the crystal to generate such strong force and perform large mechanical work. It is possible to estimate the maximum stress generated by UV irradiation from the photostimulated bending of the crystal cantilever beam. The maximum stress was estimated to be 56 MPa, which is 180 times larger than that of muscle (~0.3 MPa) and comparable to that of piezoelectric crystals, such as lead zirconate titanate (PZT, ~50 MPa). The two-component mixed crystals have fatigue-resistant and substantial mechanical properties comparable to piezoelectric crystals.

1.5 Light-Driven Bending of Rectangular Cocrystals

Although densely packed crystals have been proved to be useful to link the molecular-scale events to macroscopic movements of materials, the crystals so far examined are too small to be used in real-world applications. As a large-size

Fig. 1.10 Chemical structures of **8a** and **FN**

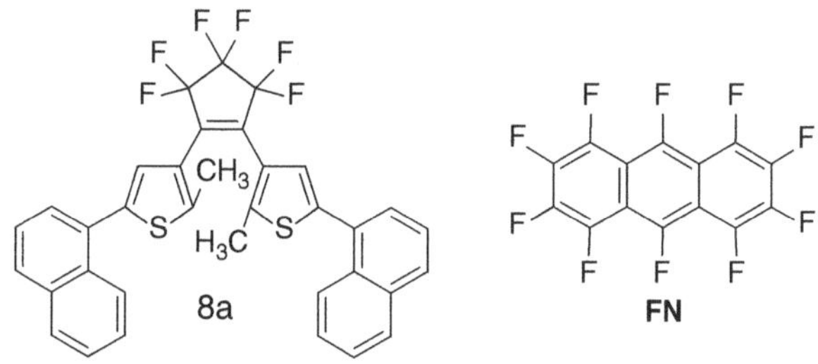

photoactive crystal, we prepared a cocrystal composed of a diarylethene derivative, 1,2-bis(2-methyl-5-(1-naphthyl)-3-thienyl)perfluorocyclopentene (**8a**), and perfluoronaphthalene (**FN**) with the size of 1–5 mm in length and examined their light-driven reversible bending (Fig. 1.10) [30].

The relatively large-size thin plate with rectangular faces (typically 1–5 mm in length, 0.2–1.5 mm in width and 10–50 μm in thickness) grew in a hexane solution of a mixture of **8a** and **FN**. The crystal is composed of **8a** and **FN** in the ratio of 1:2. The rectangular crystal changed its color from colorless to blue and bent moving away from the light source. The bending direction was opposite to that observed in the previous diarylethene crystals. The bending ceased when the illumination light was switched off and the crystal kept the bent shape in the dark. Upon visible light irradiation, the blue color disappeared and the crystal returned to the initial straight shape. As the UV light cannot penetrate deep into the crystal, the photoisomerization of the component diarylethene molecules takes place only in the surface thin layer. The bending is ascribed to the gradient of the degree of photoisomerization of the diarylethene molecules in the crystal.

X-ray crystallographic analysis was carried out to reveal the correlation between the structure change of individual molecules and the crystal shape deformation. The geometrical structures of the open- and closed-ring isomers are shown in Fig. 1.11. The two isomers were isolated by HPLC and independently recrystallized. We assumed that the geometric parameters derived from the structure of the independently recrystallized closed-ring isomers are valid except the conformation of the naphthyl side groups for the structure that is formed *in situ* in the crystal of the open-ring isomer. The two isomers differ in their geometrical structures, as shown by the triangle shape (blue broken line). As can be seen from the structures, the height H increases from 0.534 to 0.679 nm and the base width B decreases from 1.554 to 1.411 nm upon photocyclization. A similar change of the triangle shape of the diarylethene molecule was also confirmed by *in situ* X-ray crystallographic analysis. The diarylethene molecules align while maintaining their longitudinal height direction in parallel to *b*-axis in the crystal packing (see the molecular packing viewed from the (0 0 1) face in Fig. 1.11). Each diarylethene molecule reversibly changes the triangle shape upon UV and visible light irradiation. The height increase due to the photocyclization reaction forces the densely packed molecular layers to expand, resulting in the elongation of the *b*-axis. The contraction of *a*-axis is also explained by the change of the triangle shape. The geometrical structure change of

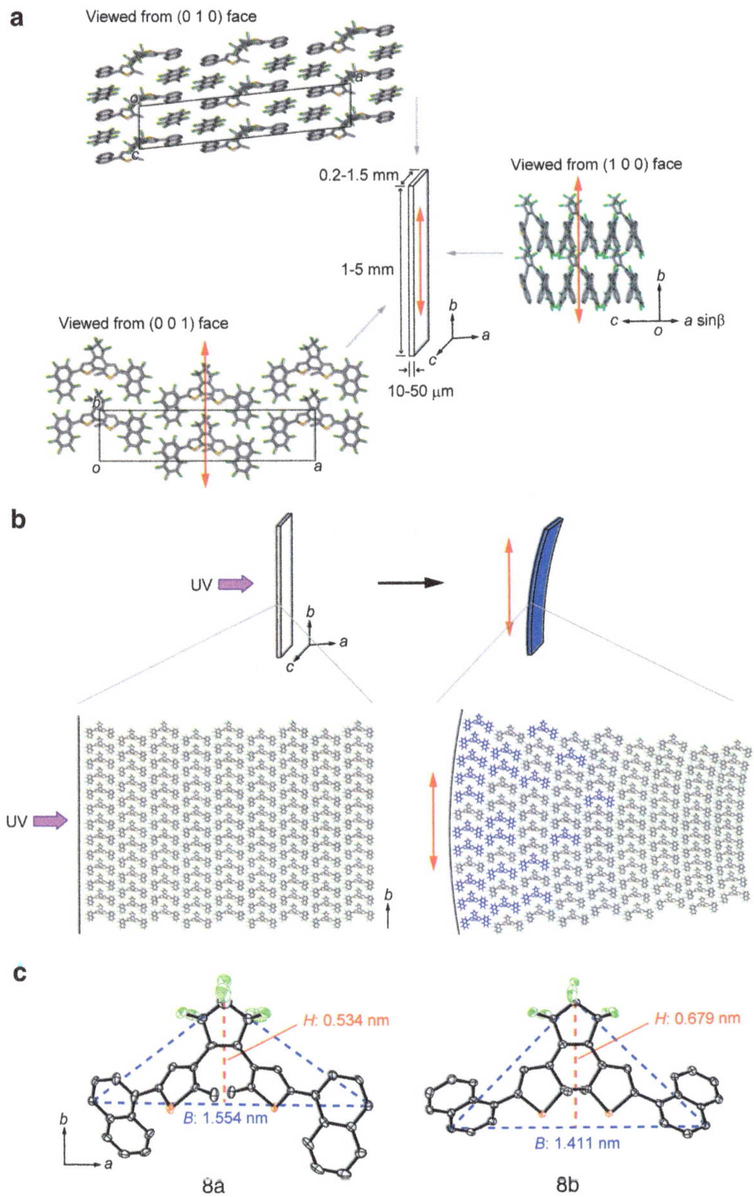

Fig. 1.11 (**a**) Illustration of crystal shape of the cocrystal (**8a–FN**) and molecular packing diagrams viewed from (1 0 0), (0 1 0) and (0 0 1) faces. The *red arrows* indicate the direction of the anisotropic elongation of the crystal, which is parallel to the *b*-axis, upon UV irradiation. (**b**) Schematic illustration of the photoinduced bending. The *blue* molecules are photogenerated closed-ring isomers in the crystal. (**c**) ORTEP drawing of the open- and closed-ring isomers in the cocrystal. The two isomers were isolated by HPLC and independently recrystallized. Reprinted with permission from [30]. Copyright (2010) Americal Chemical Society

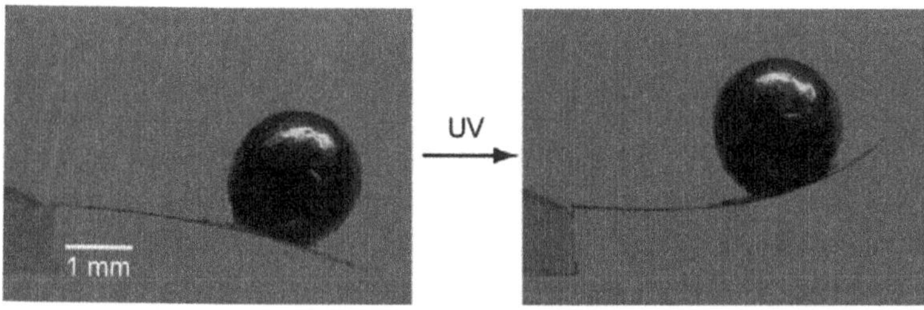

Fig. 1.12 Photomechanical work of a molecular crystal cantilever made of the cocrystal (**7a–FN**). Illumination of light was carried out below the crystal. The weights of the lead ball and the crystal cantilever were 46.77 and 0.17 mg, respectively

the individual molecules causes the macroscopic shape deformation of the crystal. The stacking mechanism proposed in the previous diarylethene crystals cannot be applied to the present cocrystal.

The photoinduced bending was applied to lift a heavy metal ball. A rectangular plate-like crystal was fixed at the edge of a glass plate as a cantilever arm and a lead ball was loaded onto the crystal, as shown in Fig. 1.12 [30]. The crystal and the ball weigh 0.17 and 46.77 mg, respectively. The weight of the lead ball is 275 times heavier than that of the crystal. Upon irradiation with UV light, the heavy ball was lifted as high as 0.95 mm. The cantilever arm performed lifting work and the amount of the work was as large as 0.43 μJ. The force generated by UV irradiation is larger than 1.1 mN. The photogenerated maximum stress was estimated to be 44 MPa. The work and force are ascribed to a large elastic modulus of the crystal. The modulus of this cocrystal was measure to be as large as 11 GPa. The large elastic modulus enables the crystal to generate a strong force and perform large mechanical work.

The bending dynamics after irradiation with pulsed laser light (355 nm, 8 ns width, 30 mJ/pulse) was followed using a high-speed camera at 295 K, as well as at 4.7 K. The photoisomerization of the diarylethene molecules is known to complete during the laser pulse (8 ns) [19]. We tried to measure how fast the mechanical bending movement of the crystal follows. At both temperatures the crystal showed bending in the first frame after irradiation with the laser pulse. The exposure time of each frame was 5 μs. Therefore, the light-driven bending took place in less than 5 μs at 295 K, as well as at 4.7 K. The very rapid movement of the crystal at 4.7 K suggests that the contribution of slow relaxation processes in the crystal lattice is negligible.

The X-ray crystallographic analysis clearly revealed that the deformation of the crystal shape is ascribed to the anisotropic deformation of the crystal lattice induced by the geometrical shape changes of the component diarylethene molecules upon photocyclization. The molecular crystal links the geometrical structure changes of molecules in the molecular world to macroscopic movement of the materials and performs mechanical work in the macroscopic real world.

Similar shape deformation of single crystals upon photoirradiation has also been reported for other diarylethene derivatives [31, 32], azobenzenes [33], salicylideneanilines [34], furylfulgide [35], and anthracene derivatives [36–38].

1.6 Conclusion

Diarylethene derivatives can undergo photoisomerization reactions even in crystals due to their rather small geometrical structure changes during the reactions. The photostimulated shape changes of individual molecules in crystals are successfully linked to macroscopic motion of the crystals and perform mechanical work in the real world. The inner stress generated by the shape changes of component molecules in densely packed crystals results in the changes of the surface morphology of the crystals, as well as the macroscopic bulk shape. Single-component and two-component diarylethene crystals and the cocrystal composed of **8a** and **FN** were found to exhibit reversible anisotropic shape deformation and bending responses upon alternate irradiation with UV and visible light. The light-driven crystal actuators made of diarylethene molecules, which remotely work without electric wire connection, potentially find various applications in micro- and nano-mechanical devices.

References

1. Whittaker M, Wilson-Kubbbalek EM, Smith JE, Faust L, Milligan RA, Sweeney HL (1995) A 35-Å movement of smooth muscle myosin on ADP release. Nature 378:748–751
2. Kay ER, Leigh DA, Zerbetto F (2007) Synthetic molecular motors and mechanical machines. Angew Chem Int Ed 46:72–191
3. Balzani V, Credi A, Venturi M (2008) Molecular devices and machines: concepts and perspectives for nanoworld. Wiley, Weinheim
4. Bissell RA, Córdova E, Kaifer AE, Stoddart JF (1994) A chemically and electrochemically switchable molecular shuttle. Nature 369:133–137
5. Jiménez MC, Dietrich-Buchecker C, Sauvage J-P (2000) Towards synthetic molecular muscles: contraction and stretching of a linear rotaxane dimer. Angew Chem Int Ed 39:3284–3287
6. Liu Y, Flood AH, Bonvallet PA, Vignon SA, Northrop BH, Tseng H-R, Jeppesen JO, Huang TJ, Brough B, Baller M, Magonov S, Solares SD, Goddard WA, Ho C-M, Stoddart JF (2005) Linear artificial molecular muscles. J Am Chem Soc 127:9745–9759
7. Badji JD, Balzani V, Credi A, Silvi S, Stoddart JF (2004) A molecular elevator. Science 303:1845–1849
8. Kelly TR, De Silva H, Silva RA (1999) Unidirectional rotary motion in a molecular system. Nature 401:150–152
9. Koumura N, Zijlstra RWJ, van Delden RA, Harada N, Feringa BL (1999) Light-driven monodirectional molecular rotor. Nature 401:152–155
10. Eisenbach CD (1980) Isomerization of aromatic azo chromophores in poly(ethyl acrylate) network and photomechanical effect. Polymer 21:1175–1179
11. Matéika L, Ilavský M, Dušek K, Wichterle O (1981) Photomechanical effects in crosslinked photochromic polymers. Polymer 22:1511–1515
12. Irie M (1990) Photoresponsive polymers. Adv Polym Sci 94:28–67
13. Finkelmann H, Nishikawa E, Pereira GG, Warner M (2001) A new opto-mechanical effect in solids. Phys Rev Lett 87:015501
14. Yu Y, Nakano M, Ikeda T (2003) Directed bending of a polymer film by light. Nature 425:45
15. Ikeda T, Mamiya J, Yu Y (2007) Photomechanics of liquid-crystalline elastomers and other polymers. Angew Chem Int Ed 46:406–528

16. Yamada M, Kondo M, Mamiya J, Yu Y, Kinoshita M, Barrett CJ, Ikeda T (2008) Photomobile polymer materials: towards light-driven plastic motors. Angew Chem Int Ed 47:4986–4988
17. Aliev AE, Oh J, Kozlov ME, Kunznetsov AA, Fang S, Fonseca AF, Ovalle R, Lima MD, Haque MH, Gartstein YN, Zhang M, Zakhidov AA, Baughman RH (2009) Giant-stroke, superelastic carbon nanotube aerogel muscles. Science 323:1575–1578
18. Irie M, Uchida K, Eriguchi T, Tsuzuki H (1995) Photochromism of single crystalline diarylethenes. Chem Lett 899–900
19. Jean-Ruel H, Cooney RR, Gao M, Lu C, Kochman M, Morrison CA, Miller RJD (2011) Femtosecond dynamics of the ring closing process of diarylethene: a case study of electrocyclic reactions in photochromic single crystals. J Phys Chem A 115:13158–13168
20. Kobatake S, Uchida K, Tsuchida E, Irie M (2002) Single-crystalline photochromism of diarylethenes: reactivity–structure relationship. Chem Commun 2804–2805
21. Yamada T, Kobatake S, Muto K, Irie M (2000) X-ray crystallographic study on single crystalline photochromism of bis(2,5-dimethyl-3-thienyl)perfluorocyclopentene. J Am Chem Soc 122:1589–1592
22. Yamada T, Kobatake S, Irie M (2000) X-ray crystallographic study on single-crystalline photochromism of 1,2-bis(2,5-dimethyl-3-thienyl)perfluorocyclopentene. Bull Chem Soc Jpn 73:2179–2184
23. Irie M, Kobatake S, Horichi M (2001) Reversible surface morphology changes of a photochromic diarylethene single crystal by photoirradiation. Science 291:1769–1772
24. Irie M (2008) Photochromism and molecular mechanical devices. Bull Chem Soc Jpn 81:917–926
25. Irie M (2010) Photochromism of diarylethene single molecules and single crystals. Photochem Photobiol Sci 9:1535–1542
26. Reddy CM, Gundakaram RC, Basavoju S, Kirchner MT, Padmanabhan KA, Desiraju GR (2005) Structural basis for bending of organic crystals. Chem Commun 3945–3947
27. Kobatake S, Takami S, Muto H, Ishikawa T, Irie M (2007) Rapid and reversible shape changes of molecular crystals on photoirradiation. Nature 446:778–781
28. Kuroki L, Takami S, Yoza K, Morimoto M, Irie M (2010) Photoinduced shape changes of diarylethene single crystals: correlation between shape and molecular packing. Photochem Photobiol Sci 9:221–225
29. Terao F, Morimoto M, Irie M (2012) Light-driven molecular crystal actuators: rapid and reversible bending of rod-like mixed crystals of diarylethene derivatives. Angew Chem Int Ed 51:901–904
30. Morimoto M, Irie M (2010) A diarylethene cocrystal that converts light into mechanical work. J Am Chem Soc 132:14172–14178
31. Uchida K, Sukata S, Matsuzawa Y, Akazawa M, deJong JJD, Katsonis N, Kojima Y, Nakamura S, Areephong J, Meetsma A, Feringa BL (2008) Photoresponsive rolling and bending of thin crystals of chiral diarylethenes. Chem Commun 326–328
32. Kobatake S, Hasegawa H, Miyaura K (2011) High-conversion photochromism of a diarylethene single crystal accompanying the crystal shape deformation. Cryst Growth Des 1223–1229
33. Koshima H, Ojima N, Uchimoto H (2009) Mechanical motion of azobenzene crystals upon photoirradiation. J Am Chem Soc 131:6890–6891
34. Koshima H, Takechi K, Uchimoto H, Shiro M, Hashizume D (2011) Photomechanical motion of salicylideneaniline microcrystals. Chem Commun 47:11423–11425
35. Koshima H, Nakaya H, Uchimoto H, Ojima N (2012) Photomechanical motion of furylfulgide crystals. Chem Lett 41:107–109
36. Al-Kaysi RO, Müller AM, Bardeen CJ (2006) Photochemically driven shape changes of crystalline organic nanorods. J Am Chem Soc 128:15938–15939
37. Al-Kaysi RO, Bardeen CJ (2007) Reversible photoinduced shape changes of crystalline organic nanorods. Adv Mater 19:1276–1280
38. Zhu L, Al-Kaysi RO, Bardeen CJ (2011) Reversible photoinduced twisting of molecular crystal microribbons. J Am Chem Soc 133:12569–12575

Chapter 2
Photomechanical Bending of Molecular Crystals

Hideko Koshima

Abstract Creation of mechanical crystals has been the focus of my study on solid-state photoreactions of molecular crystals for 20 years. Our search of photomechanical crystals, based on the operating principle that crystals should bend via reversible photochromic reactions, found that typical photochromic crystals composed of azobenzene, salicylideneaniline, and furylfulgide bend upon photoirradiation. When plate-like microcrystals of *trans*-4-(dimethylamino)azobenzene were irradiated at 365 nm, the crystals quickly bent away from the light. Thirty seconds after the light was extinguished, the crystal returned to its initial flat shape. To our knowledge, *trans–cis* photoisomerization of azobenzene chromophores has not been observed in the crystalline state, but the occurrence of photoisomerization near the crystal surface was confirmed. Plate-like microcrystals of a salicylideneaniline compound could be repeatedly bent and straightened over 200 cycles by alternately irradiating with UV and visible light, respectively. Microcrystals of furylfulgide exhibited the same behavior. X-ray crystallographic analyses before and after photoirradiation showed that the mechanism of bending in salicylideneaniline and furylfulgide crystals involved molecular-level shape changes that induced macroscale mechanical motion. Synthetic molecular machines that convert photoenergy directly into mechanical work through molecular structural changes are attractive for both basic research and many potential applications. These photomechanical crystals will contribute to the advancement of molecular machinery.

Keywords Azobenzene • Crystal machines • Furylfulgide • Photomechanical bending • Salicylideneaniline

H. Koshima (✉)
Department of Materials Science and Biotechnology,
Graduate School of Science and Engineering, Ehime University,
Matsuyama 790-8577, Japan
e-mail: koshima.hideko.mk@ehime-u.ac.jp

M. Irie et al. (eds.), *New Frontiers in Photochromism*,
DOI 10.1007/978-4-431-54291-9_2, © Springer Japan 2013

2.1 Introduction

Crystals have a well-deserved reputation for rigidity. The attractive forces that trap molecules into three-dimensional solids also keep the atoms relatively immobile. In recent decades, however, a large number of solid-state reactions have been observed in molecular crystals [1, 2]. Bimolecular photoreactions in the crystalline state have also been observed [3, 4], and the relationship between crystal structure and these reactions has been elucidated. Crystalline state photoreactions usually induce molecular motion in the crystal lattice, thereby suggesting the possibility of motion in bulk crystals. It has been my goal to make photomechanical crystals for 20 years.

Recently, molecular machines based on the conversion of molecular motion to macroscale mechanical motion have attracted interest from both basic researchers and applications engineers. Such machines are amenable to remote operation with light and do not require wire connections. Supramolecular machines such as shuttles and motors have been reported, but the motion was only spectroscopically detected [5, 6] and not linked to macroscale mechanical motion in bulk materials. Macroscale mechanical motion of molecular materials has been observed in gels [7] and nematic elastomers [8–13], in which a photoinduced order–disorder phase transition acted as the driving force. Irie and co-workers [14–18] first reported rapid, reversible bending in rod-shaped microcrystals of photochromic diarylethene upon photoirradiation, thereby linking molecular-level shape changes to macroscale mechanical motion in crystals. Bardeen and co-workers [19–23] demonstrated irreversible and reversible shape changes in rod-shaped nanocrystals of anthracene carboxylates by photodimerization. Since then, several photomechanical crystals have been reported [24–26], revealing potential opportunities for synthetic molecular machinery [27].

We started studying photomechanical crystals several years ago. Initially, our work focused on known, typical photochromic crystals in order to identify the characteristics required for photomechanical motion in crystals. Since then, several photomechanical crystals have been found, based on the operating principle that crystals should bend via reversible, photochromic reactions. This chapter introduces recent studies of photomechanical bending in crystals of azobenzene [28, 29], salicylideneaniline [30], and furylfulgide [31].

2.2 Photomechanical Bending of Azobenzene Crystals

Azobenzenes are typical chromophores that undergo *trans–cis* photoisomerization and are incorporated into various photofunctional materials. Azobenzene-containing elastomer films bend upon photoirradiation, and this effect has been applied in the fabrication of a light-driven plastic motor [8–12] and a high-frequency, light-driven polymer oscillator [13]. This section describes plate-like microcrystals of *trans*-4-(dimethylamino)azobenzene (*trans*-**1a**) [28] and *trans*-4-aminoazobenzene (*trans*-**1b**) [29] that exhibit reversible bending upon irradiation with UV light (Fig. 2.1).

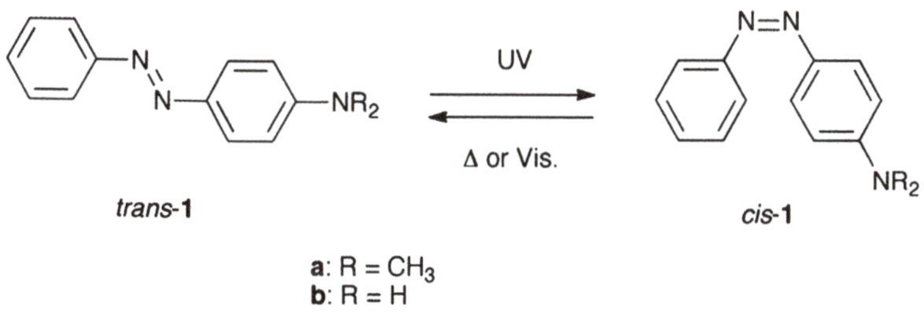

Fig. 2.1 Photoisomerization of azobenzene **1**

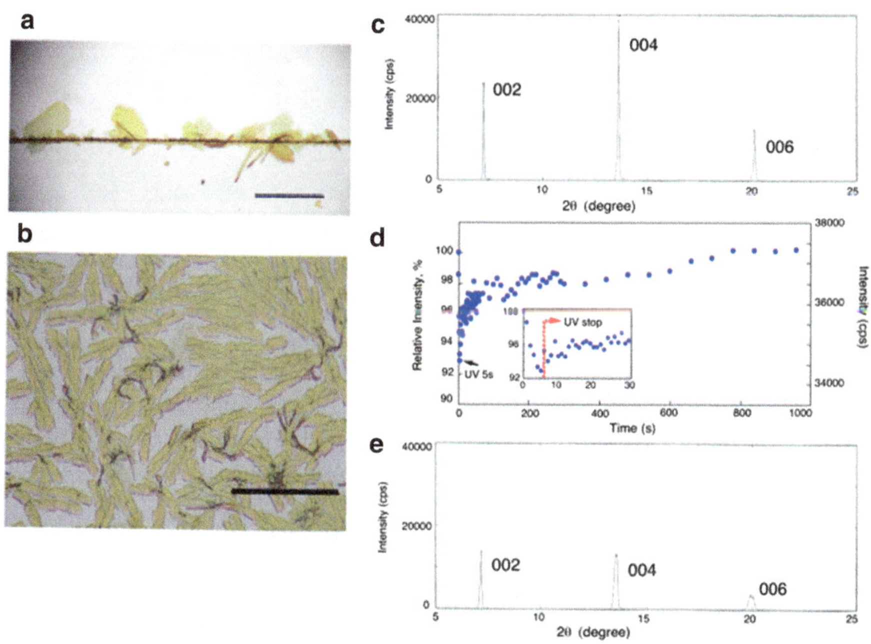

Fig. 2.2 Plate-like microcrystals of *trans*-**1a** grew on the edges (**a**) and surface (**b**) of a glass plate by sublimation and condensation. The scale bars are 500 μm. (**c**) X-ray diffractogram of *trans*-**1a** microcrystals before irradiation. (**d**) The 004 peak intensity decreased to 93% upon irradiation at 365 nm for 5 s and then increased when the irradiation was stopped. (**e**) X-ray diffractogram of *trans*-**1a** microcrystals were measured under constant UV irradiation. Peak heights decreased and peak widths expanded, but no new peaks appeared

2.2.1 4-(Dimethylamino)azobenzene

Microcrystals of *trans*-**1a** were prepared by sublimating crystalline powders in a glass tube oven under vacuum or in air. Plate-like microcrystals (several hundred micrometers in length) grew on the edges and surface of the glass plate after heating to a temperature near the melting point (114 °C) for several hours (Fig. 2.2a, b). X-ray diffraction (XRD) measurements revealed three sharp peaks (Fig. 2.2c),

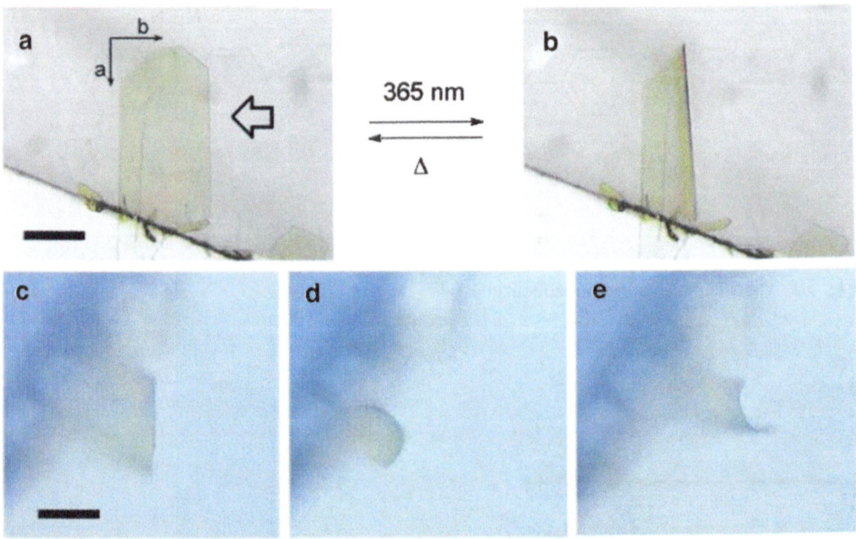

Fig. 2.3 The (0 0 1) face of a plate-like *trans*-**1a** microcrystal (**a**) before and (**b**) after UV irradiation from the right rear. The same crystal is shown (**c**) before and after UV irradiation from the (**d**) right and (**e**) left sides. The scale bars are 200 μm

which were assigned to the 0 0 2, 0 0 4, and 0 0 6 peaks on the basis of consistencies with crystallographic data [32]. The top surface of the plate-like microcrystals was identified as the (0 0 1) face with its longitudinal direction along the *a*-axis, based on comparisons with plate-like bulk crystals, which exhibited a (0 0 1) face along the *a*-axis.

Figure 2.3a shows a piece of a plate-like microcrystal ($525 \times 280 \times 5$ μm^3) whose lower portion was fixed to the glass surface and whose upper portion was free. Upon UV irradiation (365 nm, 5 mW/cm^2) with a light-emitting diode (LED) at the (0 0 1) surface from the right rear, the crystal quickly bent away from the light source, reaching a maximum deflection after 0.5 s (Fig. 2.3b). The maximum deflection angle was quite large, leading to the formation of a ~180° semicircle along the *b*-axis. When illumination was stopped, the crystal returned to its initial flat shape after 30 s (Fig. 2.3a). Figure 2.3c shows the same crystal when irradiated from the right (Fig. 2.3d) and left sides (Fig. 2.3e), which induced bending in opposite directions.

A narrow, plate-like crystal ($70 \times 5 \times 1$ μm^3) with one end fixed to another crystal was subjected to repeated bending (Fig. 2.4a). As described above, comparisons with bulk crystals showed that the (0 0 1) plane was oriented perpendicular to the surface of the glass plate with the longitudinal direction along the *a*-axis. When subjected to UV irradiation at the (0 0 1) surface, the crystal bent away from the light source, reaching a maximum deflection of 5 μm after 0.2 s, as measured at the end of the crystal (Fig. 2.4b). The degree of deflection along the *a*-axis was considerably smaller than that along the *b*-axis (Fig. 2.3b, d, e). The bent crystal returned to its

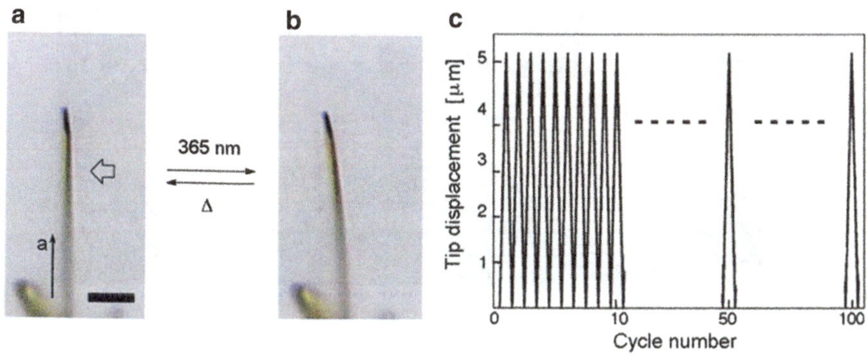

Fig. 2.4 Bending of a narrow, plate-like crystal of *trans*-**1a**, as shown by a comparison of images acquired (**a**) with and (**b**) without UV irradiation. The scale bar is 20 μm. (**c**) The reversible bending was repeatable over as many as 100 cycles

initial straight form 3.8 s after the illumination was stopped. This reversible bending was observed over 100 cycles of alternating UV irradiation (2 s) and darkness (5 s) (Fig. 2.4c).

The crystals bend away from the light source due to a gradient in the extent of *trans* to *cis* photoisomerization as a function of light penetration. Photoisomerization and consequent elongation of the crystal at the irradiated surface along the *a*- and *b*-axes, with no changes at the non-illuminated rear surface, resulted in a bent macrostructure. However, only slight differences were observed in the absorption spectra (λ_{max}=400 nm) of the *trans*-**1a** microcrystals before and after UV irradiation. Therefore, the occurrence of photoisomerization could not be confirmed by absorption data. Direct evidence of the formation of *cis*-**1a** was obtained from nuclear magnetic resonance (NMR) measurements. The ^{1}H-NMR spectrum of *trans*-**1a** microcrystals that had been irradiated for 1 min was measured in benzene-d_6. Singlet peaks were observed at 2.35 and 2.23 ppm, corresponding to the $N(CH_3)_2$ protons of the *trans* and *cis* isomers, respectively, with a 160:1 relative intensity [33]. This provided unambiguous evidence of *trans*-to-*cis* photoisomerization in the crystalline state. The estimated yield of the *cis* isomer was ~1% based on the decay of absorbance in benzene. With no UV irradiation, the *cis* isomer almost disappeared after 30 min as a result of *cis*-to-*trans* thermal isomerization. During UV irradiation, the melting point of *trans*-**1a** microcrystals (113–114 °C) decreased to 66–80 °C because of the coexistence of the *cis* isomer and returned to 113–114 °C when the irradiation was extinguished, further demonstrating the reversibility of isomerization. To our knowledge, *trans*–*cis* photoisomerization of azobenzene chromophores has not been observed in the bulk crystalline state because of the large geometric changes that would be required in such densely packed crystal lattices. However, we confirmed the occurrence of photoisomerization near the surface of *trans*-**1a** microcrystals.

The (0 0 1) top surface of the *trans*-**1a** microcrystal was smooth prior to irradiation, as observed by atomic force microscopy (AFM) (Fig. 2.5a). After UV

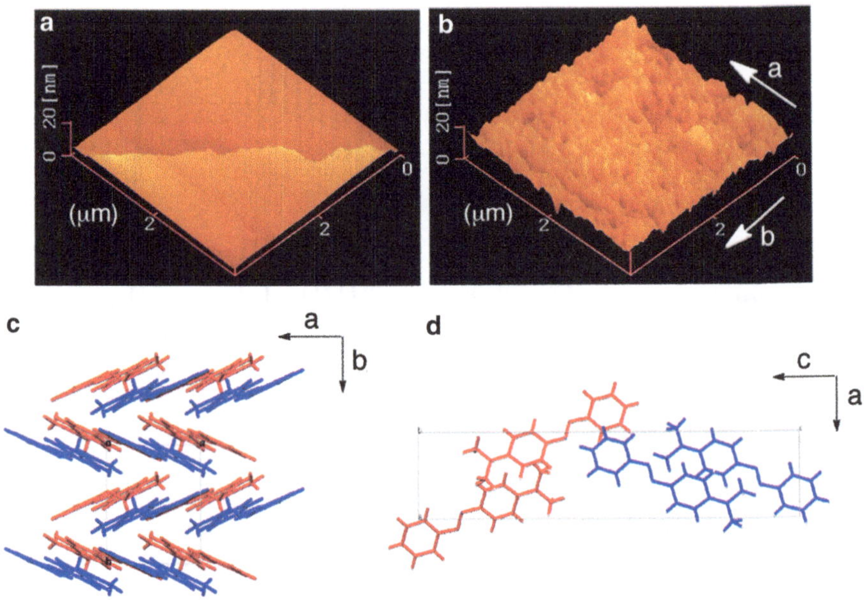

Fig. 2.5 AFM images of the (0 0 1) top surface of a *trans*-**1a** microcrystal (**a**) before and (**b**) after UV irradiation and illustrations of the molecular arrangements of the (**c**) (0 0 1) and (**d**) (0 1 0) faces before irradiation

irradiation for 5 s, uneven features appeared along the *a*-axis with a height of 10 nm and a relative roughness of 1% of the crystal thickness (900 nm) (Fig. 2.5b). The uneven features receded slightly after 1 h in the dark, but the initial smooth surface was not recovered. The relative XRD peak intensities for the microcrystals quickly decreased to 93% upon UV illumination for 5 s, likely because of both *trans*-to-*cis* photoisomerization and the deterioration of crystallinity (Fig. 2.2d). After the irradiation was stopped, the diffraction peaks increased to 97% of their initial intensities after 30 s and recovered completely after 15 min as a result of the *cis*-to-*trans* thermal isomerization and the recovery of crystallinity. However, no new peaks were observed in the diffractogram, even after continued irradiation (Fig. 2.2e). This suggests that the *cis*-**1a** molecules did not form a new crystalline phase, but may have instead adopted a fine polycrystalline or amorphous phase.

In the *trans*-**1a** crystal, the planar molecules are arranged almost perpendicularly at the (0 0 1) face to form a herringbone structure along the *a*-axis (Fig. 2.5c) [32]. Upon UV irradiation, the planar *trans*-**1a** molecules undergo photoisomerization to *cis*-**1a** at the (0 0 1) crystal surface, resulting in an increased torsional conformation due to repulsion between the two phenyl planes. Crystalline *cis*-azobenzene molecules normally exhibit a dihedral angle of 64.26° between the two phenyl planes [34]. The *trans*-to-*cis* photoisomerization elongates the unit cell length along the *b*- and *a*-axes near the (0 0 1) crystal surface, giving rise to the uneven features. In contrast, since there is no photoisomerization in the absence of light, the unit cell dimensions remain constant at the non-illuminated surface, resulting in bending of

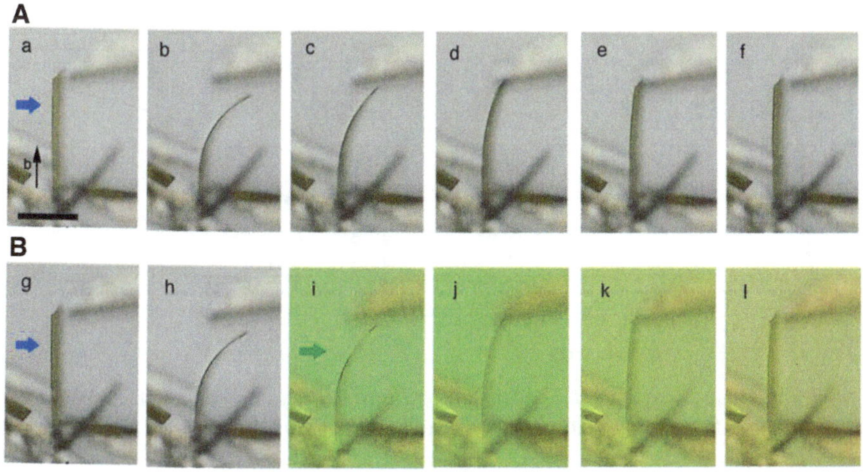

Fig. 2.6 Photomechanical bending of the *trans*-**1b** microcrystal: (**A**) (*a*) before irradiation, (*b*) after UV irradiation from the left side for 0.5 s, and no irradiation for (*c*) 10 s, (*d*) 60 s, (*e*) 120 s, (*f*) 240 s; (**B**) (*g*) before irradiation, (*h*) after UV irradiation from the left side for 0.5 s, and after visible light (530 nm) irradiation from the left side for (*i*) 2 s, (*j*) 14 s, (*k*) 30 s, (*l*) 60 s. The scale bar is 100 μm (from [29] with permission. © 2012 Elsevier)

the microcrystal. Furthermore, molecular interactions between the neighboring herringbone structures are weak, as seen in the packing arrangements at the (0 0 1) and (0 1 0) faces (Fig. 2.5c, d). Therefore, the uneven features along the *b*-axis are sharper than those along the *a*-axis (Fig. 2.5b). This leads to a drastic bending motion along the *b*-axis (Fig. 2.3) relative to a small crystal bending observed along the *a*-axis (Fig. 2.4).

2.2.2 4-Aminoazobenzene

Plate-like microcrystals of *trans*-4-aminoazobenzene *trans*-**1b** were also obtained by sublimation and condensation. The top surface was identified as the (1 0 −1) face by XRD, and the longitudinal direction was assigned to *b*-axis based on comparisons with plate-like bulk crystals. Figure 2.6a shows a piece of a plate-like microcrystal ($200 \times 25 \times 1.2\ \mu m^3$) in which the lower portion was adhered to the glass surface and the upper portion was free. When the (1 0 −1) surface was irradiated at 365 nm (40 mW/cm^2) from the left side, the crystal quickly bent away from the light, reaching a displacement angle of 34° after 0.5 s (Fig. 2.6b). Stopping the illumination resulted in the gradual return of the crystal to its initial linear shape after 4 min (Fig. 2.6c–f).

The microcrystal bending speed and displacement angle increased in proportion to the UV light intensity (Fig. 2.7). When the microcrystals were irradiated with 2.5 mW/cm^2, 8 s was required to reach the maximum displacement angle of 27°.

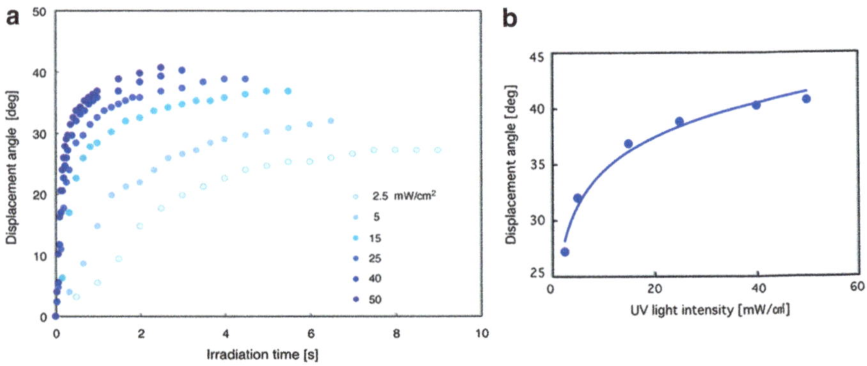

Fig. 2.7 Dependence of (**a**) UV irradiation time and (**b**) UV light intensity on the displacement angle of a plate-like microcrystal ($200 \times 25 \times 1.2\ \mu m^3$) of *trans*-**1b**

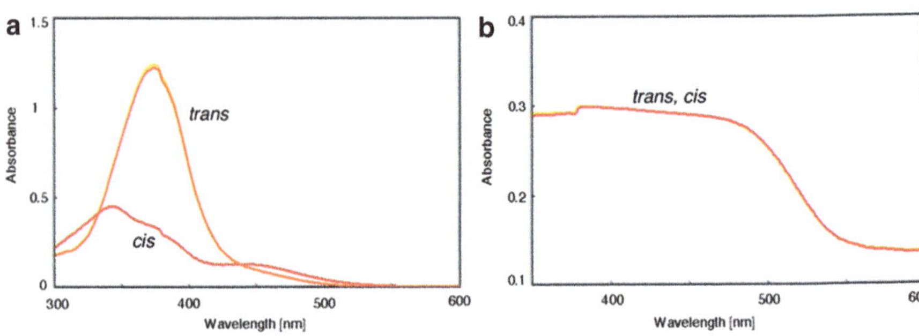

Fig. 2.8 (**a**) Absorption spectra of *trans*-**1b** (0.05 M) in benzene before (*yellow*) and after UV irradiation at 365 nm for 30 s (*red*). Stopping irradiation for 10 min recovered the initial spectrum in benzene (*orange*). (**b**) Absorption spectra of the *trans*-**1b** microcrystals before (*yellow*) and after UV irradiation for 30 s (*red*)

When 50 mW/cm² was used, 2 s was required to reach the maximum displacement angle of 41°.

After 30 s of UV irradiation at 365 nm, the absorption spectrum ($\lambda_{max}=375$ nm) of *trans*-**1b** in benzene [35] changed to that of *cis*-**1b** (photostationary state) with two bands at $\lambda_{max}=344$ and 442 nm, corresponding to (π, π*) and (n, π*) excitations, respectively (Fig. 2.8a). When irradiation was stopped for 10 min, the initial spectrum of *trans*-**1b** was recovered, demonstrating thermal *cis*-to-*trans* isomerization. Visible light irradiation at 530 nm led to a faster recovery (1 min) of the initial *trans*-**1b** spectrum. Thus, 4-aminoazobenzene **1b** in benzene underwent reversible photoisomerization between the *trans* and *cis* isomers via alternate irradiation with UV (365 nm) and visible (530 nm) light. In contrast, the absorption spectrum of *trans*-**1b** microcrystals did not change after UV irradiation at 365 nm for 30 s (Fig. 2.8b).

However, the changes in *trans*-**1b** microcrystals upon UV irradiation were detectable by XRD (Fig. 2.9). The 20−2 peak intensity decreased to 87% of the original

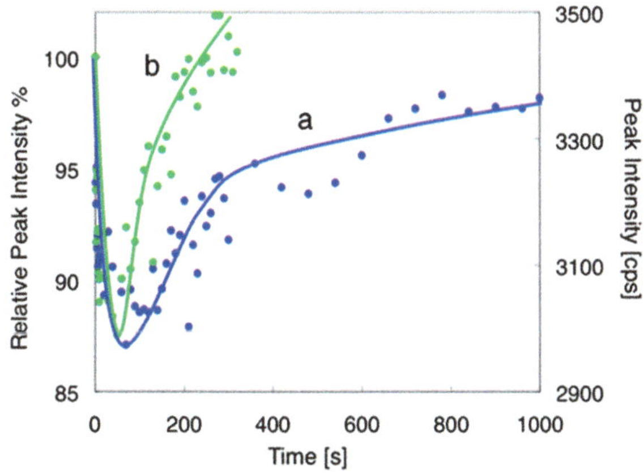

Fig. 2.9 Dependence of irradiation time on the XRD 2θ–2 peak intensity of *trans*-**1b** microcrystals. (**a**) *Blue*: UV light was applied for 60 s and then stopped. (**b**) *Green*: UV light was applied for 60 s and then visible light was applied at 530 nm (from [29] with permission. © 2012 Elsevier)

value after UV irradiation (40 mW/cm^2) for 60 s because of the *trans*-to-*cis* photoisomerization and deterioration of crystallinity. However, no shift in the diffraction angle 2θ or new peaks were observed in the XRD profile, even after continued irradiation. This suggests that the *cis*-**1b** molecules did not form a new crystalline phase. The XRD peak intensity almost recovered 15 min after irradiation was stopped due to *cis*-to-*trans* thermal isomerization (Fig. 2.9a). In contrast, irradiation with visible light at 530 nm (10 mW/cm^2) recovered the initial intensity after 4 min due to *cis*-to-*trans* photoisomerization induced by (n, π*) excitation (Fig. 2.9b). As discussed above, the crystal returned to its initial, linear shape 4 min after irradiation was extinguished (Fig. 2.6f). In contrast, visible light irradiation recovered the initial straight crystal in 1 min (Fig. 2.6l), revealing that both bending and straightening of **1b** microcrystals could be controlled by irradiation with UV and visible light, respectively.

2.3 Photomechanical Bending of Salicylideneaniline Crystals

Salicylideneanilines are typical photochromic compounds in the crystalline state [36–38]. Their photochromicity is based on the formation of colored species by photoinduced proton transfer. This section describes plate-like microcrystals of *N*-3,5-di-*tert*-butylsalicylidene-3-nitroaniline (**2**) in their enol form that exhibit reversible bending upon alternate irradiation with UV and visible light (Fig. 2.10). The bending mechanism was determined by changes in crystal structure before and after photoirradiation.

2.3.1 Bending Motion

Plate-like microcrystals of enol-**2** were obtained on a silanized glass plate by sublimation and condensation at approximately 10 °C below the melting point (132 °C).

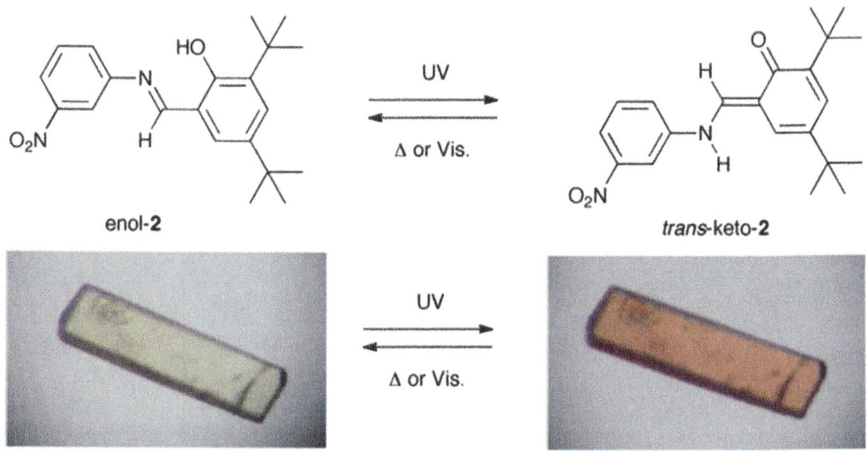

Fig. 2.10 Photochromic reaction of salicylideneaniline **2**

The top surface of the plate-like microcrystals was identified as the (0 0 1) face based on consistencies with existing crystallographic data [39], and its longitudinal direction was assigned to the *a*-axis based on comparisons with plate-like bulk crystals, which exhibit a (0 0 1) face along the *a*-axis. Figure 2.11a shows the frontal (0 0 1) face of a plate-like enol-**2** crystal ($73 \times 4.5 \times 1.1\ \mu m^3$) with one end fixed to an adjacent crystal. When the (0 0 1) face was irradiated from the right at 365 nm (40 mW/cm^2, Fig. 2.11b), the crystal bent away from the light to reach a maximum tip displacement angle of 45° after 5 s (Fig. 2.11c). Subsequent illumination with a halogen lamp equipped with a color filter (530 nm, 10 mW/cm^2) returned the bend to its initial straight shape after 10 s (Fig. 2.11b).

A 2-s pulse of UV irradiation, with simultaneous and continuous illumination with visible light (>390 nm) from the above halogen lamp, bent the crystal to a displacement angle of 37°. When the UV source was blocked, the crystal returned to its initial straight form after 4.6 s. This reversible bending was observed over 200 cycles of alternating UV irradiation (2 s on, 5 s off) under continuous illumination with visible light (Fig. 2.11d). The bending motion was accompanied by a color change from pale yellow to reddish orange due to the formation of *trans*-keto-**2** (Fig. 2.12a). The reddish-orange crystal also returned to the initial pale-yellow color due to photochemical fading to the enol-**2** isomer (Fig. 2.12b).

The bending motion of the microcrystals depended on the intensity of UV and visible light. When a plate-like microcrystal ($50 \times 6.3 \times 2.1\ \mu m^3$, Fig. 2.13a) was irradiated at 365 nm at 5–40 mW/cm^2 for 30 s, the tip displacement angle increased from 9° to 25°, respectively, in proportion to the UV intensity (Fig. 2.13b, c). In the absence of UV radiation, the crystals straightened to their initial form after 500 to 1,400 s. The lifetime of thermal fading from the *trans*-keto-**2** to the enol-**2** isomer is very long (42 days, $k = 2.8 \times 10^{-7}$/s at 25 °C, Fig. 2.12c). This is due to stabilization of the *trans*-keto-**2** form by intermolecular hydrogen bonding and dimer formation between the N–H group of the *trans*-keto imine and the oxygen atom of the nitro

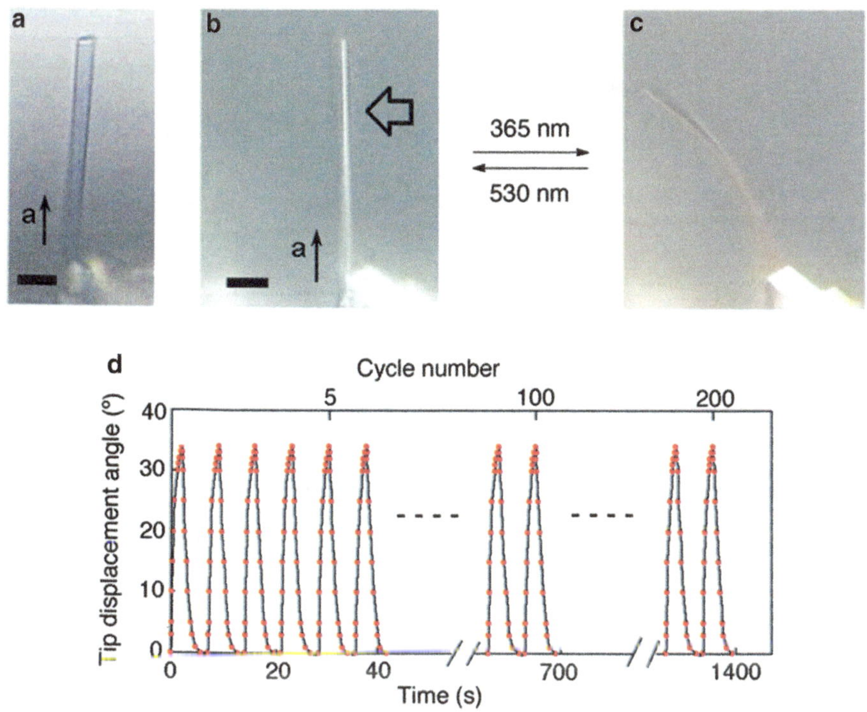

Fig. 2.11 A plate-like enol-**2** microcrystal is shown (**a**) at the frontal (0 0 1) face (**b**) before and (**c**) after UV irradiation from the right side. The scale bars are 10 mm. (**d**) The repeatability of the reversible bending can be observed over as many as 200 cycles (from [30] with permission. © 2011 Royal Society of Chemistry)

Fig. 2.12 Absorption spectra of enol-**2** microcrystals upon (**a**) UV irradiation at 365 nm, and then (**b**) visible light irradiation at 530 nm, or (**c**) in the dark. The lifetimes estimated from the changes of absorbance at 480 nm are (**a**) 2.6 s, (**b**) 63 s, and (**c**) 42 days

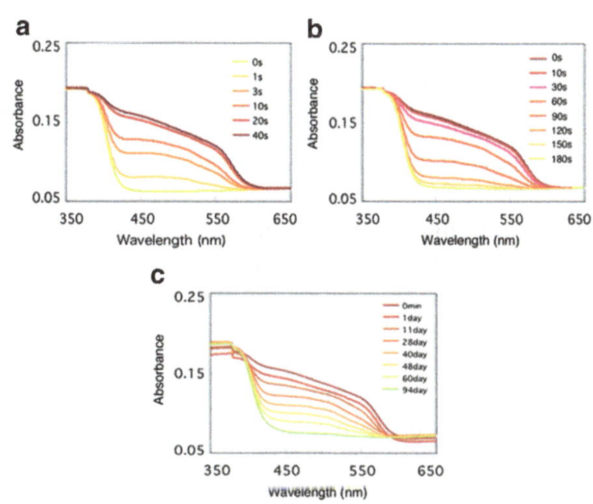

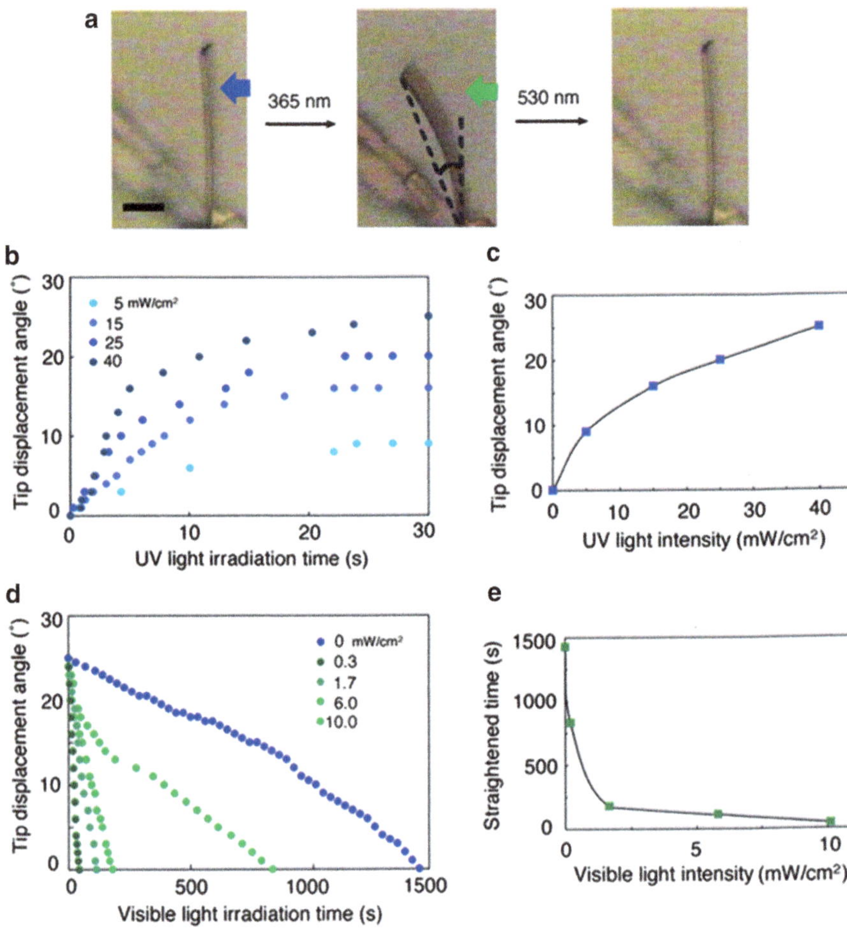

Fig. 2.13 (**a**) A microcrystal ($50 \times 6.3 \times 2.1\ \mu m^3$) is shown before and after irradiation with UV (365 nm) and visible light (530 nm). The scale bar is 10 mm. (**b**) The bending motion is dependent on the UV irradiation time. (**c**) Tip displacement angle is shown as a function of UV intensity. (**d**) The straightening motion is dependent on the visible irradiation time, and (**e**) the time to recover the initial straightness is shown as a function of visible light intensity (from [30] with permission. © 2011 Royal Society of Chemistry)

group [40]. However, the lifetime was dramatically shortened to 63 s upon illumination with 530 nm light at 10 mW/cm^2 due to photochemical fading (Fig. 2.12b). The time required to recover the initial straight form was also dramatically shortened from 1,400 to 43 s with increasing visible light intensity up to 10 mW/cm^2 (Fig. 2.13d, e). This recovery time is comparable to the 63 s lifetime observed with 530 nm light when the intensity was increased to 10 mW/cm^2.

The degree of tip displacement with plate-like microcrystals increased with increasing aspect ratio (length/thickness). For example, maximum displacement angles of 25 °C, 52 °C, and 70 °C were observed with microcrystals having aspect

a

	UV irradiation time (s)				
	0	900	1800	3600 (relative change)	
a (Å)	6.09139(17)	6.1150(4)	6.1146(4)	6.1172(4)	(+0.42%)
b (Å)	10.2994(2)	10.3353(8)	10.3347(7)	10.3417(7)	(+0.41%)
c (Å)	16.2874(4)	16.2230(11)	16.2172(9)	16.2127(10)	(−0.46%)
α (deg)	102.1852(14)	102.055(4)	102.050(4)	102.017(4)	(−0.16%)
β (deg)	92.2252(16)	92.705(4)	92.728(3)	92.797(3)	(+0.62%)
γ (deg)	102.9942(16)	103.264(5)	103.264(4)	103.296(4)	(+0.29%)
V (Å^3)	969.23(4)	971.06(12)	970.57(10)	971.23(11)	(+0.21%)

b

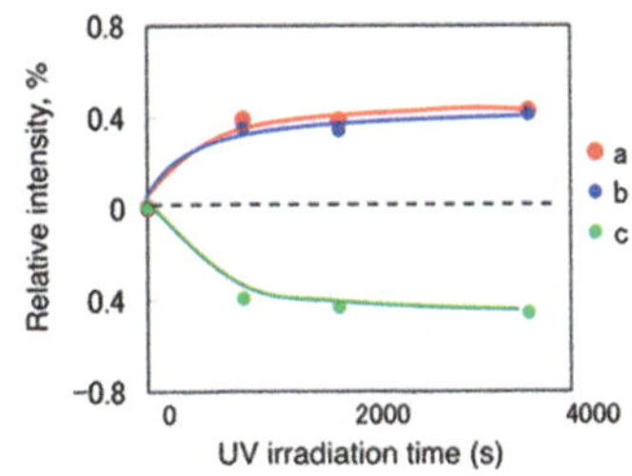

Fig. 2.14 (**a**) Changes in the unit cell parameters and (**b**) the relative changes in unit cell lengths of enol-**2** crystal upon UV irradiation

ratios of 25 (50 × 6.3 × 2.1 μm^3), 71 (120 × 7.1 × 1.7 μm^3), and 132 (159 × 6.8 × 1.2 μm^3), respectively.

Notably, upon UV irradiation, a plate-like crystal (120 × 7.1 × 1.7 μm^3) was able to flip a silica gel particle that was approximately 50-fold heavier than the microcrystal itself, suggesting that the mechanical force generated in these crystals is relatively large. Most of the plate-like crystals prepared by sublimation adhered to the glass surface and were unable to bend freely. UV irradiation over the silanized glass plate, however, bent most of the crystals away from the light source, stripping the stuck crystals from the glass surface. In contrast, crystals prepared on bare glass plates did not bend, indicating that silanization decreased the degree of interaction between the crystals and the glass.

2.3.2 *Mechanism of Crystal Bending*

Crystallographic analyses of a thin enol-**2** microcrystal (200 × 30 × 20 μm^3) were performed at 123 K before and after successive UV irradiation at room temperature. Here, the thinness of the crystal allows for complete light penetration. A nearly constant unit cell size was attained after 1 h (Fig. 2.14). The resulting crystal exhibited a disordered structure with contributions from both the enol-**2** and the *trans-keto*-**2** isomers at a ratio of 0.905(2):0.095(2). Figure 2.15a shows the ORTEP drawing of this structure. This is identical to structures obtained following irradiation at 360 nm [40] and following two-photon excitation at 730 nm [39].

In the *trans*-keto-**2** molecule, the oxygen atom O(1′) and the nitrogen atom N(1′) are *trans* with respect to the C(1′)–C(7′) bond. The bond lengths of C(2′)–O(1′), C(1′)–C(7′), and C(7′)–N(1′) were 1.305(9), 1.415(9), and 1.349(9) Å, respectively, which differed significantly from those (1.355(4), 1.441(5), and 1.273(4) Å) of the enol-**2** molecule. The C(8′)–N(1′)–C(7′)–C(1′) torsion angle of the *trans*-keto-**2** was −172.4(18)°, and the corresponding angle of the enol-**2** isomer was 177.2(3)°. The phototransformation likely occurs through the motion of a pair of benzene rings,

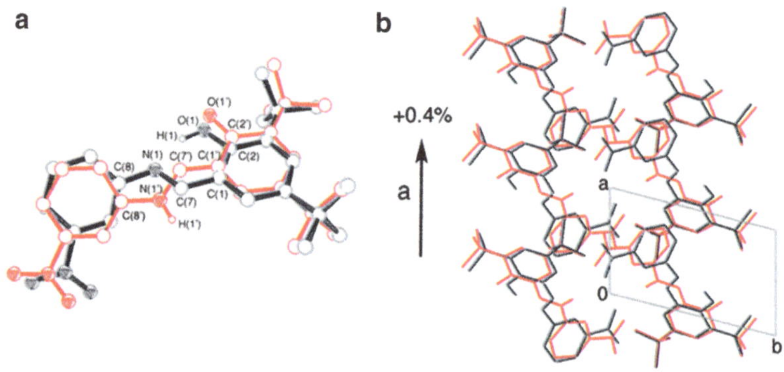

Fig. 2.15 (**a**) ORTEP drawings show the disordered structures of the enol-**2** (*black*) and *trans*-*keto*-**2** (*red*) isomers after irradiation at 365 nm at the 25% probability level. Hydrogen atoms, except H(1) and H(10), are omitted for clarity. (**b**) Molecular arrangements on the (0 0 1) face

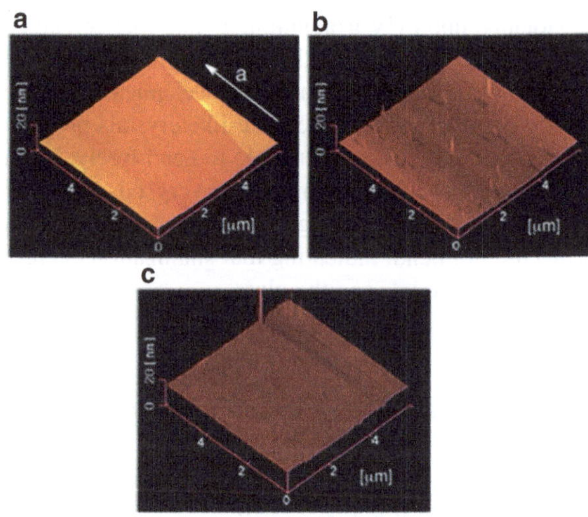

Fig. 2.16 AFM images show the (0 0 1) top surface of an enol-**2** microcrystal (**a**) before and after irradiation with (**b**) UV and (**c**) visible light

analogous to the bicycle pedal motion of stilbenes and azobenzenes, without requiring major structural changes [41–43].

The length of the *a*-axis in the unit cell increased by 0.42% from 6.0913(2) to 6.1171(4) Å (Figs. 2.14 and 2.15b). Hence, photoisomerization elongates the crystal length along the *a*-axis near the (0 0 1) crystal surface. In contrast, since no photoisomerization occurs in the absence of light, the unit cell dimensions remain constant at the non-illuminated surface, resulting in bending of the microcrystal away from the light source. When UV or visible (530 nm) irradiation is stopped, the reverse isomerization from *trans*-keto-**2** to enol-**2** brings the crystal back to its initial straight conformation. These molecular-level shape changes therefore result in bending of the bulk crystal.

The (0 0 1) surface of the enol-**2** microcrystal was smooth before irradiation, as observed by AFM (Fig. 2.16a). Small dents with a depth of 1.5 nm (0.3% of the total crystal thickness) appeared after UV irradiation for 40 s (Fig. 2.16b). The overall surface morphology, however, remained constant. Subsequent irradiation with visible light for 180 s did not alter the surface morphology (Fig. 2.16c). These results suggest minimal molecular movement, such as a pedaling motion, during photoisomerization. The minimal movement affords excellent repeatability and reversibility over as many as 200 cycles (Fig. 2.11d).

2.4 Photomechanical Bending of Furylfulgide Crystals

Fulgides constitute a class of photochromic compounds that undergo reversibly electrocyclic ring-closure and ring-opening reactions [44]. The furylfulgide (*E*)-2-[1-(2,5-dimethyl-3-furyl)ethylidene]-3-isopropylidenesuccinic anhydride (**E3**) in its *E*-form exhibits photochromism in the crystalline state (Fig. 2.17) [45–48]. This section describes plate-like microcrystals of **E3** with reversible bending upon alternate irradiation with UV and visible light. The bending mechanism is based on changes in crystal structure before and after photoirradiation.

Microcrystals of **E1** were prepared by sublimation and condensation of crystalline powders in a small platinum pan covered with a glass plate. Plate-like microcrystals grew on the surface of the glass plate after heating to approximately 20 °C below the melting point (126 °C) and holding for several hours. X-ray diffractograms of the microcrystals contained two peaks, assigned to the 1 0 1 and 2 0 2 reflections based on consistencies with existing crystallographic data [45, 46, 48].

Figure 2.18a shows the hexagonal face of a plate-like **E3** microcrystal ($115 \times 60 \times 2\ \mu m^3$) with the left end fixed to an adjacent crystal and the remaining portion free. The top surface of the plate-like microcrystal was identified as the (1 0 1) face with its longitudinal direction along the *b*-axis, based on comparisons with bulk crystals having a hexagonal surface, which exhibit a (1 0 1) face along the *b*-axis. When the (1 0 1) face of the microcrystal was irradiated from the diagonal underside at 365 nm (10 mW/cm^2) for 1 s, the crystal curled from the right upper corner toward the light, reaching a maximum twisted curl after 2 s with a color change from pale yellow to red due to the formation of the closed **C3** isomer (Fig. 2.18a–c).

When the (1 0 1) face of a narrow, plate-like microcrystal ($109 \times 6 \times 2\ \mu m^3$) was irradiated at 365 nm from the lower side, the crystal bent toward the light, reaching a maximum tip displacement angle of 9° after 2 s (Fig. 2.19a, b). Subsequent illumination with a halogen lamp equipped with a filter (>390 nm, 10 mW/cm^2) returned the crystal to its initial straight shape after 30 s (Fig. 2.19a). This reversible bending was observed over 200 cycles of alternating irradiation with UV (2 s) and visible light (30 s) (Fig. 2.19c). The bending motion was accompanied by a color change from pale yellow to red ($\lambda_{max} = 512$ nm) due to the formation of the closed **C3** isomer in the crystals. The red crystal also returned to the initial pale-yellow color due to the photochemical ring-opening reaction, generating the **E3** isomer.

The bending effect was ascribed to a gradient in the extent of electrocyclic ring closure as a function of light penetration, such that shrinkage of the irradiated

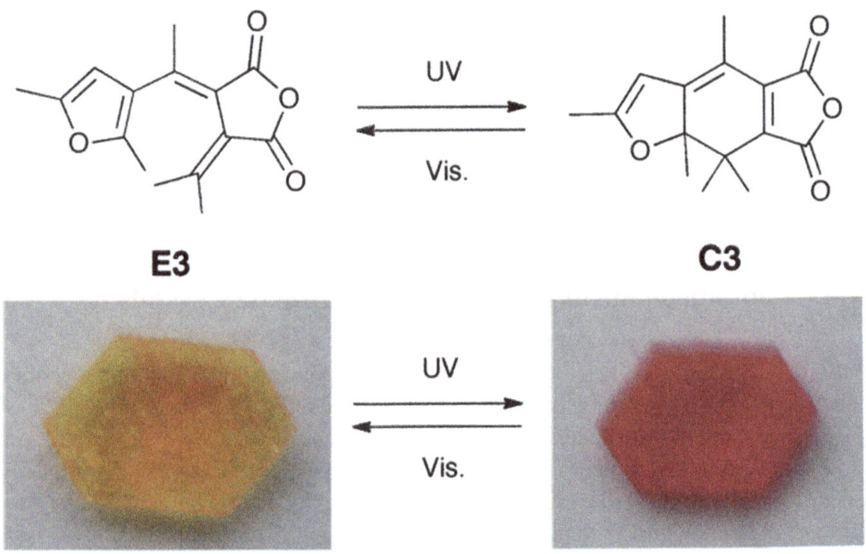

Fig. 2.17 Photochromic reaction of furylfulgide **3**

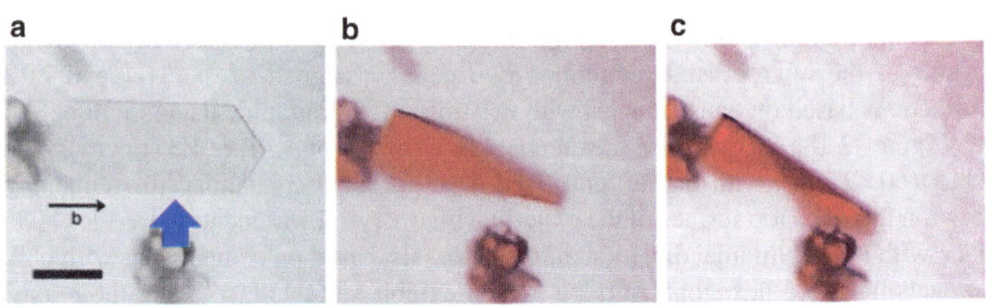

Fig. 2.18 The (1 0 1) face of a plate-like **E3** microcrystal before (**a**) and after UV irradiation for (**b**) 1 s and (**c**) 2 s. The scale bar is 50 μm

crystal surface along the *b*-axis resulted in a bent macrostructure. In the **E3** crystal, the **E3** molecules exhibit a torsional conformation with a dihedral angle of 38.3° between the succinic anhydride portion and the furan ring (Fig. 2.20a) and are arranged at the (1 0 1) face in a twofold screw along the *b*-axis [45, 46]. Upon UV irradiation, the torsional **E3** molecules underwent electrocyclic ring closure to **C3** molecules at the (1 0 1) crystal surface. The crystalline **C3** isomer is nearly planar with a dihedral angle of 13.8° between the succinic anhydride portion and the furan rings (Fig. 2.20b) [47]. Therefore, the ring-closure photoisomerization from the torsional **E3** isomer to the nearly planar **C3** isomer shrank the length of the *b*-axis of the unit cell. By contrast, because photoisomerization does not occur in the absence of light, the unit cell dimensions remained constant at the non-illuminated surface, causing the microcrystals to bend toward the light source.

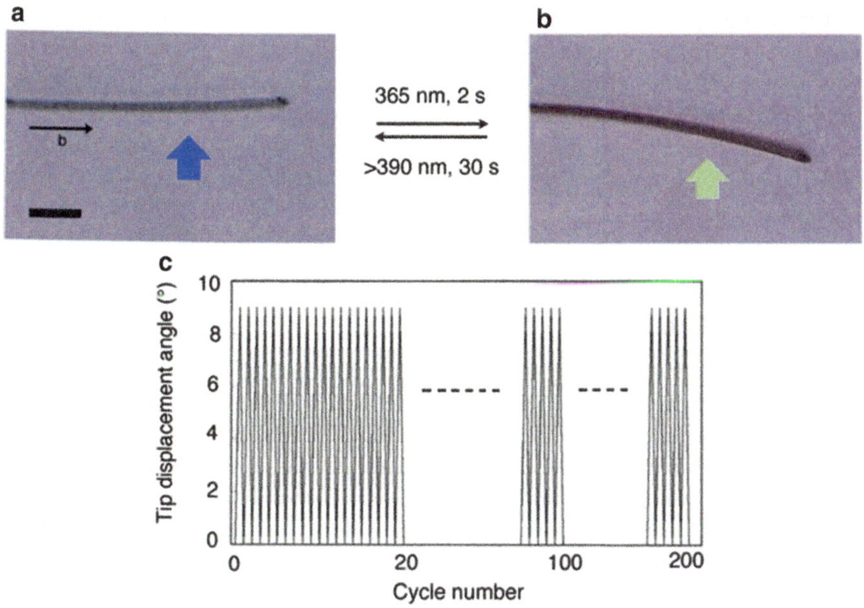

Fig. 2.19 Bending of a narrow, plate-like crystal of **E3** (**a**) before and (**b**) after UV irradiation from the lower side. Scale bar is 20 μm. (**c**) Reversible bending was repeatable over as many as 200 cycles (from [31] with permission. © 2011 Chemical Society of Japan)

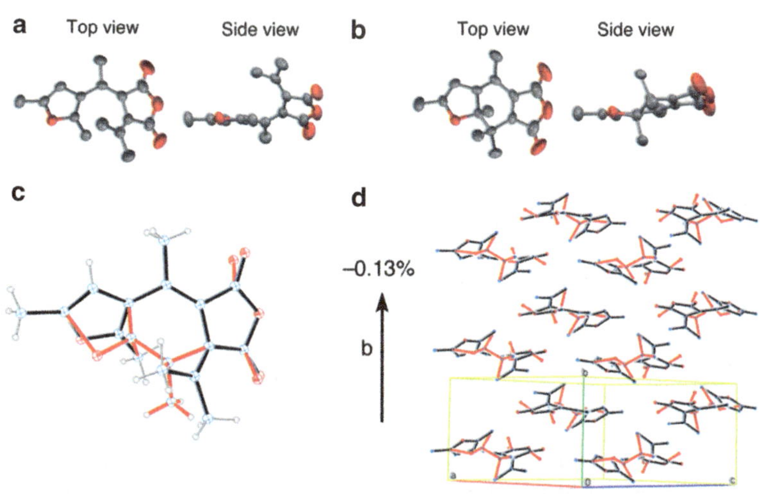

Fig. 2.20 Molecular structures of (**a**) the **E3** isomer and (**b**) the **C3** isomer in crystals of **E3** and **C3**, respectively. (**c**) ORTEP drawings show the disordered structures of the **E3** (*black*) and the **C3** (*red*) isomers after irradiation at 742 nm at the 25% probability level. (**d**) Molecular arrangement on the (1 0 1) face. Hydrogen atoms are omitted for clarity

Crystallographic analysis of **E3** after two-photon excitation at 742 nm by pulsed laser light provides more direct evidence for the mechanism of photomechanical bending [48]. The resulting crystal exhibited a disordered structure with contributions from both the **E3** and the **C3** isomers at a ratio of 0.945:0.055 (Fig. 2.20c). The length of the *b*-axis of the unit cell decreased 0.13% from 9.6203(5) to 7.6101(8) Å after two-photon excitation (Fig. 2.20d), bending the plate-like microcrystal toward the light source.

2.5 Conclusions

Several typical photochromic crystals composed of azobenzene, salicylideneaniline, and furylfulgide exhibit mechanical bending upon photoirradiation. The bend motion is reversible and repeatable by tuning the illuminating wavelength. The mechanism of bending in salicylideneaniline and furylfulgide crystals was elucidated by X-ray crystallographic analyses before and after photoirradiation. Molecular-level structural changes resulted in macroscale mechanical motion of the bulk crystals. Synthetic molecular machines have attracted a great deal of interest from the perspective of basic research and offer a wide range of potential applications. Further studies are necessary for the development of molecular machinery.

Acknowledgments This work was supported by the Grant-in-Aid for Scientific Research on Priority Area "New Frontiers in Photochromism" (471) of the Ministry of Education, Culture, Sports, Science and Technology and the Grant-in-Aid for Scientific Research (B) (22350063) of the Japan Society for the Promotion of Sciences.

References

1. Desiraju GR (1987) Organic solid state chemistry. Elsevier, Amsterdam
2. Ohashi Y (1993) Reactivity in molecular crystal. Kodansha, VCH, Tokyo
3. Koshima H (2002) In: Toda F (ed) Organic solid-state reactions. Kluwer, Dordrecht, pp 189–268
4. Koshima H (2004) In: Inoue Y, Ramamurthy V (eds) Chiral photochemistry. Dekker, New York, pp 485–531
5. Balzani A, Credi A, Raymo FM, Stoddart JF (2000) Artificial molecular machines. Angew Chem Int Ed 39:3348–3391
6. Kay ER, Leigh DA, Zerbetto F (2007) Synthetic molecular motors and mechanical machines. Angew Chem Int Ed 46:72–191
7. Irie M (1990) Photoresponsive polymers. Adv Polym Sci 94:27–67
8. Yu Y, Nakano M, Ikeda T (2003) Directed bending of a polymer film by light. Nature 425:145
9. Yamada M, Kondo M, Mamiya J, Yu Y, Kinoshita M, Barrett CJ, Ikeda T (2008) Photomobile polymer materials: towards light-driven plastic motors. Angew Chem Int Ed 47:4986–4988
10. Yamada M, Kondo M, Miyasato R, Naka Y, Mamiya J, Kinoshita M, Shishido A, Yu Y, Barrett CJ, Ikeda T (2009) Photomobile polymer materials—various three-dimensional movements. J Mater Chem 19:60–62

11. Yoshino T, Kondo M, Mamiya J, Kinoshita M, Yu Y, Ikeda T (2009) Three-dimensional photomobility of crosslinked azobenzene liquid-crystalline polymer fibers. Adv Mater 22:1361–1363
12. Naka Y, Mamiya J, Shishido A, Washio M, Ikeda T (2011) Direct fabrication of photomobile polymer materials with an adhesive-free bilayer structure by electron-beam irradiation. J Mater Chem 21:1681–1683
13. White TJ, Tabiryan NV, Serak SV, Hrozhyk UA, Tondiglia VP, Koerner H, Vaia RA, Bunning TJ (2008) A high frequency photodriven polymer oscillator. Soft Matter 4:1796–1798
14. Kobatake S, Takami S, Muto H, Ishikawa T, Irie M (2007) Rapid and reversible shape changes of molecular crystals on photoirradiation. Nature 446:778–781
15. Irie M (2008) Photochromism and molecular mechanical devices. Bull Chem Soc Jpn 81:917–920
16. Kuroki L, Takami S, Yoza K, Morimoto M, Irie M (2010) Photoinduced shape changes of diarylethene single crystals: correlation between shape changes and molecular packing. Photochem Photobiol Sci 9:221–225
17. Morimoto M, Irie M (2010) A diarylethene cocrystal that converts light into mechanical work. J Am Chem Soc 132:14172–14178
18. Terao F, Morimoto M, Irie M (2012) Light-driven molecular-crystal actuators: rapid and reversible bending of rodlike mixed crystals of diarylethene derivatives. Angew Chem Int Ed 51:901–904
19. Al-Kaysi RO, Müller AM, Bardeen CJ (2006) Photochemically driven shape changes of crystalline organic nanorods. J Am Chem Soc 128:15938–15939
20. Al-Kaysi RO, Bardeen CJ (2007) Reversible photoinduced shape changes of crystalline organic nanorods. Adv Mater 19:1276–1280
21. Good JT, Burdett JJ, Bardeen CJ (2009) Using two-photon excitation to control bending motions in molecular-crystal nanorods. Small 5:2902–2909
22. Zhu L, Agarwal A, Kai J, Al-Kaysi RO, Tham FS, Ghaddar T, Mueller L, Bardeen CJ (2011) Solid-state photochemical and photomechanical properties of molecular crystals nanorods composed of anthracene ester derivatives. J Mater Chem 21:6258–6268
23. Zhu L, Al-Kaysi RO, Bardeen CJ (2011) Reversible photoinduced twisting of molecular crystal microribbons. J Am Chem Soc 133:12569–12575
24. Uchida K, Sukata, Matsuzawa Y, Akazawa M, de Jong JJD, Katsonis N, Kojim Y, Nakamura S, Areephong J, Meetsma A, Feringa BL (2008) Photoresponsive rolling and bending of thin crystals of chiral diarylethenes. Chem Commun 326–328
25. Kobatake S, Hasegawa H, Miyamura K (2011) High-convertible photochromism of a diarylethene single crystal accompanying the crystal shape deformation. Cryst Growth Des 11:1223–1229
26. Naumov P, Kowalik J, Solntsev KM, Baldridge A, Moon JS, Kranz C, Tolbert LM (2010) Topochemistry and photomechanical effects in crystals of green fluorescent protein-like chromophores: effects of hydrogen bonding and crystal packing. J Am Chem Soc 132:5845–5857
27. Garcia-Garibay MA (2007) Molecular crystals on the move: from single-crystal-to-single-crystal photoreactions to molecular machinery. Angew Chem Int Ed 46:8945–8947
28. Koshima H, Ojima N, Uchimoto H (2009) Mechanical motion of azobenzene crystals upon photoirradiation. J Am Chem Soc 131:6890–6891
29. Koshima H, Ojima N (2012) Photomechanical bending of 4-aminoazobenzene crystals. Dyes Pigm 92:798–801
30. Koshima H, Takechi K, Uchimoto H, Shiro M, Hashizume D (2011) Photomechanical bending of salicylideneaniline crystals. Chem Commun 47:11423–11425
31. Koshima H, Nakaya H, Uchimoto H, Ojima N (2011) Photomechanical motion of furylfulgide crystals. Chem Lett 41:107–109
32. Whitaker A (1992) Crystal and molecular structure of C.I. Solvent Yellow 2, 1-phenylazo-4-(N,N-dimethylamino)benzene. J Cryst Spec Res 22:151–155
33. Tait KM, Parkinson JA, Bates SP, Ebenezer WJ, Jones AC (2003) The novel use of NMR spectroscopy with in situ laser irradiation to study azo photoisomerization. J Photochem Photobiol A 154:179–188

34. Mostad A, Rømming C (1971) A refinement of the crystal structure of cis-azobenzene. Acta Chem Scand 25:3561–3568
35. Nishimura N, Sueyoshi T, Yamanaka H, Imai E, Yamamoto S, Hasegawa S (1976) Thermal cis-to-trans isomerization of substituted azobenzenes II. Substituent and solvent effects. Bull Chem Soc Jpn 49:1381–1387
36. Kawato T, Koyama H, Kanatomi H, Isshiki M (1985) Photoisomerization and thermoisomerization I: unusual photochromism of N-(3,5-di-tert-butyl-salicylidene)anilines. J Photochem 28:103–110
37. Amimoto K, Kawato T (2005) Photochromism of organic compounds in the crystal state. J Photochem Photobiol C 6:207–226
38. Hadjoudis E, Vittorakis M, Moustakali-Mavridis I (1987) Photochromism and thermochromism of Schiff bases in the solid state and in rigid glasses. Tetrahedron 43:1345–1360
39. Harada H, Uekusa H, Ohashi Y (1999) X-ray analysis of structural changes in photochromic salicylideneaniline crystals. Solid-state reaction induced by two-photon excitation. J Am Chem Soc 121:5809–5810
40. Johmoto K, Sekine A, Uekusa H, Ohashi Y (2009) Elongated lifetime of unstable species by intermolecular hydrogen bond formation in photochromic crystals. Bull Chem Soc Jpn 82:50–57
41. Harada J, Ogawa K, Tomoda S (1997) Molecular motion and conformational interconversion of azobenzenes in crystals as studied by X-ray diffraction. Acta Cryst B 53:662–672
42. Harada J, Ogawa K (2001) Invisible but common motion in organic crystals: a pedal motion in stilbenes and azobenzenes. J Am Chem Soc 123:10884–10888
43. Harada J, Ogawa K (2009) Pedal motion in crystals. Chem Soc Rev 38:2244–2252
44. Yokoyama Y (2000) Fulgides for memories and switches. Chem Rev 100:1717–1739
45. Yoshioka Y, Tanaka T, Sawada M, Irie M (1989) Molecular and crystal structures of E- and Z-isomers of 2,5-dimethyl-3-furylethylidene(isopropylidene)succinic anhydride. Chem Lett 19–22
46. Ulrich K, Port H (1990) Optical investigations of the valence isomerizations of fulgides. J Mol Struct 218:45–50
47. Khedhiri L, Corval A, Casalegno R, Rzaigui M (2004) Molecular conformation an structural changes in crystalline photochromism of 3-furylfulgide. J Phys Chem A 108:7473–7478
48. Harada J, Nakajima R, Ogawa K (2008) X-ray diffraction analysis of photochromic reaction of fulgides: crystalline state reaction induced by two-photon excitation. J Am Chem Soc 130:7085–7091

Chapter 3
Photoinduced Reversible Topographical Changes on Diarylethene Microcrystalline Surface with Wetting Property

Kingo Uchida

Abstract Reversible topographical changes are observed on a photochromic diarylethene microcrystalline film surface by alternate irradiation with UV and visible light. The crystals of closed- or open-ring isomers of the diarylethenes were grown from the eutectic mixtures of both isomers by UV or visible light irradiations. The photogenerated crystals form either needle shape or cubic shape. The former surfaces show a superhydrophobic wetting property, which is classified as a lotus or petal effect of wetting depending on the size of the needle-shaped crystals and fractality of the surfaces, while the latter does not show such superhydrophobicity. We further developed a method to fabricate the petal surface by controlling the surface preparation condition; the surface shows a lotus effect without the control. In addition, reversible epitaxial crystal growth was found on a 1 1 0 surface of a strontium titanate single crystal whose lattice constant is similar to that of the diarylethene crystal. These reversible topographical changes are typical phenomena for diarylethene derivatives which have specific photochromic characteristics, i.e., thermal stability of both isomers and photoreactivity in the crystalline state.

Keywords Lotus effect • Petal effect • Superhydrophobicity • Topographical change • Wettability

3.1 Introduction

Diarylethenes are well-known photochromic compounds which undergo a cyclization reaction to form colored closed-ring isomers by UV irradiation and revert to colorless open-ring isomers by visible light irradiation. Their photochromism has

K. Uchida (✉)
Department of Materials Chemistry, Ryukoku University,
Seta, Otsu, Shiga 520-2194, Japan
e-mail: uchida@rins.ryukoku.ac.jp

M. Irie et al. (eds.), *New Frontiers in Photochromism*,
DOI 10.1007/978-4-431-54291-9_3, © Springer Japan 2013

excellent properties: thermal stability of both isomers, a high fatigue resistance, rapid response, high quantum yield, and reactivity even in the crystalline state [1].

On the other hand, the polarity changes during the photochromism of diarylethenes are much smaller compared to those of azobenzene and spiropyran derivatives. Consequently, those two derivatives have been used for photocontrol of the changes in surface wettability. For example, Garcia et al. reported the contact angle (CA) changes of a water droplet on a spiropyran-coated surface [2]. The CA before UV irradiation was 74°, but it dropped to 54° after UV irradiation. This is due to the enhanced polarity by photoisomerization, which converted nonpolar spiropyran to a polar merocyanine isomer. Such wettability changes have been extensively studied not only in organic materials but also in inorganic materials.

Another approach to controlling surface wettability is to change the surface morphology. It is known that increasing surface roughness results in a superhydrophobic surface, an effect that is widely observed in natural plants such as lotus leaves [3]. Such micrometer-scale rugged [4] or fractal [5] structures have also been artificially prepared to make superhydrophobic surfaces. In 2006, we accidentally discovered reversible crystal growth of a diarylethene derivative, and this was accompanied by a reversible wetting property [6]. The finding became a breakthrough into a new surface science.

3.2 Photoinduced Reversible Topographical Changes on Diarylethene Surface

The topographical changes of diarylethene crystals were first reported by Irie et al. [7] Upon UV irradiation of a single crystalline surface, micrometer-sized steps and valleys were formed, and these then disappeared upon visible light irradiation. These phenomena were explained by the size difference between the open-ring and closed-ring isomers. However, the wettability changes were not studied in this system.

New type of photoinduced changes in morphology of a photochromic diarylethene crystalline thin film of 1,2-bis(2-methoxy-5-trimethylsilylthien-3-yl)perfluorocyclopentene (**1o**) as shown in Fig. 3.1 are appeared, and it is accompanied with wettability changes [6]. Upon UV irradiation (254 nm), the color become deep blue within 5 min, yet no change in surface morphology is observed at this moment at room temperature. After irradiation for 10 min, the crystal is stored in the dark, when small microfibrils immediately start to grow. After storage for 24 h, the surface is covered with many microfibrils as shown in Fig. 3.1b. Upon irradiation of the surface with visible light ($\lambda > 500$ nm), the blue color and the microfibrils disappear (Fig. 3.1c). Topographical changes are also observed on the surface of microcrystalline film prepared by coating the chloroform solution of **1o** on a glass substrate [6].

Measurement of the CA of a water droplet is a simple and direct method for evaluating wettability on a surface. The CA of the film surface before UV irradiation is 120° (Fig. 3.2a). Upon irradiation of the surface with UV light (254 nm), the

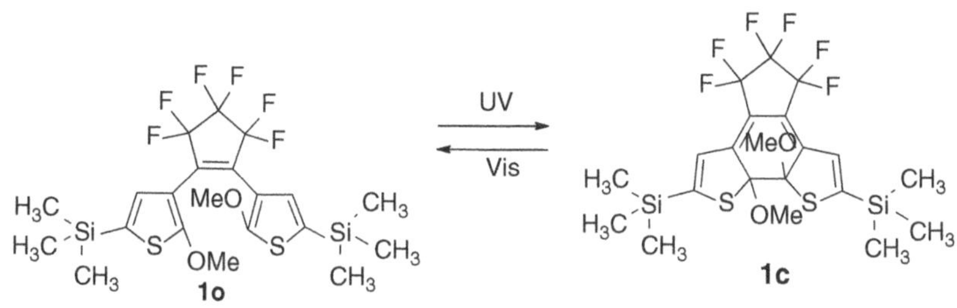

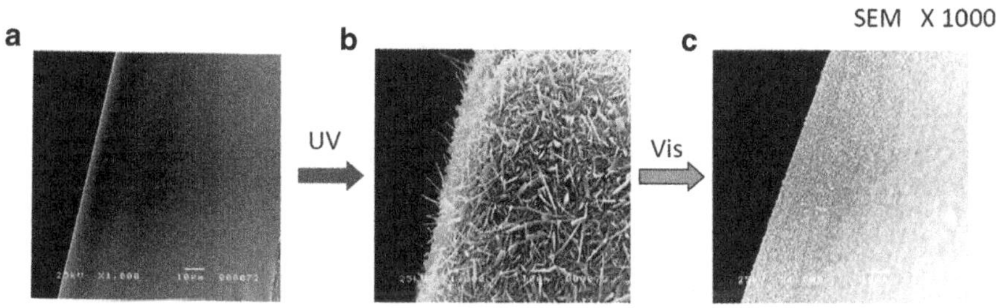

Fig. 3.1 Photoinduced topographical changes on a single crystal of **1o**. (**a**) SEM image of a crystal surface of **1o** from a side view before UV irradiation. (**b**) After UV (254 nm, 12 W) irradiation for 10 min followed by storing for 24 h at 30 °C in the dark. (**c**) SEM image from a side view of the surface after irradiation with visible light ($\lambda > 500$ nm, 500 W, 20 min) and storage in the dark for 24 h

Fig. 3.2 Contact angles (CAs) of a water droplet on the microcrystalline surface of diarylethene **1o** prepared on glass substrate by solution coating. (**a**) Water droplet (1.2 mm Ø) on the microcrystalline surface before irradiation with UV light (254 nm, 12 W). The contact angle was 120°. (**b**) Water droplet after UV irradiation for a short time (1–2 min). The contact angle changed to 92°. (**c**) Water droplet after UV irradiation (10 min) and stored in the dark at room temperature for 24 h. The contact angle increased to 163°

color becomes blue and the angle decreases to around 90° (Fig. 3.2b) after a short irradiation time (1–2 min). After 24 h, the angle gradually increases to 163° (Fig. 3.2c). The colored surface is irradiated with visible light to regenerate the open-ring isomer. After 20 min the blue color disappeared. The CA is also monitored during the fibril-disappearing process. It gradually decreased and finally recovered to the initial 120° [6].

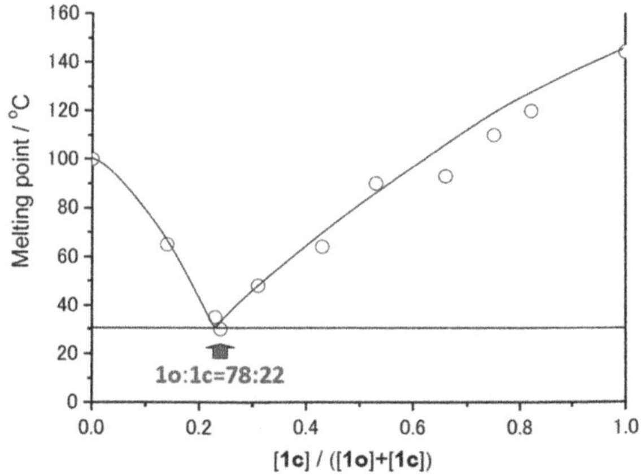

Fig. 3.3 Phase diagram of mixtures of open- (**1o**) and closed-ring (**1c**) isomers formed by irradiation of diarylethene crystals with UV light (313 nm)

Powder X-ray diffraction measurements are carried out to elucidate the composition of the microfibrils, and the results show that the XRD pattern of the fibrils is in good agreement with that of the closed-ring isomer, whose pattern is estimated from single-crystal structure analysis of **1c**. Ordinarily, closed-ring isomers formed in the crystal of open-ring isomers are strained by the lattice and have different structures from closed-ring isomer crystals obtained by recrystallization in solution [8]. The above results indicate that the fibril crystal formed on the film grew freely from the crystal lattice of **1o** [6].

The forming and disappearing mechanism of the needle-shaped crystals of **1c** by alternate irradiation with UV and visible light can be explained using a phase diagram (Fig. 3.3) [6].

Melting points of **1o** and **1c** were 100 °C and 140 °C, respectively. Furthermore, the shape of **1o** is cubic, while that of **1c** is a needle or fibril when obtained by recrystallization from hexane solution of these isomers. The needle-shaped crystals tend to grow in one direction along the *a* axis of the crystal [6]. Upon UV irradiation, **1o** is converted to **1c**. When the conversion exceeds 24 %, crystal growth of **1c** starts to form fibril structures on the film surface. To confirm this mechanism, samples were kept in air and water at different temperatures. Below around 0 °C the fibril does not grow, while fibril formation is observed in water and in air around 30 °C. These results also support the melt mechanism as depicted in the phase diagram. Fractal analysis is made by the box-counting method for the photogenerated rough surface, and this revealed that the surface is also fractal [9, 10]. The size of the roughness can be classified into three regions based on the fractal dimension: ca. 2 (roughness size smaller than 0.5 μm), 2.5 (roughness size of 0.5–20 μm), and ca. 2.0 (roughness size larger than 20 μm). The fractal dimension of ca. 2.5 was due to the fibril-like structures generated gradually by UV irradiation on diarylethene surfaces accompanied by an increase in the contact angle. The surface structure with larger fractal dimension mainly contributes to realizing the extremely high water repellency of the diarylethene surfaces. This mechanism of spontaneous formation of fractal surfaces is similar to that for triglyceride and alkyl ketene dimer (AKD) waxes [5]. Furthermore, the needle shape is important for enhancing the CA of the surface, since diarylethene derivatives without needle-shaped crystals does not show remarkable CA enhancement from crystal growth on the surface [11].

Photoinduced fibril growth of **1c** is not only observed on glass and metal substrates but also on plastic and paper. Therefore, the compound is applicable to materials used in the photocontrol of surface-water repellency.

3.3 Photoinduced Lotus and Petal Effects of Wetting

A surface whose CA of water droplets exceeds 150° is classified as a superhydrophobic surface. Actually, many plants and insects have such superhydrophobic surfaces. For example, the leaves of the plume poppy (*Macleaya cordata*) and of the lotus are covered with micrometer-sized rods or protrusions that cause water droplets to roll over the surface [3]. In particular, lotus leaves are well known for their superhydrophobic surface, where the contact angle and sliding angle (SA) of a water droplet are 161.0±2.7° and 2.0°, respectively (lotus effect) [3]. Rose petals also show super water repellency with a CA of about 152.4°; however, these droplets are pinned to the surface and cannot roll off, even when the petal is turned upside down. This phenomenon is the "pinned effect" or "petal effect" [12]. Such surfaces have attracted great interest for both basic research and practical applications in recent years. To explain the effects of surface topography on the enhancement of hydrophobicity, Wenzel [13] and Cassie and Baxter [14] have proposed models showing the roles of both the surface area and the air trapped in concavities on the surface. Jiang et al. noted how the lotus effect represents the Cassie state [12], while the petal effect represents the Cassie impregnating wetting state [12]. These two states are differentiated by the microstructures of their surfaces as shown in Fig. 3.4.

A Cassie state can be achieved in theory by reducing the microstructure scale. This type of surface is mimicked by using diarylethene, and the reversible wettability changes on the microcrystal surface of photochromic diarylethene **1o** is monitored as mentioned in the previous section. The CA of a water droplet on a surface covered with needle-shaped crystals of **1c** is 163°, and the SA of the surface is less than 2°. These angles are the same as those on a lotus leaf. Then, reproducibility of the selective formation of surfaces that show both a lotus effect and a petal effect by using a single diarylethene derivative is examined (Fig. 3.5) [10, 15].

A hint on how to produce a microcrystalline surface with the petal effect was given from experiments of monitoring the surfaces stored at different temperatures by SEM and CA. After UV irradiation for 10 min, the microcrystalline surface of **1o** is stored in the dark for 48 h at four different temperatures.

The surface shows superhydrophobicity after storage for 24 h at 30 °C in the dark, owing to the formation of needle-shaped crystals on it. The CA profiles of a water droplet at 50 °C, 70 °C, and 90 °C are also summarized in Fig. 3.6. The purple dotted line indicates the border of superhydrophobicity (CA > 150°). Before UV irradiation, the CA of the rough surface of **1o** is nearly 120°. During 10 min of UV

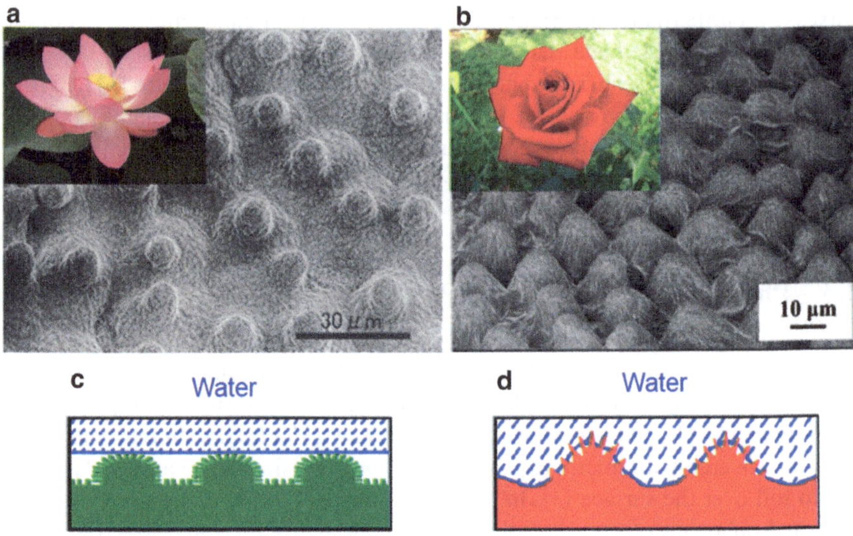

Fig. 3.4 SEM images showing the surface structures of (a) lotus leaves and (b) rose petals and schematic illustrations showing the difference in the wettability of a water droplet on the surface of (c) a lotus leaf (Cassie state) and (d) a rose petal (Cassie impregnating wetting state) [12]

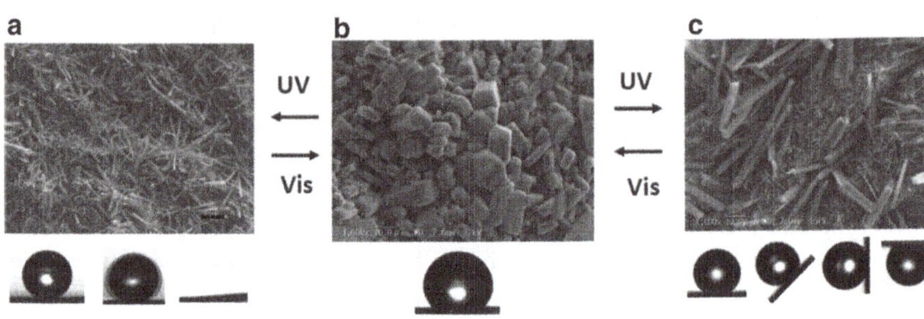

Fig. 3.5 Lotus leaf and rose petal wetting surfaces are designed and repeatedly formed on a microcrystalline photochromic diarylethene film by a sequence of photocontrol. (a) The photogenerated lotus state prepared by storing at 30 °C for 24 h in the dark. CA and SA are 163° and 2°, respectively. (b) The surface covered with cubic crystals of **1o**. (c) The photogenerated petal state. CA of the surface is 154° but a water droplet is pinned on the surface. The difference between the surfaces is characterized by the size distribution of the microcrystals. Both rough surfaces revert to the surface covered with cubic crystals of **1o** upon visible light ($\lambda > 500$ nm) irradiation (scale bars: 20 μm for all images, which are magnified 1,000×)

irradiation, the CA rapidly decreases to 92° (after an initial 1–2 min of UV irradiation) owing to the formation of a flat surface on the eutectic mixture [6]. The CA gradually increases to 150° after 6 h and finally reached 163° after 24 h of storage at 30 °C in the dark (Fig. 3.6, black squares). The surface also shows a very low SA of less than 2°. The water droplets never stay on the surface when it is slightly tilted, and the CA and SA values are the same as those on the lotus leaves. The CA of the

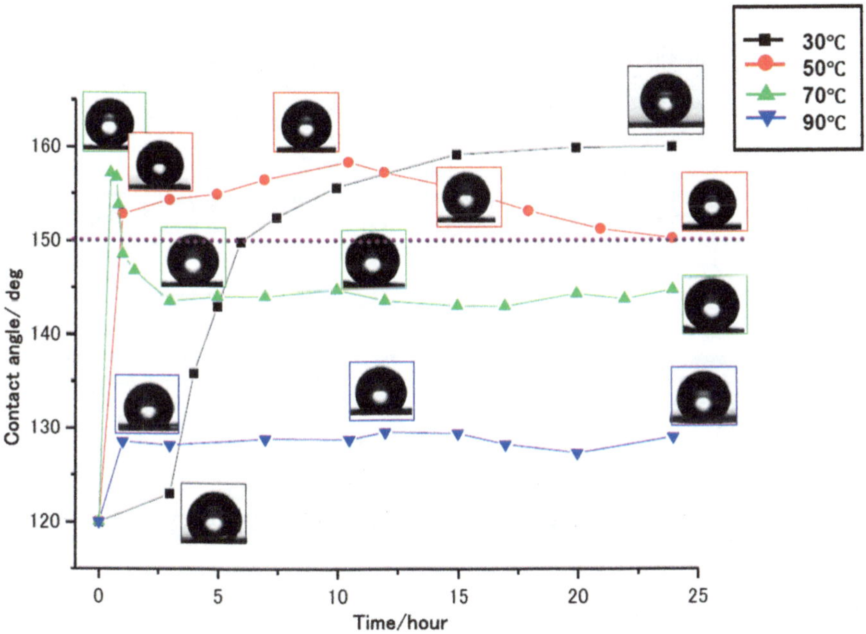

Fig. 3.6 CA time profiles at different storage temperatures (in the dark)

film stored at 50 °C rises very rapidly and reached 153° after only 1 h (Fig. 3.6, red circles).

The rough surface also shows the lotus effect with an SA of 7.5°. After 10.5 h, the CA reaches a maximum of 158° and starts to decrease gradually. The CA of the film stored at 70 °C increases dramatically after UV irradiation for the initial 30 min and reaches 157°, immediately starts to decrease and reaches about 144° 3 h later, and remains almost constant over the subsequent 23 h (Fig. 3.6, green triangles). However, the CA of the films stored at 90 °C remains at an almost constant 130° (Fig. 3.6, blue inverse triangles).

To understand the temperature dependence of the changes in CA profile, each film is monitored by SEM at different storage temperatures (Fig. 3.7). The CAs shown are the average values over 21 measurements. Smooth, gradual growth of needle-shaped crystals is observed on the film surface at 30 °C (Fig. 3.7a–d), and larger rod-shaped crystals appear after 15 h at 50 °C (Fig. 3.7g).

At 70 °C, rod formation is observed within 3 h (Fig. 3.7j), and the population increases with time. On the surface stored at 90 °C, rod-shaped crystals are observed at the initial stage without the formation of any needles (Fig. 3.7m). After prolonged heating at 90 °C, the thermal cycloreversion from **1o** to **1c** proceeds, with melting of the crystals on the surface since the ratio is closed to that of the eutectic mixture of **1o** and **1c** (Fig. 3.7p).

Powder X-ray diffraction measurements confirm that both the rod- and needle-shaped crystals are those of **1c** obtained by recrystallization from an *n*-hexane

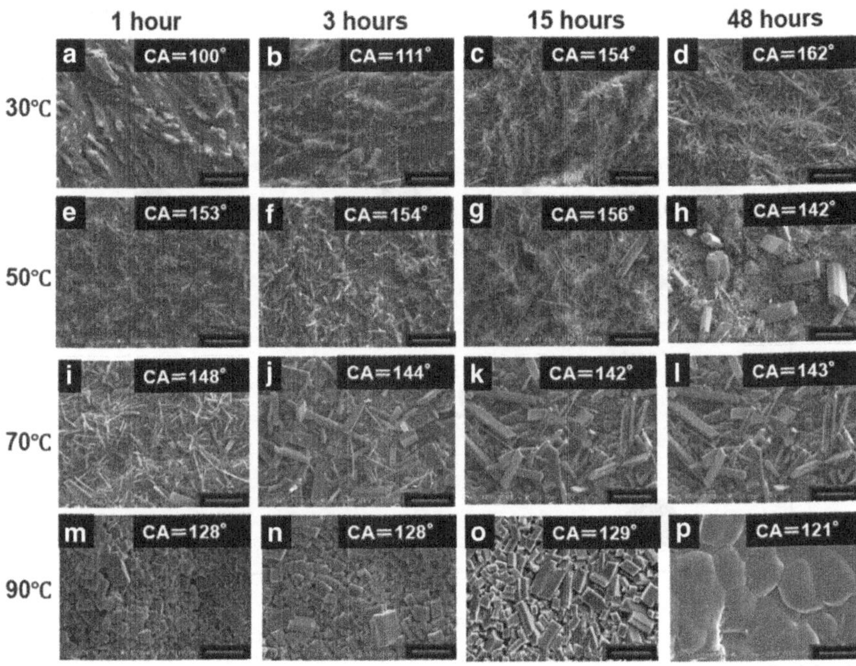

Fig. 3.7 SEM images (1,000×) of the surface of the diarylethene microcrystalline films stored at (**a**–**d**) 30, (**e**–**h**) 50, (**i**–**l**) 70, and (**m**–**p**) 90 °C in the dark after UV irradiation (scale bar: 20 μm)

solution. The samples show diffraction peaks at 5.1, 7.7, and 15.0 that are attributed to the (0 0 1), (0 1 0), and (1 −1 −1) lattice planes of the **1c** crystal [6]. These peaks are also observed for the fibril crystals formed during storage at 30 °C after UV irradiation. Therefore, the rod-shaped crystals formed at 70 °C have the same crystal structure as those of fibril crystals formed at 30 °C and the **1c** crystals obtained by recrystallization from the solution. The growth of rod-shaped crystals is observed simultaneously with the melting of the needle-shaped crystals at 50 °C. Consequently, the substitution mechanism of the needle-shaped crystals by the rod-shaped crystals is apparently Ostwald ripening [10].

The formations of needle- (Fig. 3.7d) and rod-shaped (Fig. 3.7h) crystals are observed at different temperatures and storage periods after UV irradiation. As mentioned above, the rough surface formed at 30 °C is a fascinating geometrical fractal structure characterized by self-similarity and non-integer dimensions [9]. Similar fractal surfaces have also been formed on several kinds of wax. For example, AKD forms fractal structures and provides super water repellency [5]. Preliminary research suggests that the spontaneous formation mechanism of fractal structures on the AKD surface originated in a phase transition from a metastable to a stable crystalline form [16]; this is followed by the blooming theory, which is proven by use of tristearin to involve the same mechanism [17]. The surface after UV irradiation is considered a metastable state because the photogenerated closed-ring isomer has different crystal packing compared to that of the open-ring isomer [8]. Diarylethene **1o** undergoes a photoinduced cyclization reaction to form **1c** upon UV light irradiation.

Because closed-ring isomer **1c**, formed in the crystal lattice of the open-ring isomer, has a distorted crystal packing rather than the original structure of the closed-ring isomer photogenerated in solution [8], the photogenerated **1c** is strained. By mixing the closed- and open-ring isomers, the melting point of the system is lowered, as shown in the phase diagram of **1o** and **1c** (Fig. 3.3). Therefore, the UV-light-irradiated surface melted initially, but this is followed by self aggregation and the crystal growth of **1c** that proceeded from the eutectic state.

The activation energy to form a fractal surface of an alkylketene dimer or a triglyceride has been obtained by measuring the periods required for the CA to reach 150° at different storage temperatures, followed by application of the Arrhenius equation. In this system, the CAs of a water droplet increase with the growth of the needle-shaped crystals and begin to decrease by replacing them with rod-shaped crystals by Ostwald ripening. From the periods required for the CA to reach 150° to obtain the activation energy for growing needle-shaped crystals as well as the periods in which CAs fell from 150° for growing rod-shaped crystals are measured, the activation energies for the formation of needle and rod-shaped crystals on the surface are 143 and 162 kJ/mol, respectively [10]. The activation energy for the needle-shaped crystal growth is 19 kJ/mol smaller than that for the rod-shaped crystal growth because needle-shaped crystals are formed at low temperature while rod-shaped crystals are formed at high temperature.

Tuning the size of the crystals by controlling the storage temperature enables for making the petal surface on the same material. Jiang et al. studied the surface structures of lotus leaves and rose petals and then attributed the different wetting behavior between these surfaces to the different designs in their hierarchical micro and nanostructures [12]. Because the microstructures of rose petals have larger pitches than those of lotus leaves, the liquid is allowed to penetrate their microstructure. Inspired by this analysis, attempts to make a photo-controllable system between artificial lotus and petal surfaces using photochromic diarylethene **1o** is performed. The procedure is schematically shown in Fig. 3.8.

In this fabrication, controlling the amount of roughness on the surface structures is crucial, since the lotus effect represents Cassie state, while the petal effect represents the Cassie impregnating wetting state [18]. In the latter state, the wetted surface area is smaller than that in the Wenzel state [13] but greater than that in the Cassie–Baxter state [14] (Fig. 3.9).

A surface displaying the lotus effect can be prepared by UV irradiation of the microcrystalline surface of **1** followed by storage at 30 °C for 24 h in the dark [6]. The procedure is illustrated in Fig. 3.8a–c. The CA and SA of a water droplet are 163° and less than 2°, respectively. Thus, the CA and SA of the surface in Fig 3.5a strongly resemble those of a water droplet on lotus leaves (i.e., the lotus effect). For preparation of a surface that displays the "petal effect," larger crystal sizes are desired in order to pin the water droplet. By storing the film shown in Fig. 3.8b at an elevated temperature, small fibrils sparsely appear, and then larger rod-shaped

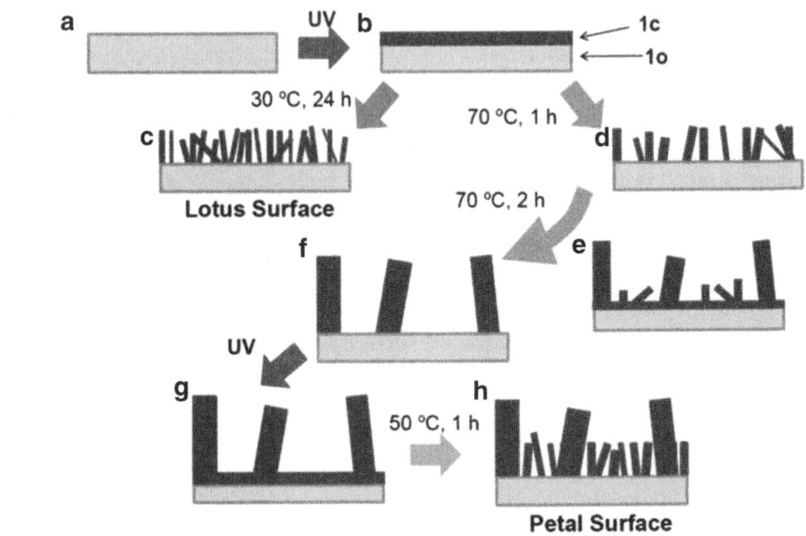

Fig. 3.8 Schematic depiction of the formation of surfaces that show lotus and petal effects. Upon UV irradiation of the surface of a film coated with diarylethene **1o** (**a**), closed-ring isomer **1c** (*dark blue layer*) is formed only on the irradiated side of the surface (**b**). After storage in the dark at 30 °C, the surface is covered with closely packed small fibrils and shows the lotus effect (**c**). After storage of the film shown in (**b**) for 1 h at 70 °C, small fibrils grow sparsely within 1 h (**d**) After additional heating in the dark, larger rod-shaped crystals grow while fine fibrils melt (**e**) and finally the surface is covered only with rod crystals (**f**). The surface is photoisomerized by a second UV light irradiation (**g**). After storing the film at 50 °C for 1 h in the dark, the surface is covered with rod crystals as well as newly formed fibrils (**h**), and the surface shows the petal effect

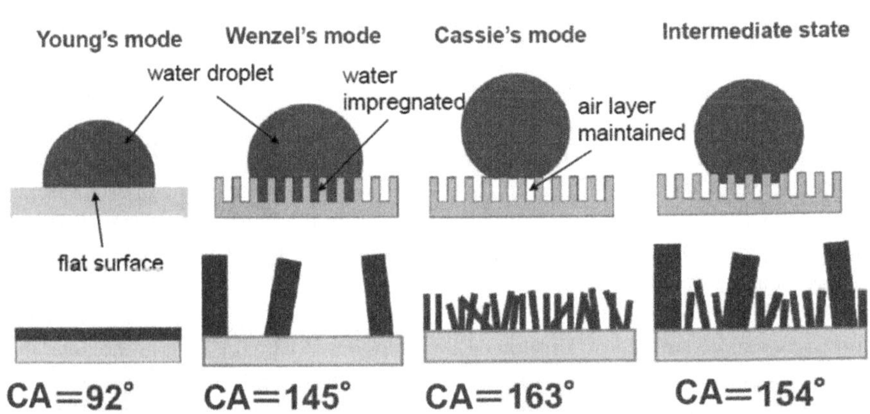

Fig. 3.9 Schematic illustration of different surface modes and corresponding microcrystalline surfaces of diarylethene **1**

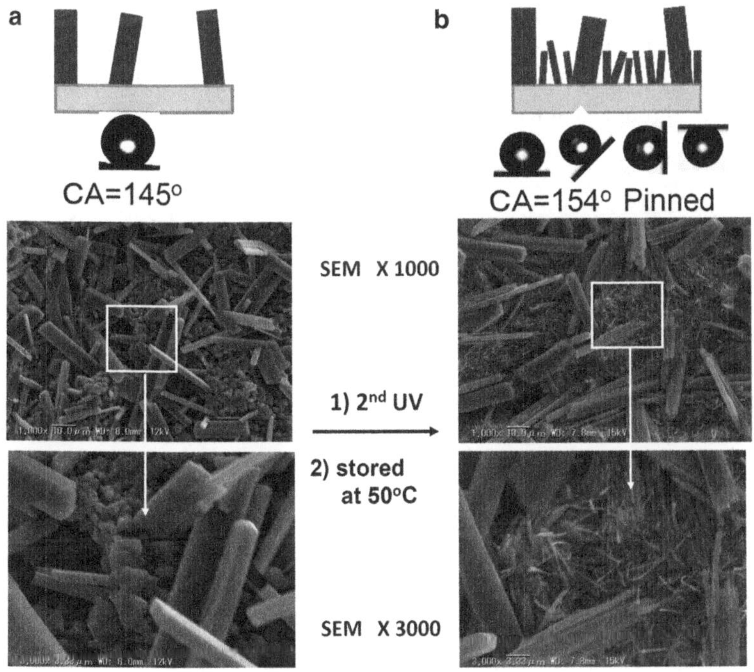

Fig. 3.10 SEM images of the microcrystalline surfaces of diarylethene **1o** with different wetting property. (**a**) Before the 2nd UV light irradiation. (**b**) After the 2nd UV light irradiation

crystals of **1c** are generated by Ostwald ripening (Fig. 3.8d–f). After the film is stored at 70 °C for 3 h in the dark, the surface is covered with rod-shaped crystals instead of fibrils. Then, the surface is irradiated with UV light for a second time, and cubic crystals of **1o** appearing at the surface (Figs. 3.8f and 3.10a) are converted to **1c** (Fig. 3.8g). By keeping the film at 50 °C for 1 h in the dark, the surface below the larger rod-shaped crystals is covered with newly formed fibrils (Figs. 3.5c, 3.8h, and 3.10b). The expanded SEM images before and after the second UV irradiation are summarized in Fig. 3.10.

Reversibility is a characteristic feature of photochromic compounds [1]. Diarylethene has a particularly excellent thermal stability for both isomers and, moreover, is fatigue resistant and reactive in the crystalline state. Therefore, this compound is a suitable material for such photoinduced topographical changes. With this material, the repetitive cycle, which shows the lotus surface followed by the petal surface, is examined. Initially, the film surface of **1o** is irradiated with UV light for 10 min, and the film is stored at 30 °C for 24 h in the dark to form a surface that showed the lotus effect (Fig. 3.8a–c). Upon visible light irradiation for 1 h at 30 °C, the needle-shaped crystals melted (Fig. 3.8c to a) and the initial surface, which is covered with cubic crystals attributable to **1o**, is regenerated. Then the recovered film is irradiated with UV light (10 min) and stored at 70 °C for 3 h in the dark to form the state shown in Fig. 3.8b, d–f. Then the surface is irradiated with UV light

(10 min) again (Fig. 3.8g) and stored at 50 °C for 1 h to form a surface with a petal effect (Fig. 3.8h). The surface is irradiated with visible light (1 h) again to regenerate the initial surface of **1o** (Fig. 3.8a). This cycle is repeated, and the lotus and petal effects are reproduced at least three times.

3.4 Photoinduced Reversible Epitaxial Crystal Growth

The photoinduced reversible growth of needle-shaped crystals of the closed-ring isomer was demonstrated in the previous section. This section describes the photoinduced reversible growth of needle-shaped crystals of the open-ring isomer. In an earlier work, a new diarylethene derivative with an asymmetric structure was prepared in order to decrease the melting point of both isomers and eutectic temperature [19]. Additionally, a surface whose wettability changes are expanded from hydrophilic, with CA of less than 90°, to superhydrophobic is desired. The melting point of **2o** is 95.2–95.8 °C which is lower than those of the parent symmetric compounds **3o** (119.0–119.5 °C) [20] and **4o** (168 °C) [21]. Asymmetric structure successfully reduces the eutectic points for ease of photocontrol of the surface. The photochromic reactions of **2o** are observed in a hexane solution, and the quantum yields of the cyclization and cycloreversion reactions of **2o** and **2c** in a hexane solution are 0.35 and 4.2×10^{-3}, respectively.

Although the compound shows polymorphism, only needle-shaped crystals are observed in the bulk film on the glass substrate. Recrystallization from the hexane solution gives platelet crystals, but the crystals from ether hexane mixtures are needles. The melting points of platelets and needles are 92.9–93.5 °C and 95.2–95.8 °C, respectively. Both crystals are monoclinic. In the crystalline states, the distances between the reactive carbon atoms of **2o** in the platelets and the needle-shaped crystals are 3.568 and 3.606 Å, respectively. It is well known that the cyclization reaction can proceed upon UV irradiation even in the crystalline state when the distance is less than 4 Å [22]. Therefore, the photocyclization can proceed in this system. To study the photoinduced topographical changes of the surface, the phase diagram of a mixture of **2o** and **2c** is prepared by DSC measurements of such mixtures with different components (Fig. 3.11).

The melting points of **2o** and **2c** are 95.2–95.8 °C and 134.5–135.0 °C, respectively. The eutectic point is 67 °C, where the ratio of **2o** and **2c** is 61:39.

The microcrystalline surface is prepared by coating the chloroform solution of **2o** onto a glass plate and subsequent evaporation of the solvent. Through SEM observation of the microcrystalline film, the surface is covered only with rod-shaped crystals. The SEM image of the surface is shown in Fig. 3.12a. The surface is covered with microcrystals of **2o**, and the CA of a water droplet on the surface is $150.4 \pm 0.3°$, showing a superhydrophobic property (Fig. 3.12d). Furthermore, a water droplet remains pinned on the surface even when the surface is turned over, showing the petal effect. Then the surface is irradiated with UV light (initially 10 min) and it becomes flat within 30 min as shown in Fig. 3.12b. The CA is drastically

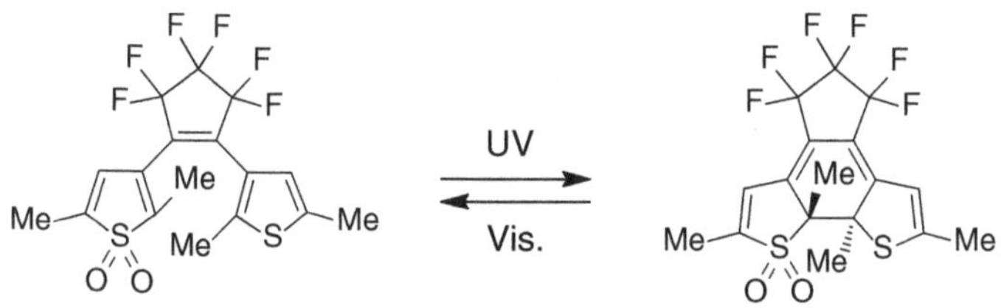

2o **2c**

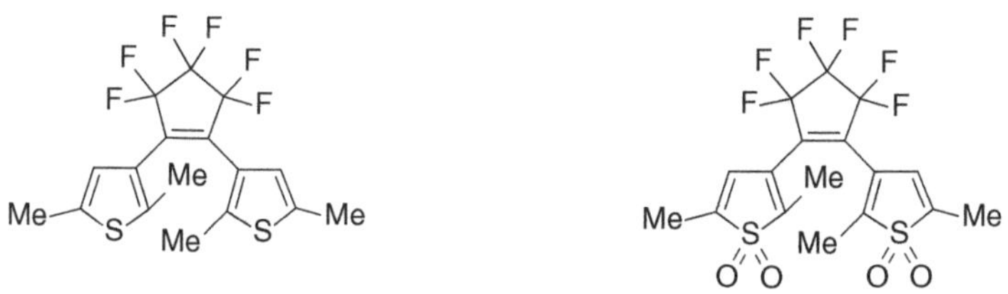

3o **4o**

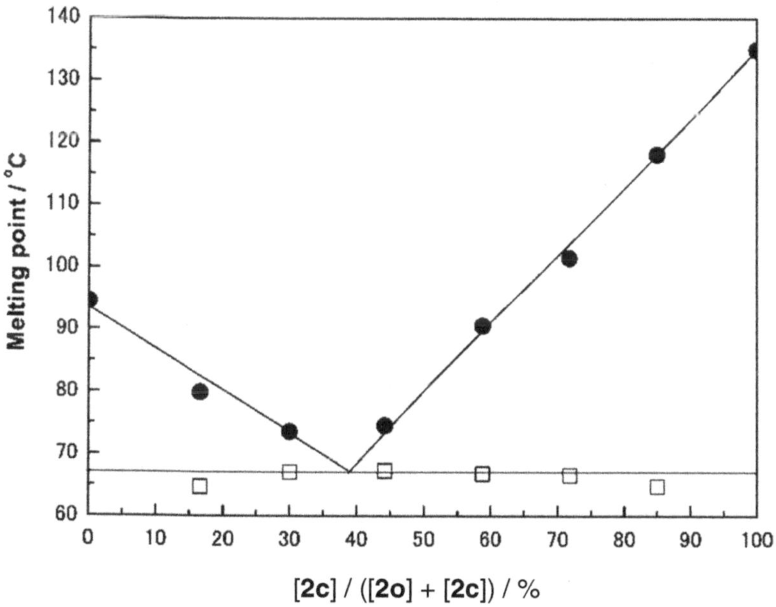

Fig. 3.11 Phase diagram of mixtures of open-ring (**2o**) and closed-ring (**2c**) isomers formed by UV light (366 nm) irradiation

reduced to 80.5 ± 0.4°, and a hydrophilic surface is generated as shown in Fig. 3.12e. Although the melting point of **2c** is high (134.5–135.0 °C), **2c** crystal does not grow on the surface because the conversion to **2c** is not high enough to provide eutectic content (39 %). By scratching the surface after UV irradiation and measuring the absorption spectra in the hexane, the content of **2c** is found to be 31 % after 10 min of UV irradiation. No remarkable increment of the content is observed by prolonged

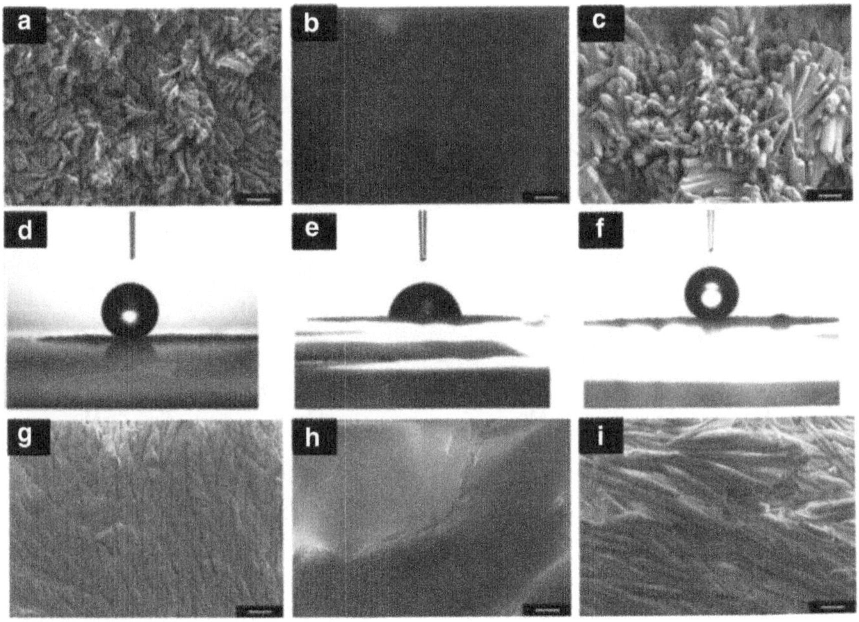

Fig. 3.12 SEM images of the reversible topographical changes in the microcrystalline surface of **2o** on a glass substrate (**a**–**c**) and STO (1 1 0) surface (**g**–**i**); water droplet shapes (**d**–**f**) on the surfaces of (**a**–**c**), respectively (scale bar: 10 μm)

UV irradiation of the surface. The flat surface reverts to a rough surface, accompanied by remarkable enhancement of the CA, by visible light irradiation within 1.5 h. During the visible light irradiation, the film is again maintained at the eutectic temperature. The SEM image of the surface is shown in Fig. 3.12c. The surface is covered with rod-shaped crystals whose diameters and lengths are around 2–5 and 20–30 μm, respectively. The CA of a water droplet is $150.9 \pm 0.6°$ (Fig. 3.12f). Alternate formations of superhydrophobic and hydrophilic surfaces are also performed as shown in Fig. 3.13 [19]. To understand the mechanism of the photoinduced crystal growth, the reversible crystal growing process in Fig. 3.13 is monitored by XRD, and the results are summarized in Fig. 3.14. The diffraction angles of the **2o** film (Fig. 3.14) correspond to those of the **2o** powder [19]. However, some of the diffraction intensities of the films are different from those of the **2o** powder, and the films are strongly oriented to the 3 0 −2 direction ($2\theta = 22.9°$).

In the previous section, needle-shaped crystals of a closed-ring isomer grew on the subphase of an open-ring isomer upon UV light irradiation; therefore, the crystals of the subphase had no effect on the growth of needle-shaped crystals [10, 15]. On the other hand, the current results show the growth of open-ring isomer crystals because the eutectic ratio is not achieved; only the crystallization of open-ring isomer **2o** is possible. Indeed, maintaining the eutectic temperature at 67 °C induces the rapid growth of the needle-shaped open-ring isomer crystals on the surface. The results demonstrate this rapid crystal growth and may explain the appearance of the

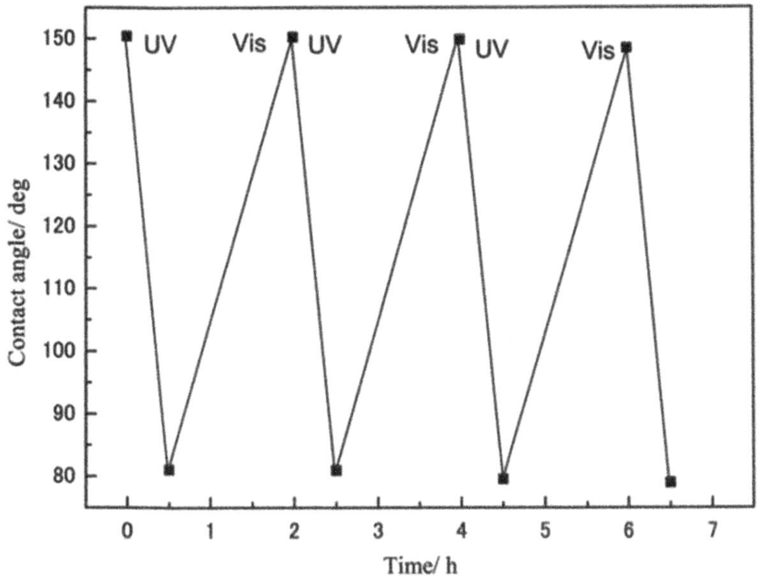

Fig. 3.13 Changes in contact angles on the surface of **2o** by alternate irradiation with UV ($\lambda = 366$ nm) and visible ($\lambda > 500$ nm) light (glass substrate)

Fig. 3.14 XRD pattern changes by alternate UV and visible light irradiation of microcrystalline surface of **2o** on glass substrate: (**a**) coated film before UV irradiation, (**b**) after UV irradiation at 67 °C for 30 min, (**c**) after next visible light irradiation while keeping at 67 °C for 1.5 h, (**d**) after second UV light irradiation for 10 min while keeping at 67 °C for 30 min

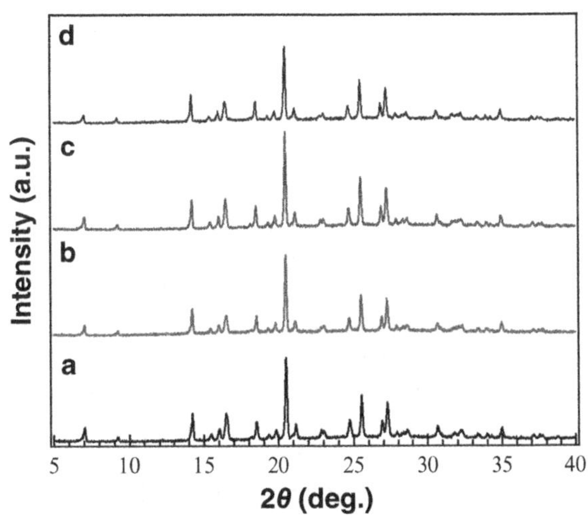

petal effect. Furthermore, this inspired to attempt epitaxial crystal growth of **2o** on a suitable substrate whose lattice parameter resembles that of **2o** [23]. For decades, the heteroepitaxial growth of organic materials has received much attention because of its importance and unique characteristics. Organic heteroepitaxial techniques are useful for the fabrication of electronic and photonic devices [24]. In device fields, single-crystal substrates of metal oxides, such as $SrTiO_3$ (STO), MgO, and sapphire,

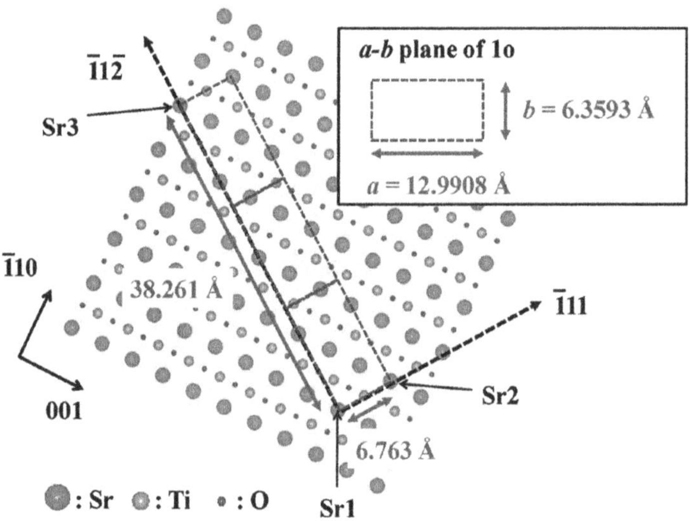

Fig. 3.15 $SrTiO_3$ (STO) 1 1 0 plane with a space group of *Pm*3–*m* (a = 3.905 Å). Distance between Sr atoms (Sr1–Sr2) on −1 1 1 axis is 6.763 Å, and this value is close to the length of the *b*-axis of rod-shaped **2o**. The −1 1 −2 axis of STO is perpendicular to the −1 1 1 axis. Distance between Sr1 and Sr3 on the −1 1 −2 axis (38.261 Å) is close to three times the length of the *a*-axis (12.9908 Å) of rod-shaped **2o**. Therefore, the *a*-axis of rod-shaped **2o** fits the −1 1 −2 axis on the 1 1 0 plane of STO

are used to obtain heteroepitaxial thin films. Among single crystal substrates, STO substrates are often used in the ferroelectric field. STO has a perovskite structure with a cubic phase (space group, *Pm*–3*m*; lattice constant, a = 3.905 Å; ICSD number, #23076). Figure 3.15 shows the 1 1 0 surface of STO, which has −1 1 −2 and −1 1 1 axes on a plane, and the −1 1 −2 axis is perpendicular to the −1 1 1 axis. The distance of Sr1–Sr2 is 6.763 Å in the −1 1 1 axis, and this value is close to lattice constant b (6.3593 Å) of rod-shaped **2o**. The Sr1–Sr3 distance on the −1 1 −2 axis (38.261 Å) is close to three times the length of the *a*-axis (12.9908 Å) of the rod-shaped **2o**. Therefore, the *a*-axis of rod-shaped **2o** fits the −1 1 −2 axis on the 1 1 0 plane of the STO. Thus, rod-shaped **1o** can grow on the 1 1 0 surface of the STO. We used an STO substrate with a 1 1 0 plane [(1 1 0) STO] to epitaxially grow **2o** on the substrate.

The film is prepared by coating a chloroform solution containing **2o** (100 mg/mL) onto the substrate (10 mm × 10 mm glass or STO crystal plates). The film is approximately 20 μm thick. The photoinduced topographical changes are carried out in the same way as in the previous case on a glass substrate. Upon UV irradiation crystals of **2o** are melted, and a flat eutectic surface is generated within 30 min. Upon visible light irradiation of the surface, the microcrystals of **2o** regenerate and cover the entire surface of the bulk film within 1.5 h. Figure 3.12g–i represents the SEM images on the surface to compare the results obtained on glass substrate. Due to differences in the substrate, the shape of the microcrystalline

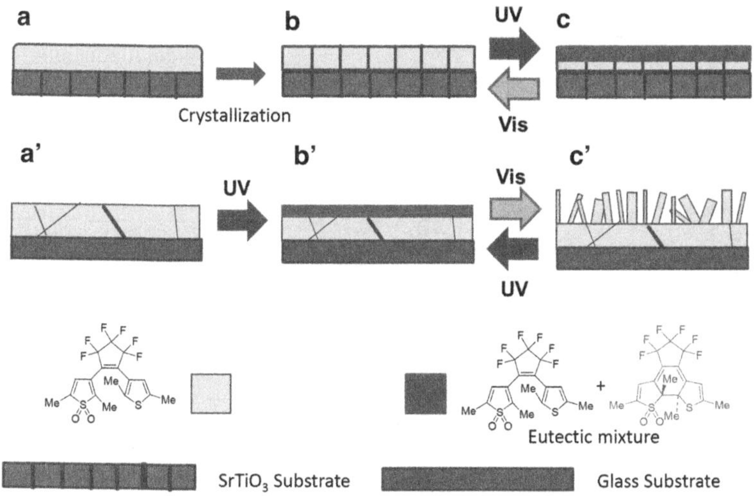

Fig. 3.16 Schematic illustration of crystal growths of **2o**. (**a**–**c**) Photoinduced topographical changes of surface on $SrTiO_3$ substrate; (**a**′–**c**′) photoinduced topographical changes of surface on glass substrate

surface is different. The surface over the STO is covered with prone rod-shaped microcrystals, while the surface on the glass substrate is covered with random standing rod-shaped microcrystals. The situations in Fig. 3.12h, i can be repeated by alternate UV and visible light irradiation. Here, the rapid crystal growth is due to the existence of seeds for the crystallization of regenerated **2o**. The results suggest that if the subphase crystals are generated by epitaxial growth on a crystal matrix, then photochemically reversible crystal growth can be performed (Fig. 3.16).

The appropriate reversible reflection pattern profiles of **2o** on the STO substrate are monitored by XRD measurements (Fig. 3.17). Diffraction intensities are normalized using the diffraction intensity of the 1 1 0 diffraction of the STO substrate. The results are compared using glass (Fig. 3.14) as a control. The diffraction peaks at 8.3°, 16.6°, and 24.9° correspond to the 0 0 2, 0 0 4, and 0 0 6 diffractions, respectively, of rod-shaped **2o**. Therefore, the **2o** film on the STO substrate with the 1 1 0 plane has a 0 0 1 orientation. In contrast, the **2o** films on (1 0 0) and (1 1 1) STO and glass had major orientations of 1 0 0, 1 0 0, and 3 0 −2, respectively, but minor orientations as well. The surface roughness of (1 0 0), (1 1 0), and (1 1 1) STO substrates are examined by AFM, and all of them are ascertained to be flat [23]. This result clearly shows that the surface lattice structure of the substrates is important for the growth orientation of the films.

The microcrystalline film of **2o** on the glass substrate shows many reflection peaks (Fig. 3.14), while that on the STO substrate shows only one reflection (Fig. 3.17). The purity of the reflections is maintained after several UV-induced crystal melting and visible light-induced crystal growing cycles. To the best of our knowledge, such a crystal growing technique is novel and will be useful in

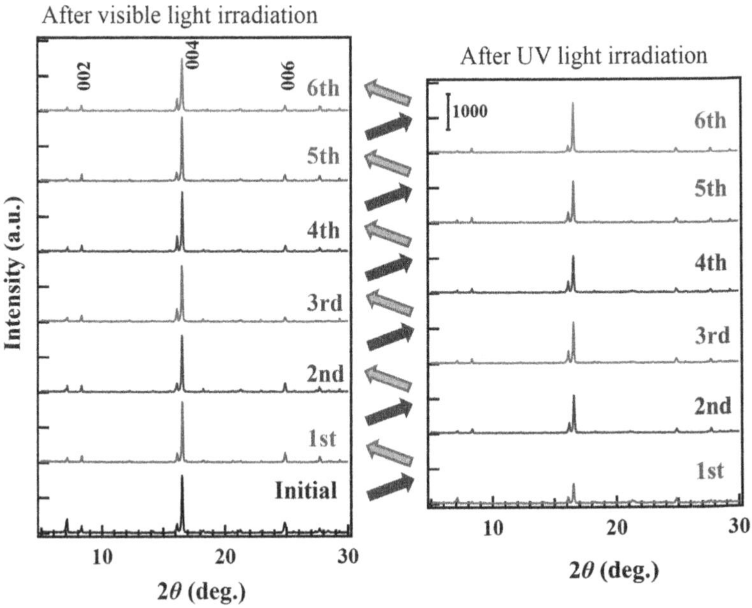

Fig. 3.17 Photoinduced reversible XRD pattern changes of microcrystalline film of **2o** on (1 1 0) surface of $SrTiO_3$ single crystal upon alternate irradiation with UV ($\lambda = 366$ nm) and visible ($\lambda > 500$ nm) light. Purple and green arrows indicate UV and visible light irradiations, respectively

controlling crystal growth for optoelectronic devices and controlling the crystal habit in a polymorphological system.

3.5 Conclusion

Diarylethenes are regarded as photochromic compounds whose polarity is not significantly changed during photochromism. However, the surface wettability of bulk diarylethene could be controlled by using photoinduced topographical changes. These changes are typical phenomena, since diarylethenes have specific properties: thermal stability of both isomers and photochromic reactivity even in the crystalline state. Of course it is necessary to have needle-shaped crystals for superhydrophobicity; however, crystal shapes are not easily estimated before the synthesis of a derivative. Furthermore, estimations of the eutectic temperature and melting points of the isomers are also obtained only after synthesis. Nevertheless, such phenomena still offer further possibilities and applications. Achieving a superhydrophilic surface is one of our remaining targets. In addition, by controlling crystal shape to mimic the natural surfaces of plants and insects, we can expect to generate additional fantastic surface functions by photoirradiation. Finally, such photocontrol of crystal growth is expected to provide important information on the selective crystal growth of organic compounds, which may lead to development of medical drugs.

References

1. Irie M (2000) Diarylethenes for memories and switches. Chem Rev 100:1685–1716
2. Rosario R, Gust D, Hayes M, Jahnke F, Springer J, Garcia AA (2002) Photon-modulated wettability changes on spiropyran-coated surfaces. Langmuir 18:8062–8069
3. Barthlott W, Neinhuis C (1997) Purity of the sacred lotus, or escape from contamination in biological surfaces. Planta 202:1–8
4. Erbil HY, Demirel AL, Avci Y, Mert O (2003) Transformation of a simple plastic into a superhydrophobic surface. Science 299:1377–1380
5. Shibuichi S, Onda T, Satoh N, Tsujii K (1996) Super water-repellent surfaces resulting from fractal structure. J Phys Chem 100:19512–19517
6. Uchida K, Izumi N, Sukata S, Kojima Y, Nakamura S, Irie M (2006) Photoinduced reversible formation of microfibrils on a photochromic diarylethene microcrystalline surface. Angew Chem Int Ed 45:6470–6473
7. Irie M, Kobatake S, Horichi M (2001) Reversible surface morphology changes of a photochromic diarylethene single crystal by photoirradiation. Science 291:1769–1772
8. Yamada T, Kobatake S, Muto K, Irie M (2000) X-ray crystallographic study on single-crystalline photochromism of bis(2,5-dimethyl-3-thienyl)perfluorocyclopentene. J Am Chem Soc 122:1589–1592
9. Izumi N, Minami T, Mayama H, Takata A, Nakamura S, Yokojima S, Tsujii K, Uchida K (2008) Superwater-repellent fractal surfaces of a photochromic diarylethene induced by UV light. Jpn J Appl Phys 47:7298–7302
10. Nishikawa N, Uyama A, Kamitanaka T, Mayama H, Kojima Y, Yokojima S, Nakamura S, Tsujii K, Uchida K (2011) Photoinduced reversible topographical changes on diarylethene microcrystalline surfaces with bio-mimetic wetting properties. Chem Asian J 6:2400–2406
11. Izumi N, Nishikawa N, Yokojima S, Kojima Y, Nakamura S, Kobatake S, Irie M, Uchida K (2009) Photo-induced reversible topographical changes of photochromic dithienylethene microcrystalline surfaces. New J Chem 33:1324–1326
12. Feng L, Zhang Y, Xi J, Zhu Y, Wang N, Xia F, Jiang L (2008) Petal effect: a superhydrophobic state with high adhesive force. Langmuir 24:4114–4119
13. Wenzel RN (1936) Resistance of solid surfaces to wetting by water. Ind Eng Chem 28:988–994
14. Cassie ABD, Baxter S (1944) Wettability of porous surfaces. Trans Faraday Soc 40:546–551
15. Uchida K, Nishikawa N, Izumi N, Yamazoe S, Mayama H, Kojima Y, Yokojima S, Nakamura S, Tsujii K, Irie M (2010) Phototunable diarylethene microcrystalline surfaces: lotus and petal effects upon wetting. Angew Chem Int Ed 49:5942–5944
16. Shibuichi S, Onda T, Satoh N, Tsujii K (1997) Super water-repellent surfaces resulting from fractal structure II. J Jpn Oil Chem Soc 46:649–659 (in Japanese)
17. Mayama H (2009) Blooming theory of tristearin. Soft Matter 5:856–859
18. Feng X, Jiang L (2006) Design and creation of superwetting/antiwetting surfaces. Adv Mater 18:3063–3078
19. Uyama A, Yamazoe S, Shigematsu S, Morimoto M, Yokojima S, Mayama H, Kojima Y, Nakamura S, Uchida K (2011) Reversible photocontrol of surface wettability between hydrophilic and superhydrophobic surfaces on an asymmetric diarylethene solid surface. Langmuir 27:6395–6400
20. Kobatake S, Yamada T, Uchida K, Kato N, Irie M (1999) Photochromism of 1,2-bis(2,5-dimethyl-3-thienyl)perfluorocyclopentene in a single crystalline phase. J Am Chem Soc 121:2380–2386
21. Jeong Y-C, Kim E, Ahn K-H, Yang SI (2005) Fatigue property of oxidized photochromic dithienylethene derivative for permanent optical recording. Bull Korean Chem Soc 26:1675–1676
22. Kobatake S, Uchida K, Tsuchida E, Irie M (2002) Single-crystalline photochromism of diarylethenes: reactivity–structure relationship. Chem Commun 38:2804–2805

23. Sakiyama S, Yamazoe S, Uyama A, Morimoto M, Yokojima S, Kojima Y, Nakamura S, Uchida K (2012) Photoinduced reversible heteroepitaxial microcrystal growth of a photochromic diarylethene on (110) surface of $SrTiO_3$. Cryst Growth Des 12:1464–1468
24. Simbrunner C, Quochi F, Hernandez-Sosa G, Oehzelt M, Resel R, Hesser G, Arndt M, Saba M, Mura A, Bongiovanni G, Sitter H (2010) Organic-organic heteroepitaxy of red-, green-, and blue-emitting nanofibers. ACS Nano 4:6244–6250

Chapter 4
Selective Metal Deposition Based on Photochromism of Diarylethenes

Tsuyoshi Tsujioka

Abstract Various novel functions on photochromic surfaces have been attracting interest. Selective metal deposition on a photochromic diarylethene surface is one such function and signifies that metal vapor atoms are deposited on the colored surface but not on the uncolored surface. This chapter introduces the phenomena, origin, extension, and applications of selective metal deposition. The origin of selective metal deposition is in a large change in the glass transition temperature (T_g) based on the photoisomerization of diarylethene. Low-T_g uncolored surface with active molecular motion causes metal atom desorption from the surface. The selective metal deposition phenomenon can be extended to various aspects including organic crystal and polymer surfaces. We demonstrate the following applications of selective metal deposition: patterned cathode preparation for organic light-emitting devices, micro thin-film fuse, multifunctional diffraction grating, and metal vapor integration. Selective metal deposition is expected for applications in various electric and optic fields.

Keywords Amorphous • Diarylethene • Electronics • Glass transition temperature • Metal deposition • Metal patterning • Vacuum evaporation

4.1 Selective Metal Deposition on Amorphous Diarylethene Surfaces

Metal film deposition with vacuum evaporation is widely used from basic academic studies to industrial mass production processes [1]. Metal vapor atoms generated by thermal evaporation are deposited on the substrate surface at room temperature.

T. Tsujioka (✉)
Department of Arts and Sciences, Osaka Kyoiku University,
Asahigaoka 4-698-2, Kashiwara, Osaka 582-8582, Japan
e-mail: tsujioka@cc.osaka-kyoiku.ac.jp

M. Irie et al. (eds.), *New Frontiers in Photochromism*,
DOI 10.1007/978-4-431-54291-9_4, © Springer Japan 2013

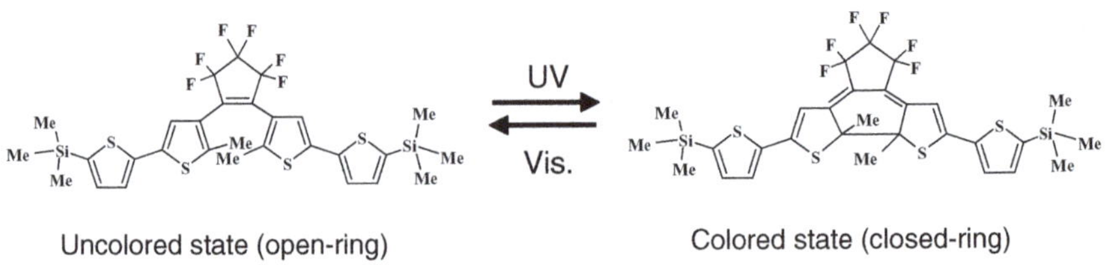

DAE1

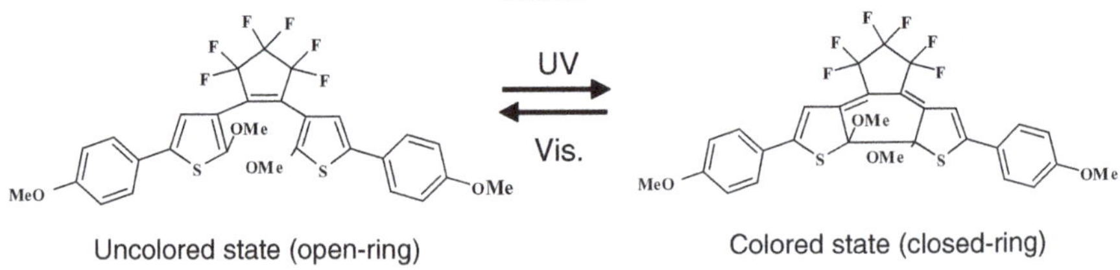

DAE2

Fig. 4.1 Photochromism of diarylethenes

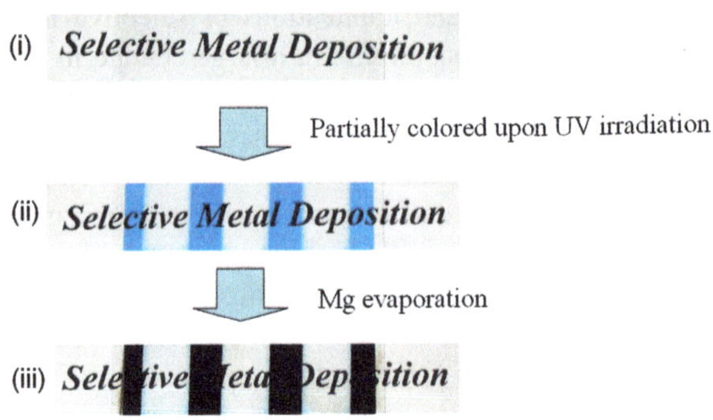

Fig. 4.2 Selective metal Mg deposition on the DAE1 surface [2]

In general, therefore, a shadow mask, which prevents metal vapor atoms from reaching a substrate, has been used to make metal-undeposited areas on the substrate. Selective metal deposition signifies that metal vapor atoms are deposited on a colored DAE surface but not on an uncolored surface [2]. The core phenomenon in selective metal deposition is an undeposition effect on the uncolored surface; this phenomenon is quite novel.

Figure 4.1 shows the photochromism of diarylethenes (1,2-bis[2-methyl-5-(5-trimethylsiliythiophen-2-yl)thiophen-3-yl]-3,3,4,4,5,5-hexafluoroccyclopentene, DAE1 and 1,2-bis[2-methoxy-5-(*p*-methoxyphenyl)thiophen-3-yl]-3,3,4,4,5,5-hexafluoroccyclopentene, DAE2), which show the selective metal deposition phenomenon. Figure 4.2 shows selective Mg deposition. (1) Amorphous DAE1 film, which consists of open-ring molecules without color, is prepared on a glass substrate using a conventional vacuum evaporation method. (2) Then the film is partially colored upon UV irradiation through a photomask, and Mg is evaporated onto the whole surface in a vacuum. (3) Mg is deposited on the colored areas but not on the

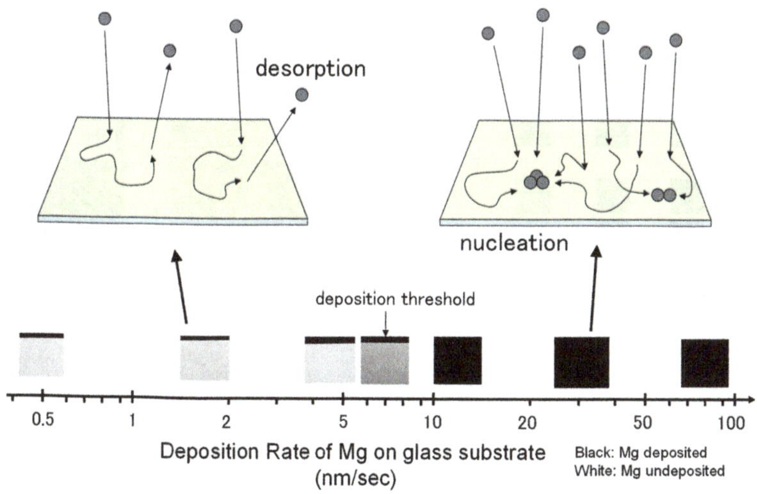

Fig. 4.3 Mg-deposition rate dependence on uncolored DAE1 surface and metal atom behavior [2]

uncolored areas. (The sample is observed by transmitted white light and the Mg-deposited areas are in black.) X-ray photoelectron spectroscopy determined no Mg atoms or Mg-containing compounds on and in the uncolored areas.

The selective Mg-deposition phenomenon originates in a large change in the glass transition temperature (T_g) of the film in response to photoreaction. The T_g of the uncolored film is around room temperature (32 °C), whereas it is 95 °C for the colored film. A core phenomenon of selective Mg deposition is the desorption of Mg atoms from the uncolored surface due to active surface-molecular motion caused by the low T_g.

Figure 4.3 shows the Mg atom behavior on the uncolored surface, and the deposition rate dependence of Mg on the uncolored surface. When the deposition rate is increased to over 7 nm/s, Mg film is formed even on the uncolored surface. This result means that Mg atoms do not rebound elastically because the elastic rebound is not directly dependent on the deposition rate: the Mg atoms are adsorbed once on the surface, migrate, and are desorbed from the surface at a low deposition rate.

Figure 4.4 shows a detailed study on Mg-deposition modulation depending on the isomerization ratio and the deposition rate [3]. The gray samples mean that Mg is deposited thinly at the Mg-deposition threshold. The threshold at a rate of 1.2 nm/s is the isomerization ratio of 70 % (b). The threshold, however, shifts to the uncolored side when the deposition rate is increased. Mg is deposited thinly at very high deposition rate (7 nm/s) (a) even in the completely uncolored state. On the other hand, the threshold shifts to the colored side when the Mg-deposition rate is decreased (c). These results indicate that the deposition rate of the metals is as important as the surface-molecular motion (or T_g) for obtaining optimum selective deposition.

The figures on the right in Fig. 4.4 show the corresponding atomic force microscope (AFM) images of the Mg nanocrystals on the DAE1 surfaces; each image

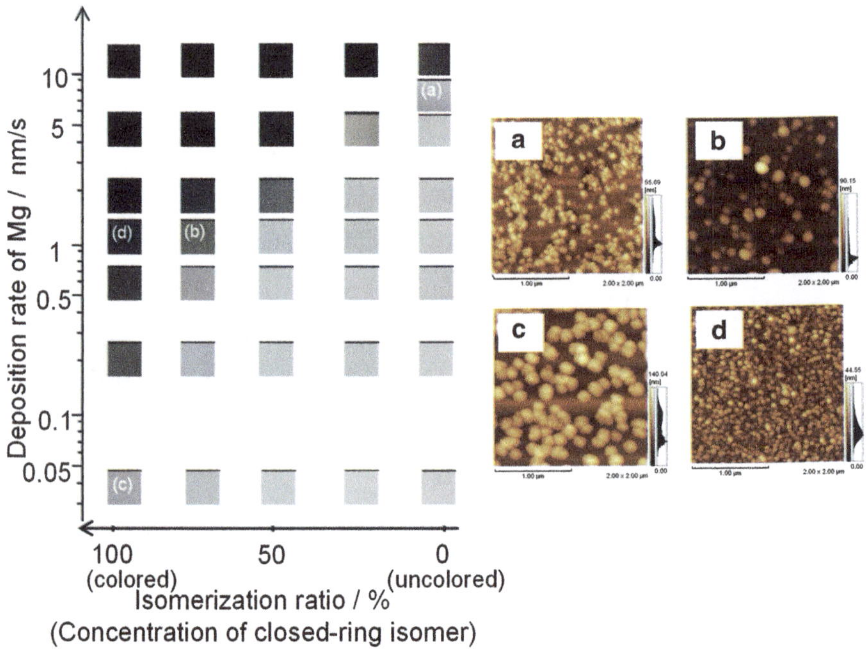

Fig. 4.4 Isomerization ratio and Mg-deposition rate dependences, and AFM images of Mg micro-nanocrystals on DAE surfaces [3]

labeled with a lower case letter corresponds to an identically labeled sample in the left figure. Relatively larger crystals are observed at the deposition threshold state in (b) than those on the colored state in (d). On the other hand, smaller Mg crystals are formed at the threshold in (a) at a high deposition rate of 7 nm/s. Large crystals of over 100 nm in diameter are formed at the threshold in (c) at a very low deposition rate of 0.04 nm/s. The dominant factors that affect the formation of the Mg film, in general, are the density of the Mg atoms, the migration length, the desorption rate, and the density of the nucleation sites on a surface. The difference in the sizes between the Mg crystals (b) and (d) means that the migration of the Mg atom on the surface in (b) is more active, and hence the crystal growth is enhanced, as in the case of the high substrate temperature.

On the other hand, the crystals formed in samples (a) and (c) are smaller and larger than those formed in (b). The small crystals in (a) result from the short migration length, which reflects the high density of the Mg atoms on the DAE surface. The enhanced crystal growth (c) at a very low deposition rate is indicative of the long migration length of the Mg atoms on the colored surface. These results indicate that the nucleation for the Mg film formation is mainly caused by the collisions among the Mg atoms, but not by the presence of the nucleation site on the surface.

These observations indicate that the Mg nucleation on the organic surface is achieved by the collision of the Mg atoms on the surface, which is strongly correlated to Mg density (or deposition rate) and the migration (isomerization ratio or T_g).

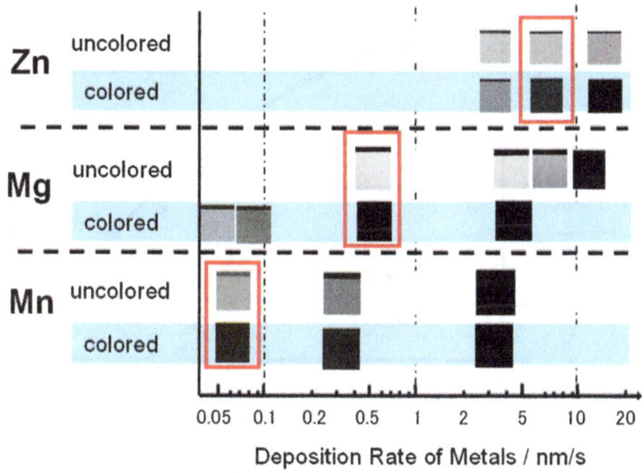

Fig. 4.5 Deposition rate dependence for selective deposition of Zn, Mg, and Mn [7]

Since the density and migration length of the Mg atoms are the most essential factors for Mg nanocrystal formation on the DAE1 surface, potential exists for these factors to govern the selective metal deposition process.

The other parameters except T_g that affect the metal deposition are the substrate temperature [2, 4], the gas pressure during the metal evaporation [5], the surface viscosity [6], and the metal species. Low temperature, high pressure, and/or high viscosity cause easy deposition.

Based on the above discussion, selective metal deposition for other metal species is possible. Figure 4.5 shows the selective deposition phenomena for Zn, Mg, and Mn that depend on the deposition rate 7 nm/s for Zn and 0.05 nm for Mn, which are higher and lower rates than those for Mg [7].

Substrate temperature (T_{sub}) during metal evaporation influences the selective metal deposition phenomenon. Pb is deposited on both the isomerization states of DAE1 at room temperature, but by controlling the temperature, selective Pb deposition is possible [4]. The substrate temperature influences the surface atom diffusion, the desorption, and, therefore, time τ between adhesion and desorption. A long τ denotes easy metal deposition. In general, time τ is given by

$$\tau = \tau_0 \exp\left(\frac{E_a}{kT}\right),$$

where E_a, k, and T are the activation energy of desorption, the Boltzmann constant, and the absolute temperature, respectively. τ_0 is around a period of surface-molecular oscillation. Selective metal deposition originates in the large difference of τ_0 and/or E_a between two isomerization states. This means the existence of a maximum ON–OFF ratio in Pb deposition with a change of temperature.

Figure 4.6a shows the T_{sub} dependence of the Pb deposition on the colored and uncolored DAE1 surfaces. The Pb evaporation quantity is the constant (thickness on the glass substrate: 55 nm). The deposited Pb thickness on the DAE1 surface decreases with increasing temperature for both the colored and uncolored states.

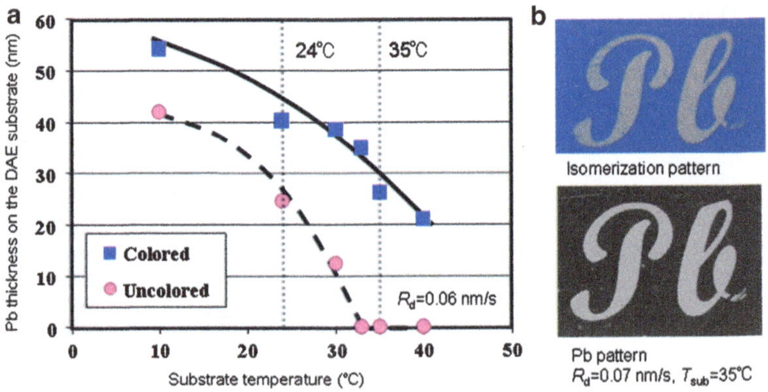

Fig. 4.6 (**a**) T_{sub} dependence of Pb deposition on DAE1 surface. (**b**) Isomerization pattern and corresponding Pb pattern based on selective Pb deposition [4]

The Pb thickness on the uncolored surface becomes zero when T_{sub} is above 32 °C, and the Pb thickness on the colored surface also decreases monotonically in the range of T_{sub} = 10–40 °C. The maximum ON–OFF ratio for Pb deposition is obtained around T_{sub} of 32–35 °C. This result indicates that a perfect light-controlled selective Pb deposition is achieved at T_{sub} = 35 °C and suggests the potential of selective deposition for other metal species by controlling T_{sub}. Figure 4.6b shows Pb patterning using selective deposition based on the above result. Pb evaporation onto the DAE1 surface with an isomerization pattern generates a corresponding clear Pb film pattern.

4.2 Selective Metal Deposition on a Diarylethene Crystal Surface

Organic crystals are essential materials in the field of organic electronics because of their high carrier mobility [8, 9]. Since diarylethenes are expected to serve as an organic semiconductor memory materials [10–12] as well as optical memory materials [13, 14], diarylethene crystals are promising candidates of organic memory. Metal patterning on the organic layers is an important process as electrodes and/or interconnection. A vacuum evaporation method with a shadow mask is the most conventional approach, but this method encounters a serious difficulty: the complex setup of a shadow mask system causes a low-resolution limit of metal patterning. The selective metal deposition method is available for metal patterning on crystal surfaces.

A soft surface with low T_g causes an undeposition effect in selective metal deposition. If a crystal surface is softened by some method, analogous undeposition is expected. Organic crystals consist of regularly arranged molecules. Doping the arranged surface molecules with different-sized molecules disarrays the original

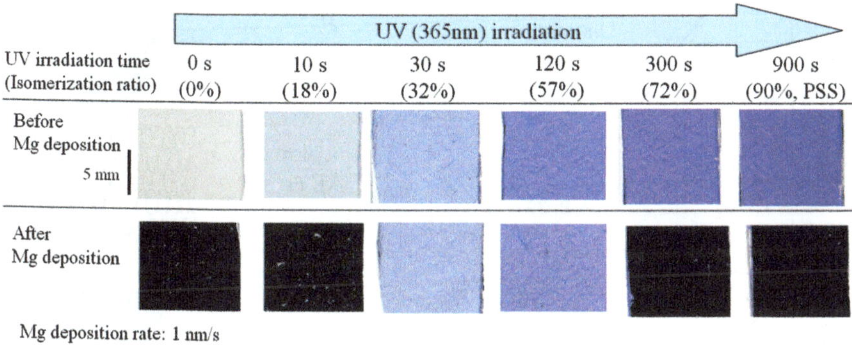

Fig. 4.7 Isomerization ratio dependence of Mg deposition on crystallized DAE1 films [15]

crystal lattice and softens the surface. Such doping can be achieved by a photoreaction from regular surface molecules to irregular photoreacted products.

A DAE1 molecule changes its molecular size based on photoreaction. UV irradiation to the uncolored DAE1 crystal generates colored molecules on the surface and softens the crystal surface, achieving selective metal deposition [15]. The upper figures in Fig. 4.7 show the DAE1 crystal film samples with various surface-isomerization ratios, which are obtained by irradiation with UV light for a variety of times. The 0-s sample is a perfectly uncolored state, and the 900-s sample is in a photostationary state (PSS) on the surface. Because of the large absorption of the DAE1 molecule at 365 nm, the isomerization depth from the surface is within several hundred nanometers; almost all of the films remain in the uncolored crystal state even after UV irradiation. The lower figures in Fig. 4.7 show the Mg-deposition property of the corresponding samples. Mg is deposited on the 0-, 10-, 300-, and 900-s samples but not on the half-colored 30- and 120-s samples (surface-isomerization ratios of 32 % and 57 %).

The Mg-undeposition effect on the uncolored amorphous diarylethene film originates in the active molecular motion on the soft surface with low T_g. The above result means the half-colored crystal surface is also in a soft surface despite the hard crystal state inside. Since the mobility of the molecules on the crystal surface is significantly different from those inside a bulk or film crystal because of their special environment, the atomic force microscopy force curve (FC) is a suitable characterization method to understand the surface softness state [16]. Figure 4.8a shows the FCs of the 0-, 120-, and 900-s sample surfaces before Mg deposition. The FCs for the 0- and 900-s samples indicate a relatively hard surface; the FC slopes (A–B) in approaching within a 10-nm depth from the surface for 0- and 900-s samples are 39 and 33 nN/nm, respectively. On the other hand, the 120-s sample surface shows the A–B slope is 24 nN/nm. This result indicates that the hard uncolored crystal surface becomes soft (not elastic but viscous) in its half-colored state.

Figure 4.8b illustrates a model of the surface states corresponding to the FC responses. The cantilever is not indented into the uncolored crystal (0-s sample)

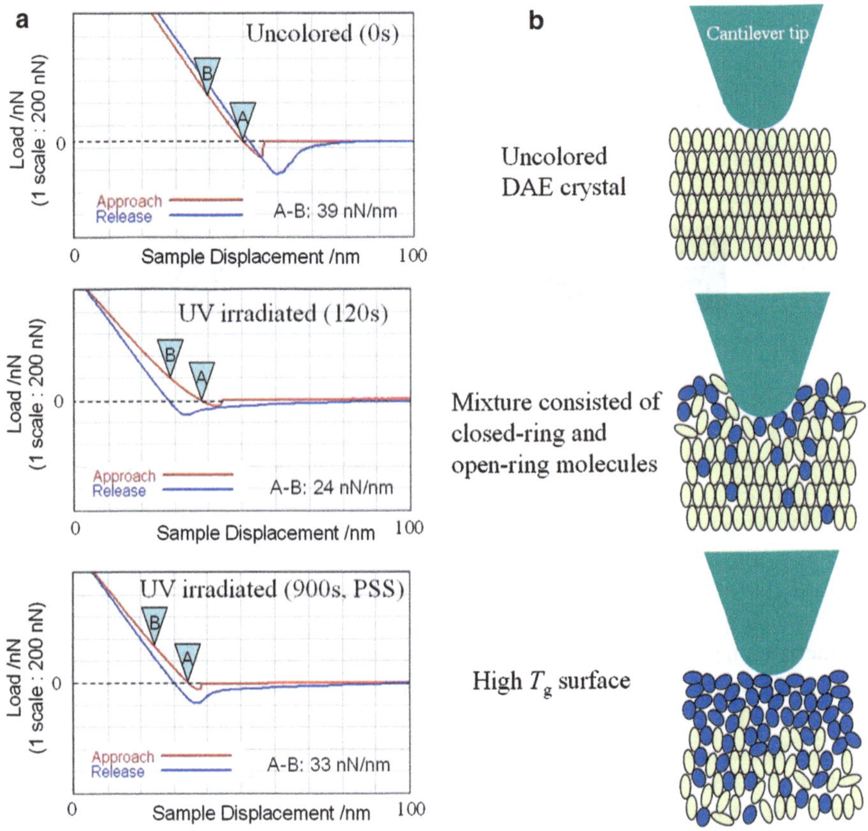

Fig. 4.8 (**a**) FC characterization of the DAE1 surfaces. (**b**) Model of the FC responses [15]

surface (upper figure). The half-colored (120-s sample) surface consists of open- and closed-ring molecules, as shown in the center figure, resulting in a disarrayed crystal lattice and a soft amorphous surface state. When the surface is further irradiated with UV light, it becomes a colored amorphous state (900-s sample) with a high T_g. This explains the Mg-deposition modulation on the crystal surfaces shown in Fig. 4.6.

Based on the above result, laser scanning and maskless metal evaporation can demonstrate fine metal patterning on the DAE1 crystal surface. Figure 4.9a shows the sample preparation process. A violet laser spot is scanned on the crystal surface, making paralleled half-colored lines on the uncolored DAE1 crystal surface. Then, Mg is evaporated onto the surface without a shadow mask. The paralleled high reflective areas are Mg-deposited areas (Fig. 4.9b). Mg is only deposited on the uncolored crystal surface but not on the half-colored lines. Either Mg-deposition or Mg-undeposition patterns can be obtained simply by adjusting the isomerization ratio by adjusting laser power or using a different laser wavelength.

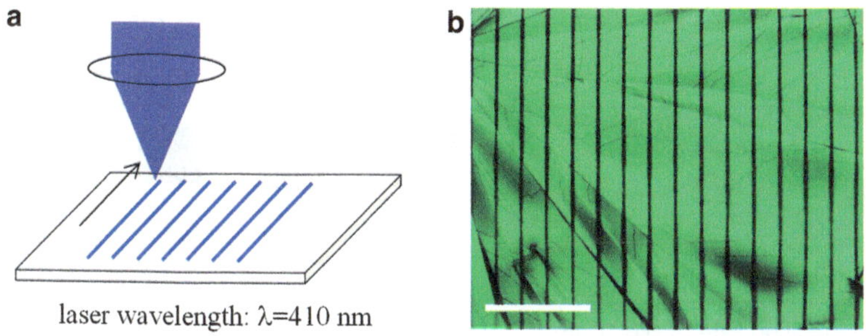

Fig. 4.9 Fine metal pattern formation on DAE1 crystal film: (**a**) parallel line formation by violet laser scanning; (**b**) corresponding parallel Mg-undeposited lines on DAE1 crystal surface. Scale bar: 200 μm

4.3 Selective Metal Deposition on Diarylethene-Doped Polymer Surfaces

A key technique to achieve selective metal deposition is controlling the surface-molecular motion. If the mobility of the surface polymer chains can be critically controlled by doping photochromic molecules into the polymer matrix, selective metal deposition is possible for polymer films [17]. Figure 4.10a shows selective Mg deposition on diarylethene-doped polystyrene (DAE-doped PS) films. In this case, Mg-deposition property on a pure PS film, which was obtained by casting a PS solution on a glass substrate, depends on its drying conditions. High-temperature and/or long-drying time causes Mg deposition, but lower temperature and shorter time cause thin Mg deposition. (See the 0 %-sample in Fig. 4.10a.) When DAE2 is doped into PS and dried under such conditions, all the DAE2-doped PS films achieve selective Mg deposition.

Figure 4.10b shows FC characterization on the DAE2-5%-doped PS surface. The FC on the uncolored surface indicates a relatively softer and more viscous response than that on the colored surface. This means that the doped DAE2 molecules can control the PS surface softness depending on its isomerization state. Analogous selective metal deposition is possible for another polymer poly[2-methoxy-5-(2-ethylhexyloxy)-1,4-phenylenevinylene], one the typical semiconducting polymer.

Metal patterning on the DAE2-doped polymer using laser scanning and maskless evaporation is possible. Figure 4.11 shows a demonstration of fine Mg patterning. An uncolored DAE2-5 %-doped PS film is prepared on a glass substrate by a casting method. (1) A laser spot with a wavelength of 410 nm and power of 0.5 mW is scanned at 2 mm/s on the film, and colored parallel lines with a 50 μm pitch are obtained. (2) Mg is evaporated to the surface at a deposition rate of 1 nm/s without a shadow mask. (3) Parallel fine 15-μm-wide Mg patterns are successfully achieved. In principle, the resolution of metal patterns is related to the diffraction limit of light: the submicron level.

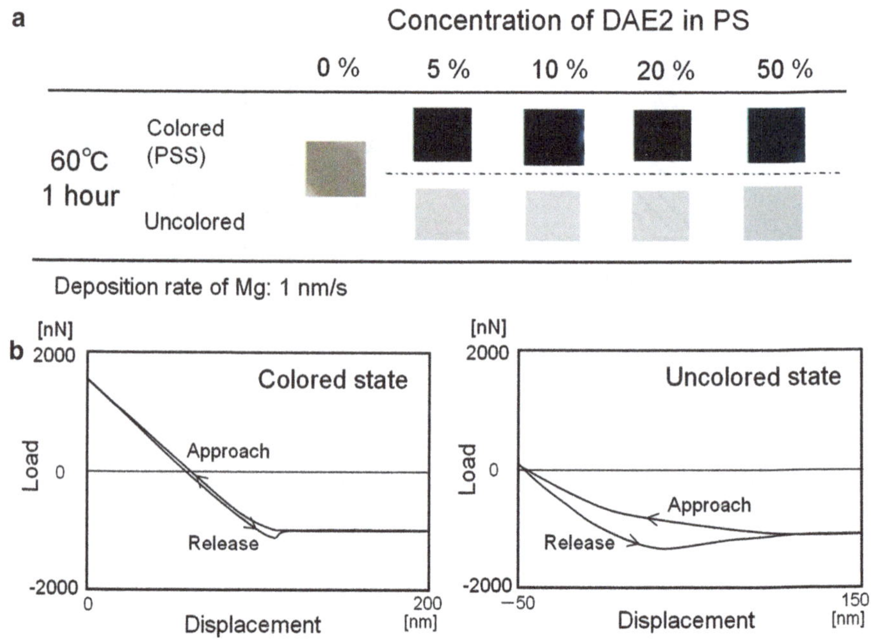

Fig. 4.10 (**a**) Selective Mg deposition on DAE2-doped PS films. (**b**) FC characterization on DAE2-doped PS films [17]

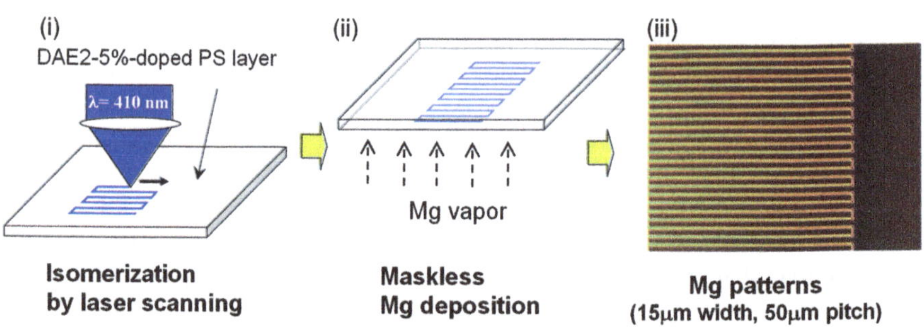

Fig. 4.11 Fine metal patterning on DAE2-doped PS film using laser scanning and maskless evaporation [17]

4.4 Applications

4.4.1 Cathode Preparation for Organic Electronic Devices

Such electronic devices as organic light-emitting devices (OLEDs) [18], organic thin-film transistors [19], and organic memories [10] have recently been attracting interest and generally require metal electrodes or wiring for their electronic

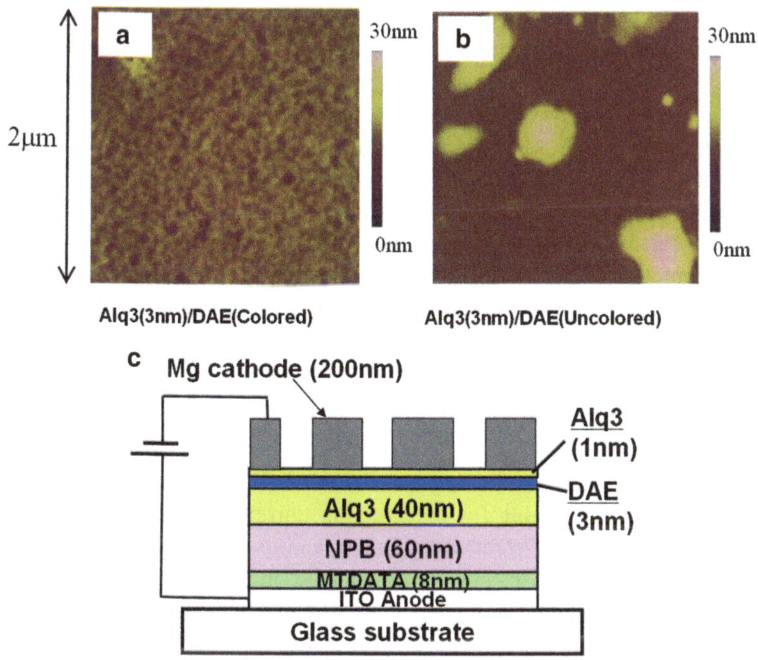

Fig. 4.12 (**a**) AFM image of thin Alq3 layer on colored surface. (**b**) AFM image of thin Alq3 layer on uncolored surface [20]. (**c**) OLED device structure with patterned Mg cathode based on selective deposition method [21]

operation. The selective metal deposition method is promising to prepare patterned metal electrodes on an organic layer.

To apply the selective deposition method to the fabrication of cathodes for organic devices, we must know the electronic characteristics at the interface between the metal and colored diarylethene layers. The diarylethene layer that neighbors the cathode metal layer suppresses electron injection and produces low performance in electronic devices. The existence of an electron injection layer, such as a thin tris(8-hydroxyquinolinato)aluminum (Alq3) intermediate layer, between the electrode and the diarylethene layer greatly improves the electron injection with holding the selective metal deposition property [20, 21].

Figure 4.12a, b shows AFM images of thin Alq3 layers with an average thickness of 3 nm on the colored and uncolored DAE1 surfaces [20]. Mesh growth of Alq3 is seen on the colored film, and large islands of Alq3 are found on the uncolored film for the samples with the 3-nm-thick intermediate layer of Alq3 on a DAE1 layer three days after the Alq3 evaporation. The development of the mesh and island structures is well known to occur at the initial stage of the film growth process [1]. These observations indicate that the deposited Alq3 molecules move more easily on the uncolored surface than on the colored surface and migrate to form island structures in the former. This results in the existence of DAE1 molecules on the surface. This is the mechanism of selective metal deposition in devices with an intermediate layer.

Based on the above results, the selective deposition method can be applied to the preparation of a cathode pattern for an OLED with typical organic materials: 4,4′,4″-tris(*N*-3-methylphenyl-*N*-phenyl-amino)-triphenylamine (MTDATA), *N,N*-*di*(naphthalene-1-yl)-*N,N*-diphenyl-benzidene (NPB), and Alq3 (Fig. 4.12c) [21]. The device has an additional 3-nm-thick DAE1 selective deposition layer and a 1-nm-thick Alq3 electron injection layer. A red laser spot scanning generates uncolored lines on the DAE1 layer, and Mg evaporation onto the surface forms a fine-patterned Mg cathode. Thus, patterned light emission corresponding to the cathode pattern is demonstrated. The cathode patterning demonstrated here is not restricted to OLEDs but can be applied to various other devices in organic electronics.

4.4.2 Thin-Film Micro-fuse

Small electronic elements including transistors, diodes, capacitors, detectors, and transducers are indispensable for constructing current and future electronics circuits. A fuse is one crucial electrical element, and its functionalizations have been studied [22]. The size of fuses has been reduced to make smaller electric circuits as well as other elements; however, fuses are much larger than such elements as transistors or diodes and prevent electric devices from becoming smaller. Pb is a typical metal species with a low melting point, and selective Pb deposition can be applied to the preparation of micro-thin-film fuses (μ-fuse) with laser scanning and maskless Pb evaporation [4]. Figure 4.13a shows the preparation of a μ-fuse array using laser scanning. Red laser spot scanning on the entire colored DAE1 surface makes uncolored lines, and colored lines are formed by scanning a violet laser spot between the uncolored lines. Pb is evaporated onto the whole area without a shadow mask under a substrate temperature of 35 °C. Figure 4.13b shows the obtained μ-fuse array. The dotted squares indicate each μ-fuse cell. This indicates the mass productivity of the selective metal deposition method in industrial applications.

The left figure in Fig. 4.13c shows the Pb pattern of the single μ-fuse. The bright areas are Pb-deposited parts. A 10-μm-wide Pb nondeposited line is generated by red laser scanning, and a 20-μm-wide Pb-bridged part is observed due to the colored line. When voltage is applied to the μ-fuse, the bridged part is broken (right figure in Fig. 4.13c). Figure 4.13d shows the current-voltage characteristics of the device. The current is interrupted around 300 μA at 2.4 V. These results indicate that the μ-fuse has a very low interrupt (highly sensitive) current level. The interrupt current level can be controlled by adjusting the cross section of the Pb-bridged part (Fig. 4.13e), indicating wide potential for various electric and electronic applications.

4.4.3 Dual-Functional Diffraction Grating

Optical elements including laser diodes, beam splitters, optical fibers, photodetectors, and/or lenses are indispensable for constructing current optical systems.

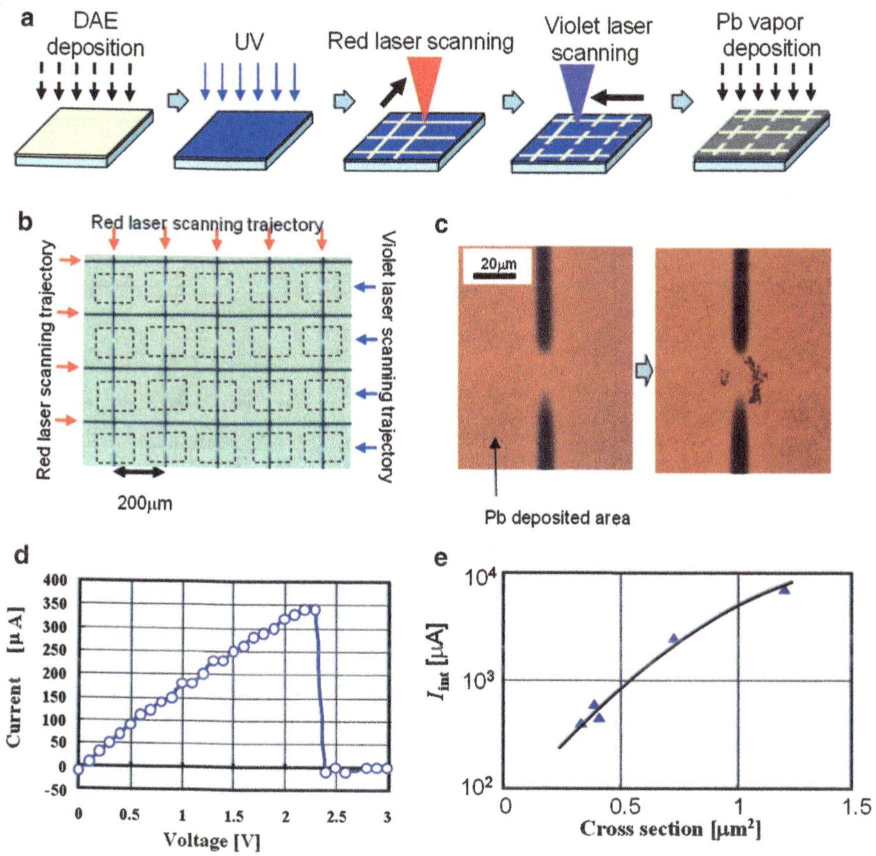

Fig. 4.13 (**a**) Preparation of μ-fuse array using selective Pb deposition. (**b**) μ-Fuse array. (**c**) μ-Fuse and its fusing operation. (**d**) Current-interrupt characteristics. (**e**) Cross-section dependence of the interrupt current level [4]

Diffraction grating, which splits and diffracts light into several beams traveling in different directions, is one crucial optical element and consists of a periodic structure with metal lines, grooves, or surface fine structures, or a periodic modulation of the refractive index on a substrate. Conventional gratings have diffraction only in transmission or in reflection: mono-functional. A dual-functional diffraction grating, however, shows different diffractions in transmission and reflection and can be realized based on a selective metal deposition method [23, 24]. Such dual-functional gratings have potential to make high-performance optical instruments and systems.

Selective metal deposition originates in a change in the T_g of amorphous film. Low T_g causes a surface deformation (flattening) effect when the DAE1 layer is prepared on a non-flat surface using vacuum evaporation (Fig. 4.14); an initial DAE1 layer is consistently formed with the original substrate surface. When part of the DAE1 film is colored, the remaining uncolored surface gradually becomes flat

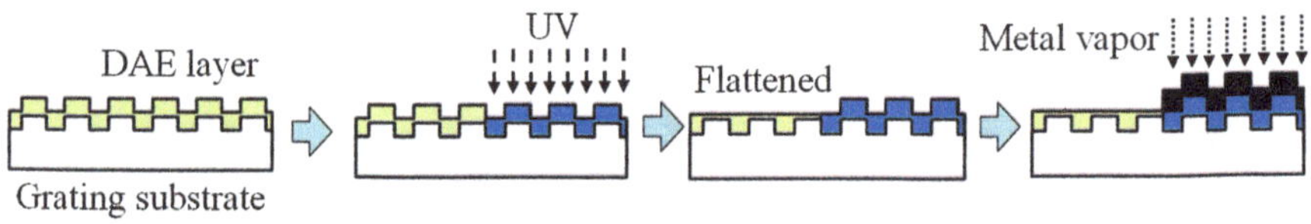

Fig. 4.14 Flattening effect of low-T_g DAE1 film on a grating substrate and selective metal deposition

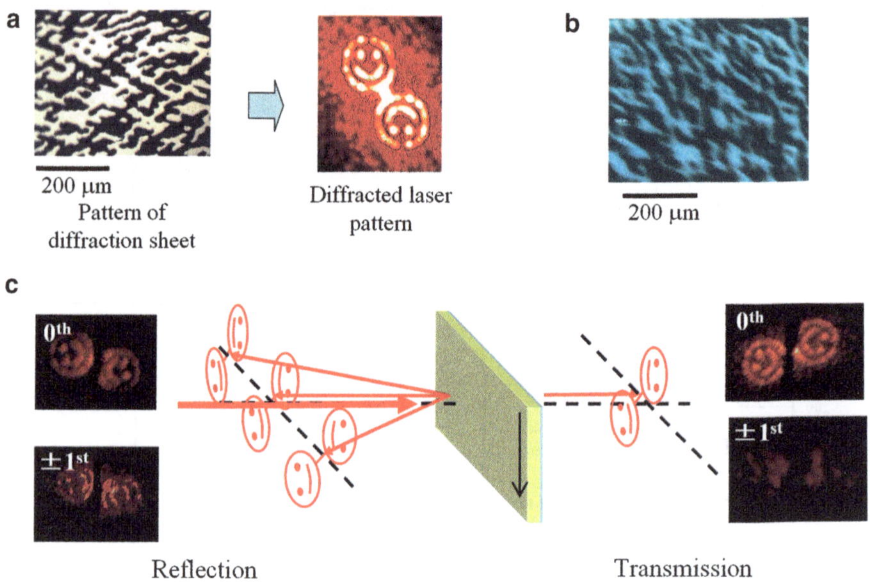

Fig. 4.15 (**a**) Diffraction sheet pattern and corresponding diffracted laser pattern. (**b**) Metal pattern on a grating substrate generated by selective metal deposition using diffraction sheet. (**c**) Obtained dual-functional grating. *Arrow* indicates groove direction of grating [24]

because it is in the low-T_g soft state, but the colored one holds the original non-flat surface shape. Metal vapor evaporation onto the surface makes reflective non-flat surface on the colored surface based on selective metal deposition.

To achieve dual-functional diffraction gratings, both selective metal deposition and the flattening effect based on a low T_g are utilized. Figure 4.15 shows a dual-functional grating obtained using a substrate with grooves and a computer-designed diffraction sheet generating a cartoon-diffraction pattern. Figure 4.15a shows a microscope image of the diffraction sheet and a corresponding laser-diffraction pattern. The uncolored DAE layer on the substrate with grooves is irradiated by UV light through the diffraction sheet, and the corresponding colored patterns to the patterns on the sheet are formed on the DAE layer. Then Mg is evaporated onto the substrate without a shadow mask to form Mg patterns that correspond to the colored pattern (Fig. 4.15b). After flattening the uncolored areas by storage at 27 °C for 500 min, explicit cartoon-diffraction patterns in the ±1st-diffraction are observed as

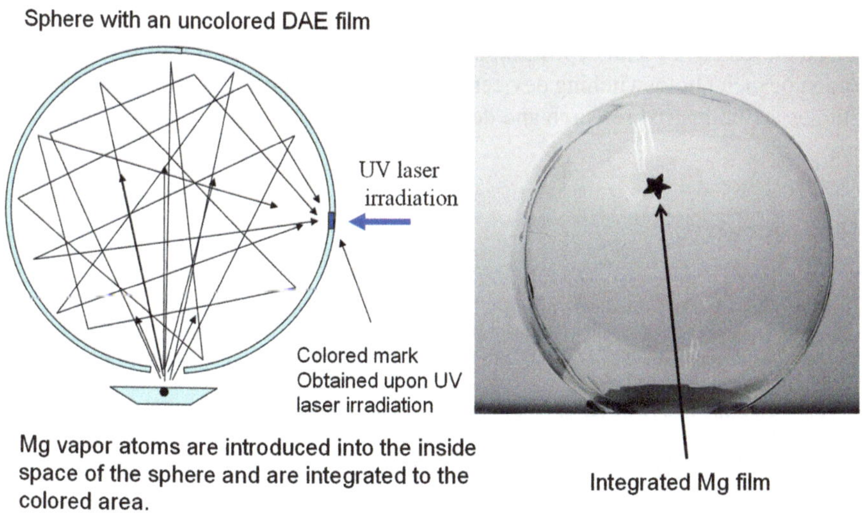

Fig. 4.16 Metal vapor integration [2]

well as in the 0th-diffraction in reflection, which are generated by a combination of the grooves and the Mg patterns. On the other hand, only one cartoon pattern in the 0th-diffraction is generated in transmission (Fig. 4.15c).

4.4.4 Metal Vapor Integration

The deposition place of metal vapor atoms can be controlled based on selective metal deposition. Figure 4.16 shows the metal vapor integration using a sphere with a DAE1 coat inside [2]. A small part of the uncolored DAE1 surface is colored by UV laser irradiation. When Mg vapor is introduced into a sphere, Mg atoms are desorbed many times from the uncolored surface and are eventually integrated onto a small colored star-shaped area. Using an analogue principle, metal vapor can be transported through a pipe at room temperature [6], like gas transformation. These functions suggest novel applications.

4.5 Conclusion

In this chapter, we highlighted a selective metal vapor deposition phenomenon, which is a novel function based on photochromism. This function is expected to be applied to metal patterning in such fields as electronics and optics. This phenomenon can also be extended to a variety of metals and organic surfaces including polymers [6, 17, 25]. Photochromic diarylethenes are expected for photoswitching in current or organic semiconductor memory materials [26, 27]. Metal patterning based on selective metal deposition is a promising method for fine electrode preparation of such photoswitching devices. These results indicate great potential in basic studies and industrial research and development.

References

1. Smith DL (1995) Thin-film deposition: principles & practice. McGraw-Hill, New York
2. Tsujioka T, Sesumi Y, Takagi R, Masui K, Yokojima S, Uchida K, Nakamura S (2008) Selective metal deposition on photo-switchable molecular surfaces. J Am Chem Soc 130:10740–10747
3. Tsujioka T, Sesumi Y, Yokojima S, Nakamura S, Uchida K (2009) Metal atom behavior on photochromic diarylethene surfaces-deposition rate dependence of selective Mg deposition. New J Chem 33:1335–1338
4. Tsujioka T, Dohi M (2012) Light-controlled selective Pb deposition on photochromic surfaces. Appl Phys Express 5:041603
5. Iwai Y, Tsujioka T (2011) Metal deposition selectivity based on photochromism of diarylethene film in intermediate vacuum. Jpn J Appl Phys 50:081602
6. Tsujioka T, Tsuji K (2012) Metal-vapor deposition modulation on soft polymer surfaces. Appl Phys Express 5:021601
7. Sesumi Y, Yokojima S, Nakamura S, Uchida K, Tsujioka T (2010) Light-controlled selective metal deposition on a photochromic diarylethene film – toward new applications in electronics and photonics. Bull Chem Soc Jpn 83:756–761
8. Mas-Torrent M, Durkut M, Hadley P, Ribas X, Rovira C (2004) High mobility of dithiophene-tetrathiafulvalene single-crystal organic field effect transistors. J Am Chem Soc 126:984–985
9. Anthony JE (2006) Functionalized acenes and heteroacenes for organic electronics. Chem Rev 106:5028–5048
10. Tsujioka T, Sasa T, Kakihara Y (2012) Nonvolatile organic memory based on isomerization of diarylethene molecules by electrical carrier injection. Org Electr 13:681–686
11. Tsujioka T, Kondo K (2003) Organic bistable molecular memory using photochromic diarylethene. Appl Phys Lett 83:937–939
12. Tsujioka T, Iefuji N, Jiapaer A, Irie M, Nakamura S (2006) Hole-injection isomerization of photochromic diarylethene for organic molecular memory. Appl Phys Lett 89:222102
13. Irie M (2000) Diarylethenes for memory and switches. Chem Rev 100:1685–1716
14. Tsujioka T, Shimizu Y, Irie M (1994) Crosstalk in photon-mode photochromic multi-wavelength recording. Jpn J Appl Phys 33:1914–1919
15. Tsujioka T (2011) Selective metal deposition on photosensitive organic crystal surfaces. J Mater Chem 21:12639–12643
16. Nakajima K, Yamaguchi H, Lee JC, Kageshima M, Ikehara T, Nishi T (1997) Nanorheology of polymer blends investigated by atomic force microscopy. Jpn J Appl Phys 36:3850–3854
17. Tsujioka T, Takagi R, Shiozawa T (2010) Light-controlled metal deposition on photochromic polymer films. J Mater Chem 20:9623–9627
18. Huang J, Su J-H, Tian H (2012) The development of anthracene derivatives for organic light-emitting diodes. J Mater Chem 22:10977–10989
19. Sun Y, Liu Y, Zhu D (2005) Advances in organic field-effect transistors. J Mater Chem 15:53–65
20. Masui K, Takagi R, Sesumi Y, Nakamura S, Tsujioka T (2009) Selective metal deposition for a structure with a thin intermediate layer on a photochromic diarylethene film. J Mater Chem 19:3176–3180
21. Takagi R, Masui K, Nakamura S, Tsujioka T (2008) Metal patterning using maskless vacuum evaporation process based on selective deposition of photochromic diarylethene. Appl Phys Lett 93:213304

22. Martin JE, Heaney MB (2000) Reversible thermal fusing model of carbon black current-limiting thermistors. Phys Rev B 62:9390–9397
23. Tsujioka T, Matsui N (2012) Photoreprogrammable dual-function diffraction grating based on photochromism and selective metal deposition. Opt Lett 37:70–72
24. Tsujioka T, Matsui N (2011) Dual-functional diffraction grating based on selective metal deposition of photochromic diarylethene. Opt Lett 36:3648–3650
25. Tsujioka T, Matsui A (2009) Light-controlled selective metal deposition on photopolymer films. Appl Phys Lett 94:013302
26. Tsujioka T, Irie M (2010) Electrical functions of photochromic molecules. J Photochem Photobiol C: Photochem Rev 11:1–14
27. Tsujioka T, Yamamoto M, Shoji K, Tani K (2010) Efficient carrier separation from a photochromic diarylethene layer. Photochem Photobiol Sci 9:157–161

Chapter 5
Photoinduced Alignment and Structuring in Thin Films of Azobenzene-Containing Liquid Crystalline Polymers

Takahiro Seki, Mitsuo Hara, and Shusaku Nagano

Abstract When an azobenzene (Az) unit is embedded in molecular assemblies such as liquid crystalline (LC) systems, the state and orientation of the surrounding LC matrix are strongly coupled with the photoisomerization reaction of Az. Due to the strong cooperative nature of LC materials, the light-driven molecular motion of Az gives rise to largely amplified photoresponses. This chapter briefly summarizes our recent activities on the photoresponsive LC thin films containing the Az unit. The motions of a wide range of size feature from nanometer levels to macroscopic levels can be driven using the side-chain Az LC polymers of almost the same structural features. The processes include light-switchable surface alignment of LC materials (nanoscopic scales at interfaces), photoalignment in a surface-grafted polymer film, photocontrol of microphase separation patterns at mesoscopic scales (10–100 nm), and mass migrations at micrometer scales. The versatility and wide applicability of the thin films of Az LC polymers are demonstrated.

Keywords Azobenzene • Block copolymers • Liquid crystalline polymers • Mass migrations • Photoalignment • Thin films

---

T. Seki (✉) • M. Hara
Department of Molecular Design and Engineering,
Graduate School of Engineering, Nagoya University,
Furo-cho, Chikusa, Nagoya 464-8603, Japan
e-mail: tseki@apchem.nagoya-u.ac.jp

S. Nagano
Venture Business Laboratory, Nagoya University,
Furo-cho, Chikusa 464-8603, Japan

PRESTO, Japan Science and Technology Agency,
5-3 Yonbancho, Chiyoda-ku, Tokyo 102-8666, Japan

M. Irie et al. (eds.), *New Frontiers in Photochromism*,
DOI 10.1007/978-4-431-54291-9_5,

5.1 Introduction

In the studies of photochromism, the fabrication of photon mode optical memories and switching of various performances have been explored [1–3]. On the other hand, recent efforts are rapidly directing to mechanical performances such as photoalignment, structure formation, mass transfer, and film/fiber deformation. The reversible photoisomerization of azobenzene (Az) is well known for nearly 80 years. However, the allure of this molecule is never lost. Due to its simple rodlike structure in the *trans*-form, the Az derivatives are highly compatible with calamitic liquid crystalline (LC) materials, and further they can be the mesogenic groups themselves. Due to these aspects, LC polymer systems containing Az derivatives provide fascinating functions and performances. In LC systems, the reaction vector and motion of an Az unit embedded are strongly coupled with the surrounding organized matrix. Such consorted motions eventually lead to uniform alignment of molecules and deformation of materials in various hierarchical levels [4–10]. The photoreaction of a molecular nanoscale event can successively be linked to larger motions in ranges of mesoscale (10–100 nm, defined as in this chapter), micro(meter)-scale, and further to macro size levels. Typical examples are the surface photoalignment of liquid crystals and photomechanical responses in LC polymer films [11–14]. When linearly polarized light (LPL) is irradiated, Az derivatives respond in directional (anisotropic) manners. The angular selective light absorption Az molecules followed by the photoisomerization causes the photoalignment to the less excitable (orthogonal) direction.

This chapter briefly introduces the alignment and structuring of thin films of Az-containing LC polymers. Structure formation and photoalignment at nanoscales (1–10 nm), mesoscales (10–100 nm), and microscale (>1,000 nm) are mentioned. A panascopic view of thin film systems covered this chapter is schematically drawn in Fig. 5.1. As exemplified here, light can be used as powerful tools for structural and orientational controls of various size features and should play important roles in thin film technologies for the future.

5.2 Surface-Assisted Photoalignment of Liquid Crystals

5.2.1 Surface Control of Nematic Liquid Crystals

A photochromic and photo-cross-linkable polymer films control the alignment of LC materials contacting with the surface. A pioneering work was reported by Ichimura et al. [11, 15, 16]. They showed that the *trans*/*cis* photoisomerization reaction of an azobenzene monolayer can reversibly switch the alignment of nematic LCs and proposed a concept of command surface effect (upper scheme of Fig. 5.2). When a photoreactive polymer film is exposed to LPL or oblique irradiation, all the LC molecules included in a several micrometer-level-thick cell align according the

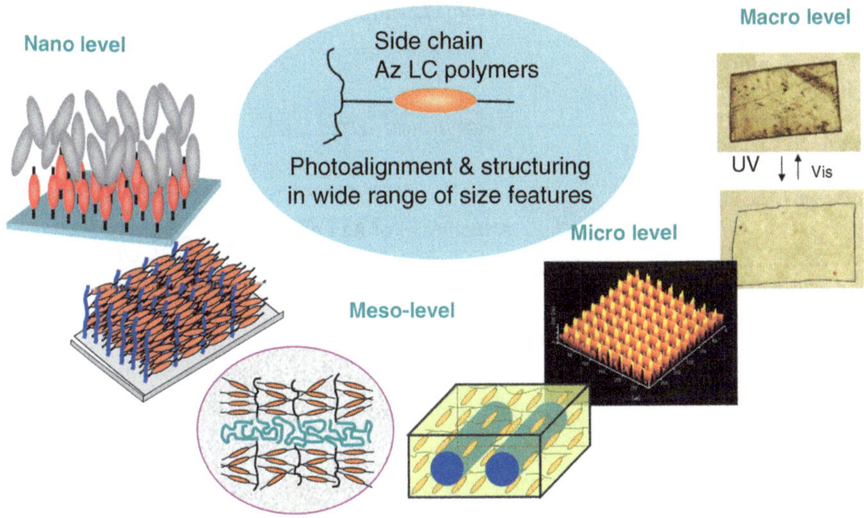

Fig. 5.1 Panascopic view (from nano to macro levels) of photoresponsive thin films consisting of Az-containing side-chain LC polymer

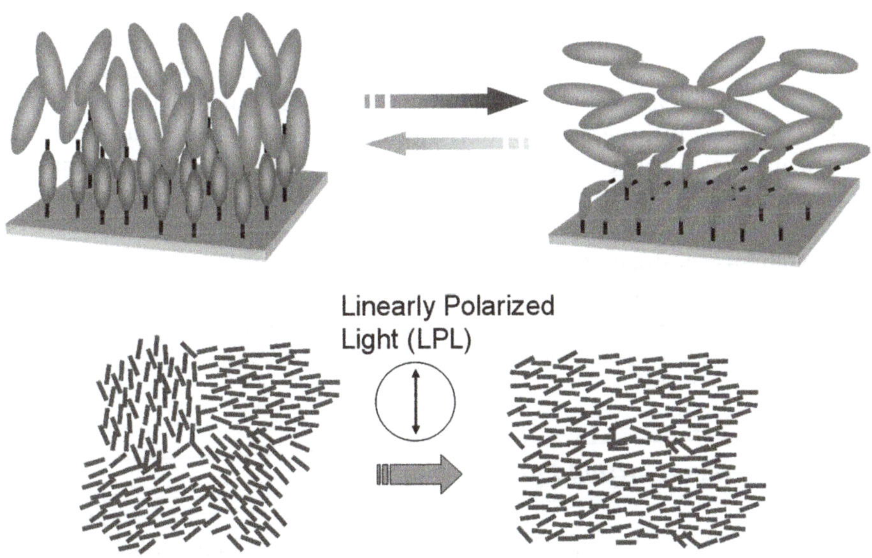

Fig. 5.2 Schematic illustration of the command surface for orientational switching of liquid crystal molecules on a photoresponsive surface of azobenzene monolayer (*upper*) and a scheme of photoinduced orientation of LC molecular assemblies by linearly polarized light (LPL)

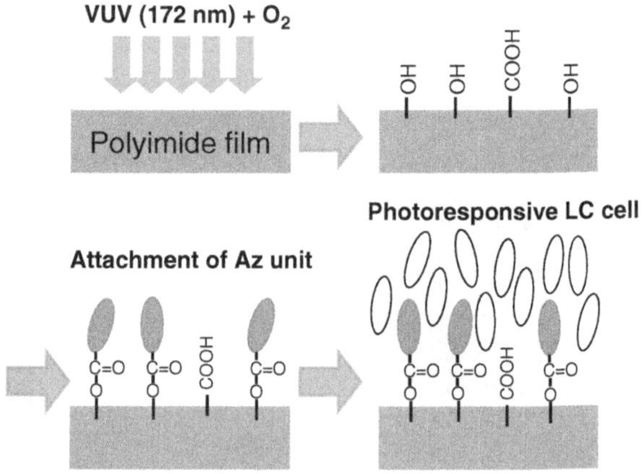

Fig. 5.3 Proposed versatile photochemical process for the direct introduction of photoresponsive units on a polyimide film. The film works as a photoresponsive command surface of a nematic LC cell [17]

angular selective photoreaction (lower scheme of Fig. 5.2). After 20 years since the discovery, photoalignment technique is now eventually adopted in the production process of LC display devices as the improved replacement of the rubbing alignment procedure.

The surface photoreactive layer is mostly prepared by spincast method, Langmuir–Blodgett technique, self-assembled monolayers (SAM), etc. In these procedures, photoreactive layers are separately prepared on the substrate. Direct covalent attachment of photoreactive groups onto a substrate via a photochemical surface modification was newly proposed by Sasaki et al. [17]. Vaccum ultraviolet (VUV) light irradiation in the presence of a trace of oxygen to a polyimide surface selectively generates hydroxyl groups on the topmost surface. This chemically functionalized surface can be utilized to attach Az derivatives or cinnamate units (Fig. 5.3). The introduced amount of Az derivative is almost a monolayer level on the polyimide film surface. When a nematic LC cell is fabricated with this substrate, the photoirradiation reversibly changes the nematic LC alignment and thus works as an effective command surface. In principle, a variety of polymer materials can be used as the substrate. This procedure should be versatile and widely applicable.

5.2.2 *Surface-Grafted Polymer Chains*

The photoalignment of Az mesogens in side-chain polymer in thin films including spincast films and LB layers is widely investigated. In these films, the smectic layers are oriented parallel to the substrate plane and the Az mesogens in the normal direction. This section introduces a system in which the orientation is just reversed by anchoring one end of the polymer to the substrate. In recent years, polymer brushes have become fascinating subjects in polymer research. They are defined as dense layers of end-grafted polymer chains confined to a solid surface or interface

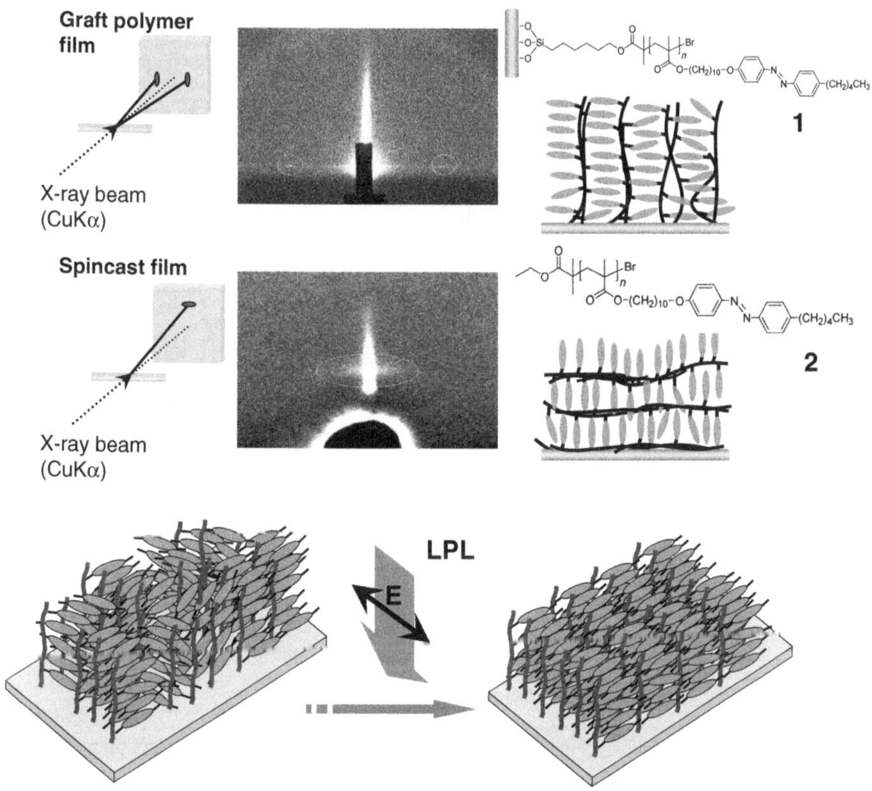

Fig. 5.4 The mesogenic group and smectic layer orientations of grafted and spincast thin films of an LC Az polymer. Surface-grafted polymer **1** and the corresponding free polymer **2** have the identical chemical structure. In the upper figures, GI-XRD patterns are shown. The scheme below indicates the illustration of LPL induced in-plane reorientation of the Az mesogenic groups [22, 23]

[18, 19]. Recently, high graft density and well-defined polymer brush structures have been developed using controlled radical polymerization, such as atom transfer radical polymerization (ATRP) [20, 21]. Since ATRP can be performed for various types of monomers, many functional polymer brushes can be designed.

Uekusa et al. [22] have synthesized LC polymer brushes bearing an Az mesogenic group in the side chains by adopting surface-initiated ATRP (**1** in Fig. 5.4). UV–visible spectroscopic and grazing incidence angle X-ray diffraction (GI-XRD) measurements reveal that the periodic layer structure of smectic LC phase orients perpendicular to the substrate plane with Az mesogens being aligned parallel to the substrate. In comparison with a spincast film of the identical homopolymer (**2** in Fig. 5.4), the molecular and layer orientations are just reversed. Since the Az mesogens are oriented parallel to the substrate for the surface-tethered chain, this orientation feature enhances the probability of light absorption, leading to a highly ordered in-plane photoalignment when exposed to LPL (Fig. 5.4) [23]. This level of high orientational order is not attained for a spincast film of **2**.

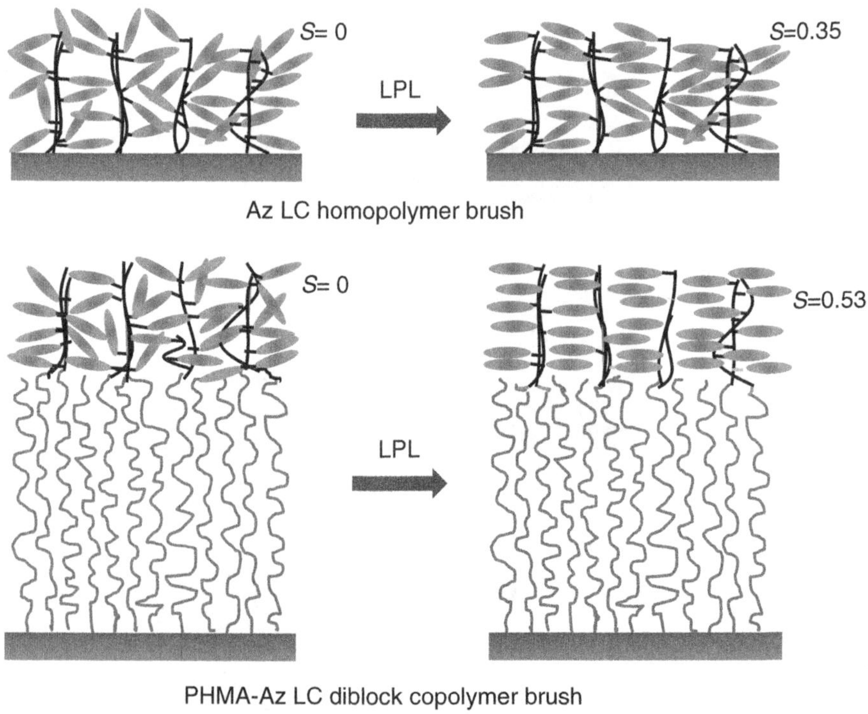

Fig. 5.5 Introduction of an amorphous flexible chain block underneath the LC Az polymer block. The flexible chain block plays a significant role as the lubricant layer for the improvement of photoalignment [24]

This approach has been extended to diblock copolymer brush systems by Haque et al. [24]. Here, an amorphous chain with a low glass transition temperature (T_g) of poly(hexyl methacrylate) (PHMA, $T_g = -20$ °C) is introduced between the Az chain block and the substrate surface. The surface-tethered diblock copolymer was synthesized by two-step surface-initiated ATRP method. In this architecture, three obvious effects are recognized. (1) Even higher optical anisotropy is obtained (Fig. 5.5). (2) The flexible chain affords more favorable packing for the LC Az mesogens. (3) The in-plane rotational motions, i.e., photoreorientation with an orthogonal LPL irradiation, are more readily achieved. These lubricant effects agree with the well-known ones in the molecular design in the side-chain LC polymers, namely, the existence of a spacer connecting the main chain and the mesogenic group plays an important role to decouple the influence of main chain in the assembling nature of mesogenic groups [25]. This approach expands this molecular design concept to macromolecular levels of block copolymers. The effect of the buffer chain length of PHMA has been further examined. The longer flexible chain leads to the higher in-plane orientational order and more efficient reorientations [26].

5.3 Photocontrol of Microphase Separation Structures of Block Copolymers

Microphase separation (MPS, regular meso-structure formation) of block copolymers in thin films has been a subject of intensive study from the viewpoint of self-assembly in a confined state [27, 28]. From practical viewpoints, such meso-structures also have been receiving considerable attention for fabrication of ever-smaller feature sizes than those obtained by the photolithography process [29–33]. The MPS morphology is generally altered by changing the block length ratio of the different polymer components. When the morphological structure and alignment of MPS is altered by irradiation procedure, new possibilities in the block copolymer technology will be developed [34].

5.3.1 Langmuir–Blodgett Monolayers

Photomechanical response is observed in monolayers of Az-containing amphiphilic polymers on a water [35–38] or mica [39, 40] surface. UV and visible light irradiation induces reversible expansion and contraction, respectively, of the monolayer. Such light-induced modulations in the monolayer have been successfully attained by Kadota et al. [41] for a triblock copolymer (**3** in Fig. 5.6). The *trans* and *cis* isomeric states of Az result in mesoscaled dot and strip morphologies, respectively, of the Az-containing domain. The light-induced morphology changes are essentially reversible [42] (Fig. 5.6).

Alignment of MPS structure in the monolayer state is a highly challenging subject because, in such 2D state, the conformational freedom of the chains is highly restricted. Aoki et al. [43] unexpectedly came across a phenomenon in which the MPS structure is vanished by UV irradiation in a block copolymer monolayer comprising an Az polymer and poly(dimethyl siloxane) (PDMS) (**4** in Fig. 5.7). In the *trans* state of Az, the monolayer exhibits a stipe morphology of Langmuir monolayer on the water surface. In contrast, the stripe morphology is fully lost upon UV light illumination. When LPL of 436 nm light is successively irradiated, clear appearance of stripe morphology with the anisotropic morphology is observed, the stripe patterns being perpendicular to the electric vector of LPL. Thus, by way of the UV-induced morphology "reset," the morphology alignment becomes possible. In Fig. 5.7, light irradiation is performed on a hydrophilic substrate surface (clean quartz plate) at high humidity conditions. UV irradiation brings about an expansion of the monolayer, resulting in the generation of larger-scale protrusions. After the successive LPL at 436 nm, the monolayer shrinks and anisotropic MPS pattern is generated.

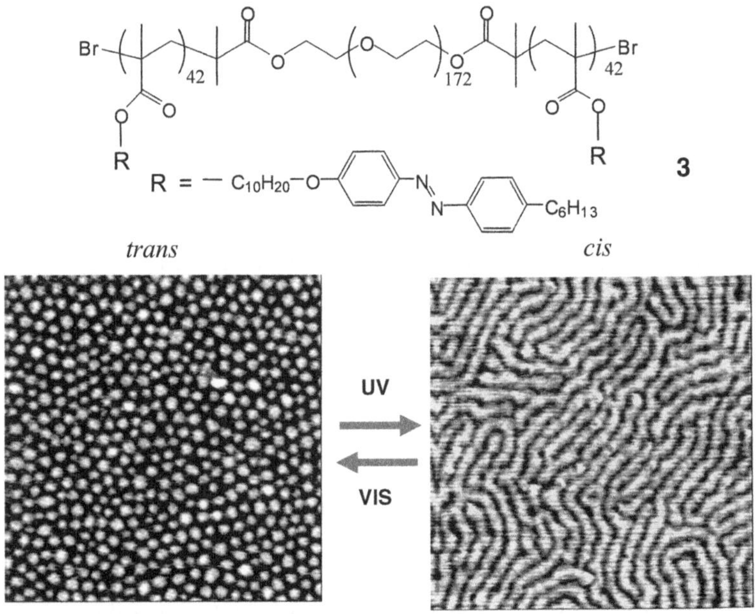

Fig. 5.6 Photocontrolled MPS structure of block copolymer **3** on water. 3 μm × 3 μm topological AFM images are indicated. *Trans*- and *cis*-Az isomers provide the dot and stripe meso-structure, respectively. The morphological change is essentially reversible [41]

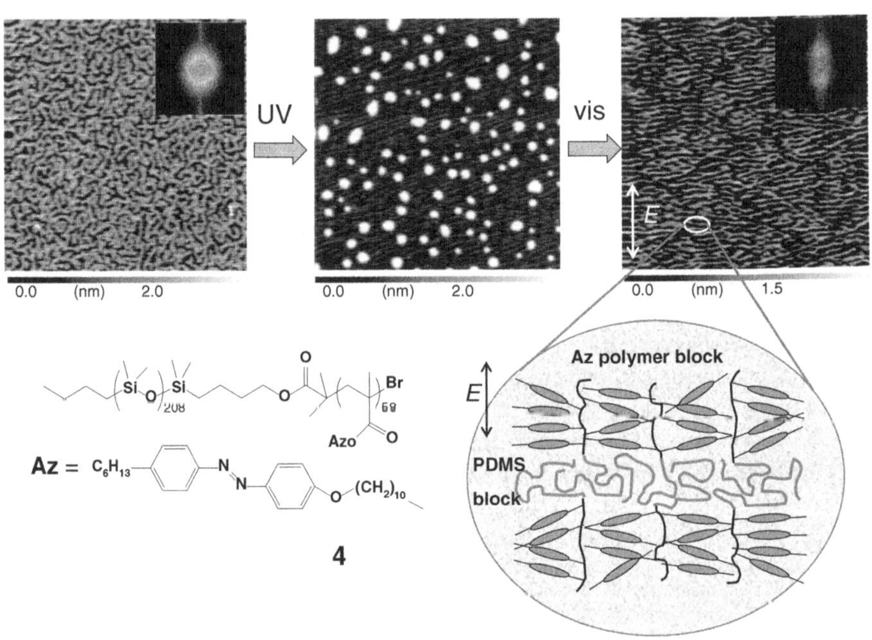

Fig. 5.7 AFM images (2 μm × 2 μm) indicating the photoaligning process of monolayered MPS stripe structure of **4** on a mica surface at a high humidity. The scheme illustrates the molecular packing model in the photoaligned monolayer [43]

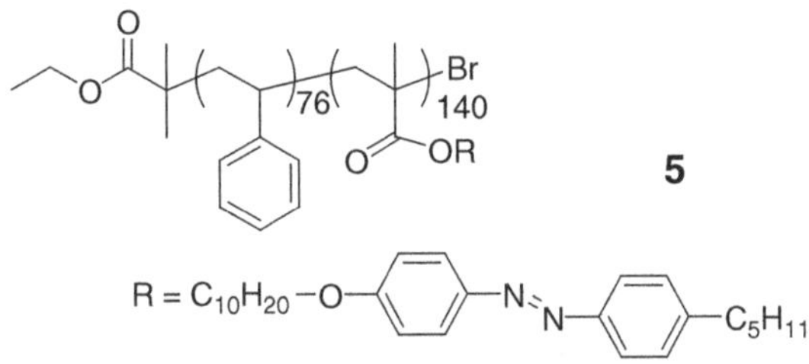

5

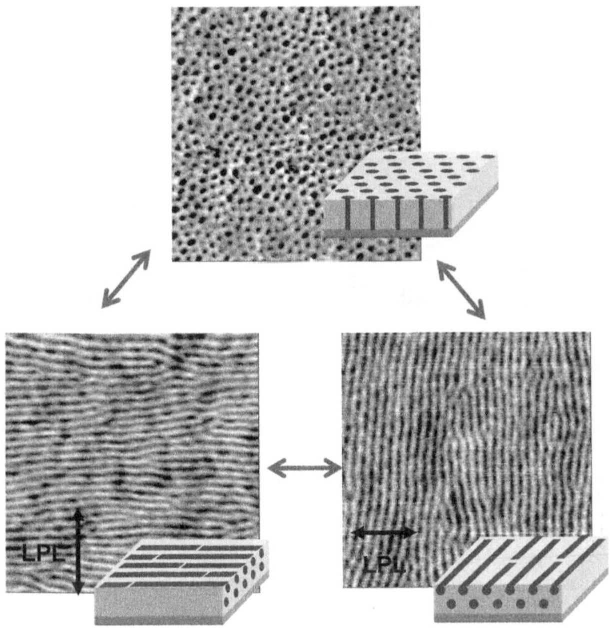

Fig. 5.8 Light-directed orientation changes of MPS nanocylinders of polystyrene in a thin film of **5**. The figures indicate 3 μm × 3 μm phase mode AFM images. The out-of-plane and in-plane directions are convertible by photoirradiation with each other [50]

5.3.2 Spincast Thin Films

Many efforts have been made to induce macroscopic alignment of the MRS patterns from viewpoints of new nanotechnologies [44]. When block copolymers with LC nature is employed, the regular MPS structure is formed over large areas by the cooperative effect. The importance of liquid crystallinity for large-scale alignment of MPS structure is demonstrated by Iyoda et al. [45, 46] by using a series of poly(ethylene oxide) (PEO)-*block*-Az LC polymers.

The obvious effect of LPL irradiation to align MPS in block copolymer films in the in-plane direction is demonstrated for soft PEO-based Az-containing block copolymers [47–49] and a polystyrene-based block copolymer [50] (**5** in Fig. 5.8). The MPS structure of mesocylinders of the polystyrene block aligned orthogonal to

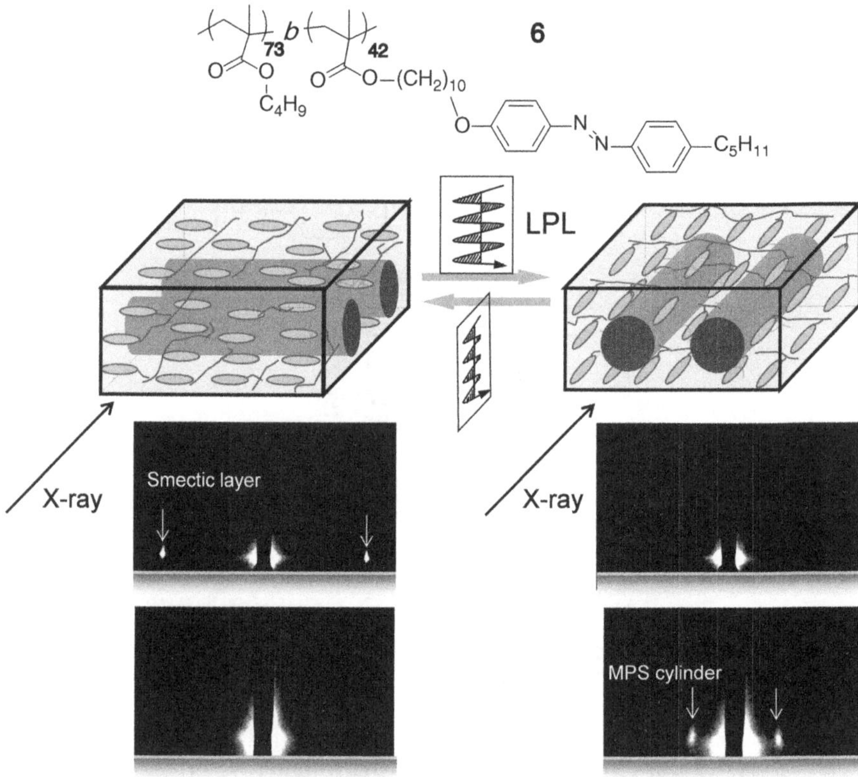

Fig. 5.9 Scheme of repeated reorientation of the MPS cylinder structure in a thin film of **6** via alternative irradiation with LPL orthogonal with each other. The images below indicate synchrotron GI-SAXS profiles for the corresponding orientation taken with the X-ray shown by an arrow [51]

the direction of the electric field vector of the irradiated LPL. This direction can be further altered to another in-plane direction corresponding to the subsequent illumination of LPL. Irradiation with a nonpolarized light in the normal incidence leads to the out-of-plane orientation of the cylinders again. These independent alignments are convertible with each other (Fig. 5.8). This is a unique example that a single material can generate three independent stable states, which derive from the memories of directions, not from the state of the matter. For a rigid segment polymer of **5**, the control of annealing temperature is an important factor for the successful alignment. The film is initially heated to 130 °C, the isotropic temperature of the Az polymer, where the erasure of the MPS cylinder structure is attained ("reset" process). Then, the film is irradiated with continuous LPL on a slow-cooling process at LC temperatures where the MPS cylinder evolves in the controlled direction.

Nagano et al. [51] have recently succeeded in conducting the real-time observations of the in-plane reorientation of MPS structure by using a homologous block copolymer possessing a low T_g poly(butyl methacrylate) (PBMA) block (**6** in Fig. 5.9). Here, grazing incidence small-angle X-ray scattering (GI-SAXS) measurements have been achieved at the BL-15A beam line of the Photon Factory (KEK-PF, synchrotron orbital radiation) in Tsukuba. The MPS cylinder structure of PBMA in a film of **6** is initially aligned uniaxially by LPL, and successive LPL with the direction rotated by 90° is performed at 90 °C. At this temperature, the LC Az block adopts the smectic A phase. The initial oriented structures of smectic LC and MPS cylinder diminish quickly within 30 s, and the structures in the orthogonal direction appear rather slowly in 300 s. These experiments reveal that strong cooperative and synchronized motions occur at the two different hierarchical structures of the LC layer (molecular level) and the MPS domains (mesoscopic level). In this dynamic reorientation process, there remain open questions on the mechanism as to how such meso-structures can be reoriented isothermally simply by irradiating with LPL at LC temperature. The experiments are conducted always at the smectic LC temperature without going through a heating "reset" process at isotropic temperatures. Schmidt et al. [52] conducted in situ synchrotron SAXS measurements and performed a computer simulation of data for the mechanism of the reorientation of a solvent-swollen block copolymer driven by an electric field. They suggested that the reorientation mechanism involves a domain nucleation and growth process or a grain rotation, which seems to also be the case for the present light-driven system.

5.4 Mass Migrations for Surface Relief Formation

5.4.1 Features of Liquid Crystalline Polymer Materials

Surface relief gratings (SRG, regular topological surface modification) formed via irradiation with an interference pattern of coherent argon ion laser beam (488 or 512 nm) were found in 1995 [53, 54] and are perhaps the most interesting target in the current research in Az polymers. A great deal of data has been accumulated rapidly due to its basic phenomenological interest and its technological applications [55, 56]. Another type of photoactivated mass migration systems has been proposed by our group using soft LC polymer systems such as binary component materials [57] and random copolymers containing an oligo(ethylene oxide) segment [58, 59]. The mass migration is completed at surprisingly small dose levels (<100 mJ/cm^2), which is three orders of magnitude smaller than those required for the conventional amorphous polymer systems. Recent explorations have revealed that the photochemical phase change of the smectic LC to isotropic phase is essential for the migration (Fig. 5.10) [60–62]. The mass transfer occurs at the boundary between the smectic phase (*trans*-Az rich) and isotropic one (*cis*-Az rich). The motions direct from the smectic to isotropic regions. The driving factor is thermal process in essence and seems to be initiated from the disparity in the surface tension and fluidity of the two phases.

Isayama et al. [62] have synthesized a series of random block copolymers consisting the Az mesogenic and hexyl chains (**7** in Fig. 5.11) systematically changing

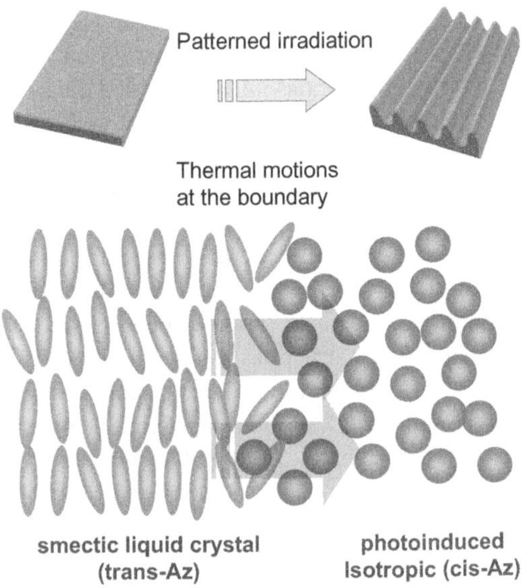

Fig. 5.10 Schematic illustration of photoinduced mass transport to form a surface relief structure (*upper*) and the transport motions at the boundary between the LC (*trans*-Az rich) and isotropic (*cis*-Az rich) phases

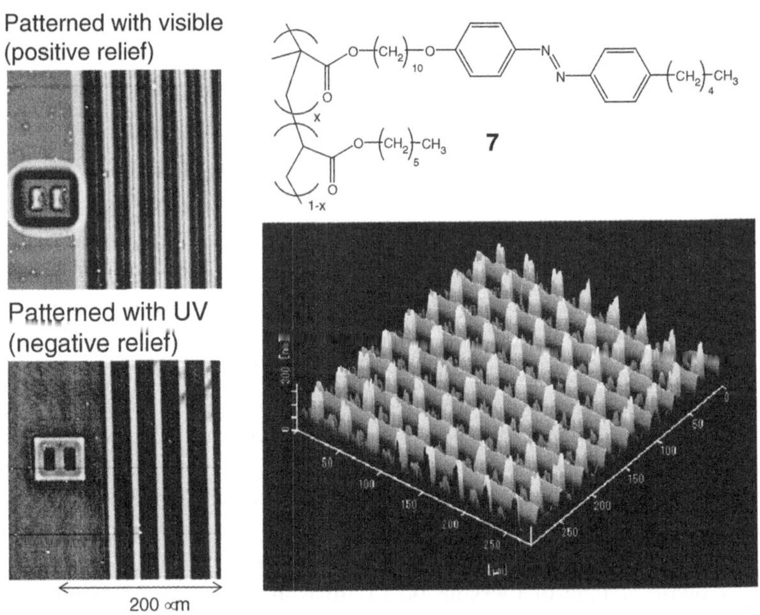

Fig. 5.11 A series of random copolymers composed of Az side group and hexyl group (**7**) [62]. Topographical AFM images display examples of relief structures. For details, see the text

the copolymerization ratio. This approach has revealed that mass migration behavior is strongly coupled with the thermal properties of the film material. The optimal mass migration occurs at temperatures above T_g and of smectic A of the polymer. This finding is of practical significance in that the mass migration can be induced for almost any Az LC polymers or materials when the experimental temperature is properly adjusted. Until this study, experiments had been achieved at room temperature where the phototriggered mass migration takes place only for limited materials. Having the above knowledge, many types of migration processes have become feasible. The migrating direction with respect to the photomask pattern can be reversed by selecting the wavelength of irradiation. The positive and negative relief structures with respect to the mask pattern are obtained by using visible and UV light, respectively (Fig. 5.11, *left*). In UV irradiation process, more complicated relief structure can be fabricated by dual irradiation (Fig. 5.11, *right*).

5.4.2 Supramolecular Strategy

The Az unit is essential for the photoinduced mass migration, but after the relief formation the existence of this strongly light-absorbing chromophore can be a drawback for optical applications such as waveguide coupler and LC alignment layer. To overcome this issue, the supramolecular strategy employing hydrogen bonding proposed by Kato [63, 64] would be a promising approach.

Zettsu et al. [65] have synthesized a supramolecular polymer in which an Az-imidazole base is introduced via hydrogen bonding (**8** in Fig. 5.12). The complexation is attained with a random copolymer consisting of poly(acrylate) with 4-oxybenzoic acid moiety as the hydrogen-bonding donor part and poly(methacrylate) with a flexible oligo(ethylene oxide) chain as a plasticizing and cross-linking part. The selective extraction of the Azo-imidazole from the film is performed after the chemical cross-linking of the polymer under mild conditions in a vapor phase [58, 59]. The Az unit attached via hydrogen bonding can be readily extracted selectively by ethanol or tetrahydrofuran from the film with retention of the periodic relief feature (AFM images of Fig. 5.12). For the purpose of cross-linking, the photochemical procedure employing [2+2] dimerization of cinnamate units is an alternative effective approach [66]. After the photo-cross-linking, the relief structure is maintained up to 300 °C.

The strategy of non-covalent attachment of Az side chain has also been reported by Kulikovska et al. [67]. They adopted the ion complexation between an Az compound having carboxylic acid and a cationic polyelectrolyte. This group has proposed another system for detaching Az unit after SRG formation [68]. The film is prepared by simply spin coating an isocyanate oligomer with 4-aminoazobenzene in the presence of a catalyst, which leads to the formation of urea unit between the oligomer and the Az unit. After the SRG is formed, the decoloration (detachment of Az unit) is readily achieved by simply heating at 270 °C. The structure remains unchanged at temperatures up to 300 °C.

5.4.3 Organic–Inorganic Hybrid Materials

The organic–inorganic hybrid materials have superior features possessing both flexibility and functionality of organic components, and chemical, thermal, and mechanical stability of inorganic ones. Little effort has been so far made to fabricate hybrid

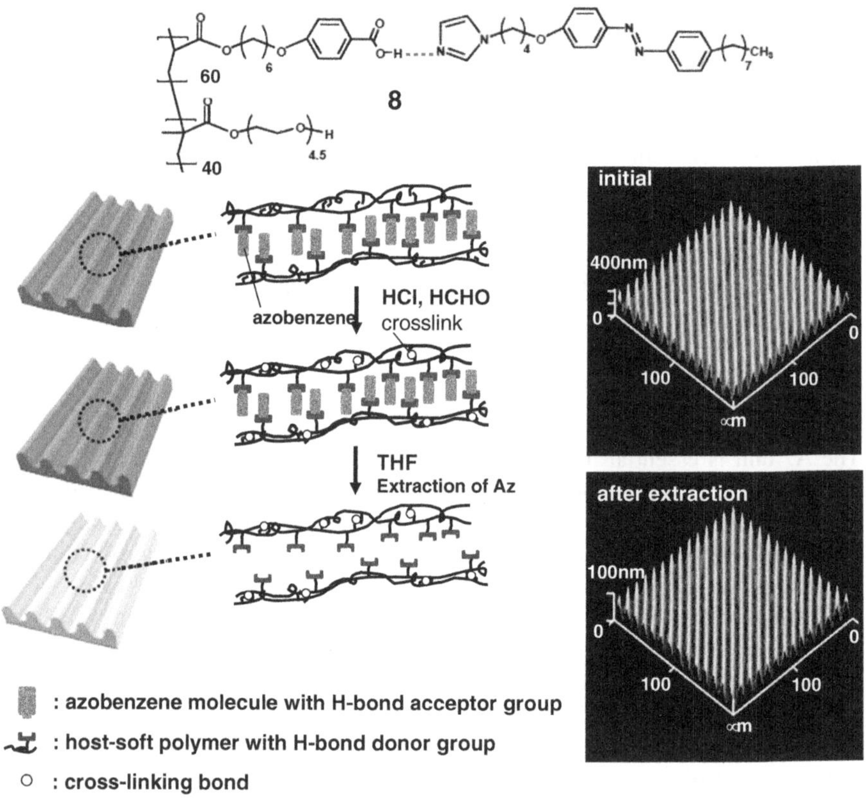

Fig. 5.12 A supramolecular strategy to remove Az unit from the relief film of **8**. Topographical AFM images display examples of relief structures before and after extraction of Az units. For details, see the text

systems that show photoinduced mass migration. In this respect, Darracq et al. [69] reported a pioneering work using Az-containing silica-based sol–gel system. They succeeded in fabricating the SRGs on the hybrid film by holographic irradiation of argon ion laser beam, but in this amorphous-type hybrid, a vast amount of recording energy as 60–140 J/cm^2 (0.6–1.4 J/mm^2) is required. More recently, Kulikovska et al. [70] reported a supramolecular sol–gel material system based on the ionic interaction between an oppositely charged Az unit and silica for optical generation of surface relief structures. The modulation depths of the recorded gratings are large, but also in this type hybrid material, a large amount of recording energy as 30–900 J/cm^2 is required. The mass transfer based on the LC framework is expected to lead to substantial reduction of light dose required.

Nishizawa et al. [71] have developed a new type of Az-containing hybrid material that shows the LC nature. In this case, titanium oxide (titania) is adopted as the inorganic component which is anticipated to be converted to photofunctional materials such as photocatalyst via removal of the organic component. An LC titania–Az

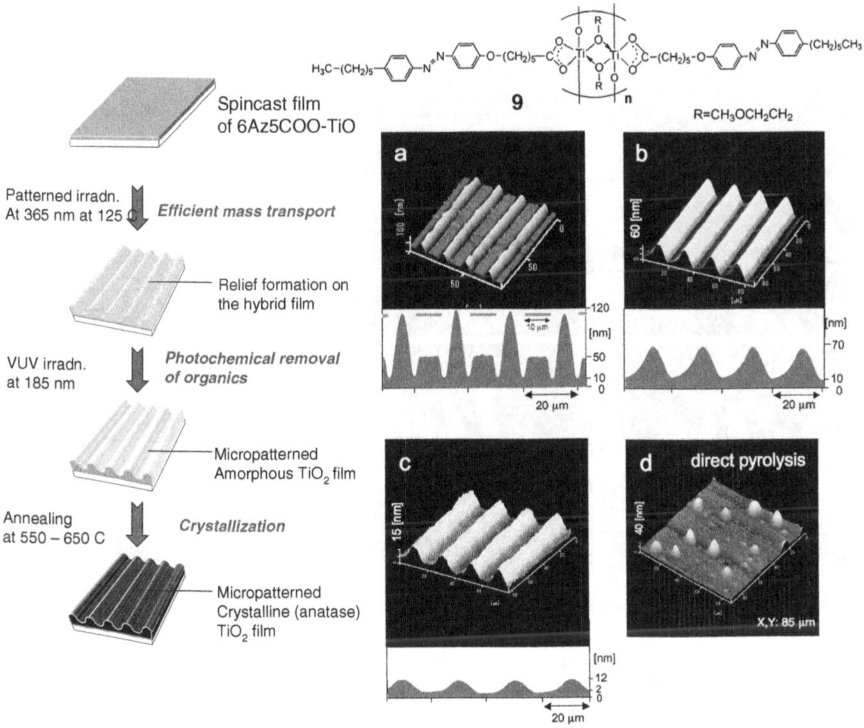

Fig. 5.13 Flowchart of the process for removal of Az from the relief film of a hybrid material of **9**. In the right side, topographical AFM images and cross-sectional height profiles are shown for the photoinscribed film of **9 at** 125 °C through a photomask irradiation (**a**), the photochemically decomposed film using a low-pressure Hg lamp for 20 min (**b**), the pyrolized film at 650 °C (**c**), and directly pyrolyzed film at 550 °C without going through the photodecomposition (**d**) [71, 72]

hybrid material (9 in Fig. 5.13) shows a smectic phase (probably smectic A) in the *trans*-Az form, in which organic Az and titania layers are alternatively piled up in the thin film at a period of ca. 5 nm. The irradiation with 365 nm light brings about the phase transition to the isotropic, just as observed for polymer materials. The mass transfer can be performed by patterned UV irradiation through a photomask of very low energy doses (ca. 10^3–10^5 times less than hitherto known sol–gel materials) at 125 °C. The topological pattern of the hybrid obtained above can be converted to pure titanium oxide [72]. The process for removal of organics involves two-step procedures. At the initial step of decomposition of organic components, the relief film (a) is exposed to UV light at 185 nm of a low-pressure Hg lamp for 20 min (b). The photochemical decomposition process modified the morphological characteristics from the two-level-height relief to a simply undulated one with ca. 70 nm modulation depth (b). Next, this film is annealed at 550 °C for 1 h and subsequently at 650 °C for 1 h to achieve full pyrolysis. The periodical feature of the relief structure is essentially retained after the annealing. A direct pyrolysis procedure without the

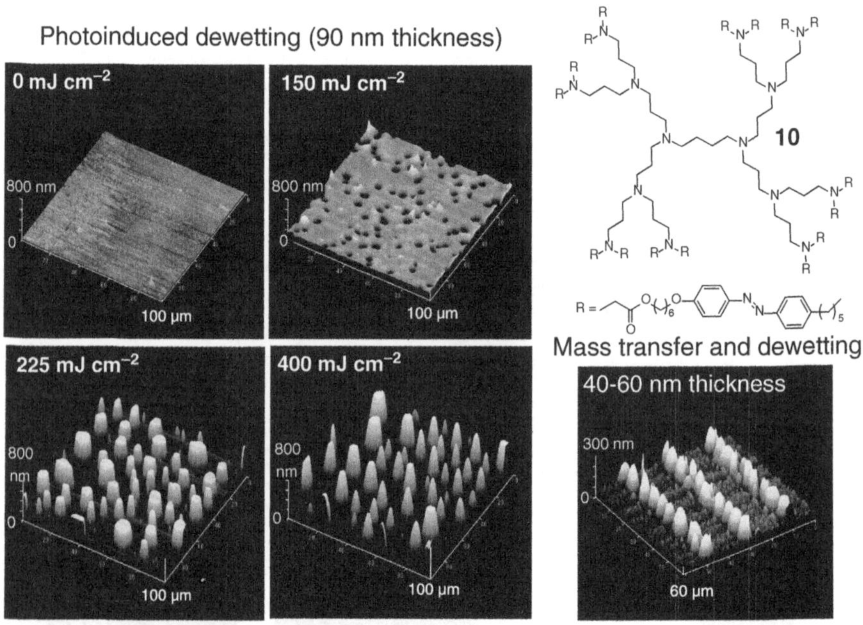

Fig. 5.14 Photoinduced dewetting behavior (morphological change) of a film of **10** on a hydrophilic surface by uniform UV irradiation (*right*). AFM image at the bottom right shows an example of a hierarchical relief structure produced via a line-and-space patterned UV irradiation consorted with the surface dewetting [75]

photochemical treatment collapses the relief structure (d). X-ray analysis and TEM observation reveal that the anatase-type crystal grains are obtained.

5.4.4 *Dendritic Materials*

Phototriggered mass migrations of LC materials have been achieved only for linear polymers. The dendritic scaffold of constituting material in the thin film would exhibit some characteristic effects in phototriggered migrating behavior, because it has been pointed out that the dendritic topology eliminates chain entanglement and interlacing with the neighboring molecules. Yonetake et al. [73] synthesized and evaluated the LC properties of poly(propylene imine) dendrimers peripherally modified with cyanobiphenyl mesogens. The Az dendrimer employed here adopts the identical scaffold of such dendrimers with the second generation (G2) (16 mesogens, **10** in Fig. 5.14).

Li et al. [74, 75] have shown that, with this dendritic architecture, particular photoinduced dewetting takes place on a hydrophilic surface upon UV light irradiation. Figure 5.14 shows the topographical AFM images observed in the photoinduced dewetting process of a film of **10** with an initial thickness of 90 nm. Upon uniform UV irradiation, the film exhibited drastic morphological changes. The initial flat state (0 mJ/cm^2) starts to exhibit dewetting at 150 mJ/cm^2. At this stage, holes of 3–5 μm diameter with the film depth were formed. Around 200 mJ/cm^2, the continuous film became separated and the morphology is characterized by formation of separated dome structures (225 mJ/cm^2). The domes possessed round bottoms of diameters of several micrometers and this basal bottom is not virtually changed upon further irradiation. At the early stage the top of domes is flat, which may reflect that the layer structure is involved. Prolonged irradiation leads to a significant growth of dome height, and the flat top is lost. At 400 mJ/cm^2, the average height of domes reaches 770 nm, which corresponds to approximately eightfold of the initial thickness.

The dendrimer film is then subjected to the mass migration by UV irradiation with using a photomask at room temperature instead of uniform irradiation. The right bottom AFM image in Fig. 5.14 shows an example of the resulting morphology after a stripe patterned irradiation. The domes are arrayed as an ordered single-column lines. In this way, the phototriggered mass transport consorted with photoinduced surface dewetting to form hierarchical surface morphologies by single patterned irradiation. A new aspect on the photofabrication process is proposed.

5.5 Conclusion

Due to the strong cooperativity in LC materials, the introduction of a photoresponsive unit (typically an Az moiety) into molecular or polymer systems gives rise to large amplification photoresponses. Such effects can be exemplified at a wide range of size features, including the molecular nanoscales (photoresponsive surface alignment of liquid crystalline materials, effective molecular photoalignment in a surface-grafted polymer film), mesoscales (10–100 nm) (photoalignment of microphase separation structures of block copolymers), and micrometer scales (mass migrations). Az derivatives are typical, simple, and classical photochromic molecules; however, there seem to remain broad ranges of new applications when they are coupled with molecular assembly systems. The topics dealt with in this chapter are mostly dealt with quasi-2D film systems, but the ultimate goal would be to create sophisticated dynamic 3D systems similar to the biological ones. It is anticipated that accumulations on the various types of light-directed motions at various scales of photoresponsive films in 2D state in ordered media will provide proper clues to future directions of materials processing and applications.

Acknowledgements We would like to acknowledge our students and collaborators for their great efforts in this research project. The synchrotron X-ray scattering experiments were performed at KEK-Photon Factory (proposal No. 2007G047) in collaboration with Prof. Y. Amemiya and Prof. Shinohara at the University of Tokyo. We also thank Mr. T. Hikage of Nagoya University for technical assistance with the X-ray measurements. The Az dendrimers were kindly provided from Prof. K. Yonetake of Yamagata University. The work is mostly supported by the Grant-in-Aid of Scientific Research on Priority Areas, "New Frontiers in Photochromism (No. 471)."

References

1. Irie M (2000) Diarylethenes for memories and switches. Chem Rev 100:1685–1716
2. Sekkat Z, Knoll W (eds) (2002) Photoreactive organic thin films. Academic, San Diego
3. Zhao Y, Ikeda T (eds) (2009) Smart light-responsive materials—azobenzene-containing polymers and liquid crystals. Wiley, Hoboken
4. Natansohn A, Rochon P (2002) Photoinduced motions in azo-containing polymers. Chem Rev 102:4139–4175
5. Seki T (2004) Dynamic photoresponsive functions in organized layer systems comprised of azobenzene-containing polymers. Polym J 36:435–454
6. Seki T (2007) Smart photoresponsive polymer systems organized in two dimensions. Bull Chem Soc Jpn 80:2084–2109
7. Ikeda T (2003) Photomodulation of liquid crystal orientations for photonic applications. J Mater Chem 13:2037–2057
8. Ikeda T, Mamiya J, Yu Y (2007) Photomechanics of liquid-crystalline elastomers and other polymers. Angew Chem Int Ed 46:506–528
9. Yu H, Ikeda T (2011) Photocontrollable liquid–crystalline actuators. Adv Mater 23:2149–2180
10. Wang Y, Li Q (2012) Photo-stimulated phase transformations in liquid crystals and their non-display applications. In: Li Q (ed) Liquid crystals beyond displays. Wiley, Hoboken, pp 157–212 (Chap. 5)
11. Ichimura K (2000) Photoalignment of liquid crystal systems. Chem Rev 100:1847–1873
12. Schadt M, Schmitt K, Kozinkov V, Chigrinov V (1992) Surface-induced parallel alignment of liquid crystals by linearly polymerized photopolymer. Jpn J Appl Phys 31:2155–2164
13. Gibbons WM, Shannon PJ, Sun ST, Swetlin BJ (1991) Surface-mediated alignment of nematic liquid crystals with polarized laser light. Nature 351:49–50
14. Zebger I, Rutloh M, Hoffmann U, Stumpe J, Siesler HW, Hvilsted S (2003) Photoorientation of a liquid–crystalline polyester with azobenzene side groups: effects of irradiation with linearly polarized red light after photochemical pretreatment. Macromolecules 36:9373–9382
15. Ichimura K, Suzuki Y, Seki T, Hosoki A, Aoki K (1988) Reversible change in alignment mode of nematic liquid crystals regulated photochemically by "command surfaces" modified with an azobenzene monolayer. Langmuir 4:1214–1216
16. Seki T, Sakuragi M, Kawanishi Y, Tamaki T, Fukuda R, Ichimura K (1993) "Command surfaces" of Langmuir–Blodgett films. Photoregulation of liquid crystal alignment by molecularly tailored surface azobenzene layers. Langmuir 9:211–218
17. Sasaki A, Aoshima H, Nagano S, Seki T (2012) A versatile photochemical procedure to introduce a photoreactive molecular layer onto a polyimide film for liquid crystal alignment. Polym J 44:639–645
18. Advincula RC, Brittain WJ, Caster KC, Rühe J (eds) (2004) Polymer brushes. Weinheim, Wiley
19. Milner ST (1991) Polymer brushes. Science 251:905–914
20. Tsujii Y, Ohno K, Yamamoto S, Goto A, Fukuda T (2006) Structure and properties of high-density polymer brushes prepared by surface-initiated living radical polymerization. Adv Polym Sci 197:1–45
21. Edmondoson S, Osborne VL, Huck WTS (2004) Polymer brushes via surface initiated polymerizations. Chem Soc Rev 33:14–22
22. Uekusa T, Nagano S, Seki T (2007) Unique molecular orientation in a smectic liquid crystalline film attained by surface-initiated graft polymerization. Langmuir 23:4642–4645
23. Uekusa T, Nagano S, Seki T (2009) Highly ordered in-plane photoalignment attained by the brush architecture of liquid crystalline azobenzene polymer. Macromolecules 42:312–318
24. Haque HA, Nagano S, Seki T (2012) Lubricant effect of flexible chain in the photoinduced motions of surface-grafted liquid crystalline azobenzene polymer brush. Macromolecules 45:6095–6103

25. Finkelmann H, Ringsdorf H, Wendorff JH (1978) Model considerations and examples of enantiotropic liquid crystalline polymers. Polyreactions in ordered systems. Macromol Chem 14(179):273–276
26. Haque HA, Nagano S, Seki T (2012) Effect of flexible chain length on the photoorientation behavior of surface-grafted liquid crystalline azobenzene block copolymer brush. Mol Cryst Liq Cryst (in press)
27. Green PF, Limary R (2001) Block copolymer thin films: pattern formation and phase behavior. Adv Colloid Interface Sci 94:53–81
28. Krausch G (1995) Surface induced self-assembly in thin polymer films. Mater Sci Eng R14:1–94
29. Park C, Yoon J, Thomas EL (2003) Enabling nanotechnology with self-assembled block copolymer patterns. Polymer 44:6725–6760
30. Park M, Harrison C, Chaikin PM, Register RA, Adamson DH (1997) Block copolymer lithography: periodic arras of 10^{11} holes in 1 square centimeter. Science 276:1401–1404
31. Widawski G, Rawiso M, François B (1994) Self-organized honeycomb morphology of star-polymer polystyrene films. Nature 369:387–389
32. Shidorenko A, Tokaref I, Minko S, Stamm M (2003) Ordered reactive nanomembranes/nanotemplates from thin films of block copolymer supramolecular assembly. J Am Chem Soc 125:12211–12216
33. Turn-Albrecht T, Schotter J, Kästle GA, Emley N, Shibauchi T, Krusin-Elbaum L, Guarini K, Black CT, Tuominen MT, Russell TP (2000) Ultrahigh-density nanowire arrays grown in self-assembled diblock copolymer templates. Science 290:2126–2129
34. Tong X, Zhao Y (2012) Functional liquid crystalline block copolymers: order meets self-assembled nanostructures. In: Li Q (ed) Liquid crystals beyond displays. Wiley, Hoboken, pp 285–302 (Chap. 8)
35. Seki T, Sekizawa H, Morino S, Ichimura K (1998) Inherent and cooperative photomechanical motions in monolayers of an azobenzene containing polymer at the air-water interface. J Phys Chem B 102:5313–5321
36. Kago K, Fürst M, Matsuoka H, Yamaoka H, Seki T (1999) Direct observation of photoisomerization of polymer monolayer on water surface by X-ray reflectometry. Langmuir 15:2237–2240
37. Ohe C, Kamijyo H, Arai M, Adachi M, Miyazawa H, Ito K, Seki T (2008) Sum frequency generation spectroscopic study on photoinduced isomerization of a poly(vinyl alcohol) containing azobenzene side chain at the air water Interface. J Phys Chem C 112:172–178
38. Ichimura K, Seki T (2011) Photoinduced motion associated with monolayers. In: Feringa BL, Browne WR (eds) Molecular switches, 2nd edn. Wiley, Weinheim, pp 629–668 (Chap. 17)
39. Seki T, Kojima J, Ichimura K (2000) Multifarious photoinduced morphologies in monomolecular films of azobenzene side chain polymer on mica. Macromolecules 33:2709–2717
40. Seki T, Kojima J, Ichimura K (1999) Light-driven dot films consisting of single polymer chain. J Phys Chem B 103:10338–10340
41. Kadota S, Aoki K, Nagano S, Seki T (2005) Photocontrolled microphase separation of a block copolymer in two dimensions. J Am Chem Soc 127:8266–8267
42. Kadota S, Aoki K, Nagano S, Seki T (2006) Morphological conversions of nanostructures in monolayers of an ABA triblock copolymer having azobenzene moiety. Colloids Surf A 284(285):535–541
43. Aoki K, Iwata T, Nagano S, Seki T (2010) Light-directed anisotropic reorientation of mesopatterns in block copolymer monolayers. Macromol Chem Phys 211:2484–2489
44. Lazzari M, De Rosa C (2008) Methods for the alignment and the large-scale ordering of block copolymer morphologies. In: Block copolymers in nanoscience. Wiley-VCH, Weinheim, pp 191–231 (Chap. 9)
45. Tian Y, Watanabe K, Kong X, Abe J, Iyoda T (2002) Synthesis, nanostructures, and functionality of amphiphilic liquid crystalline block copolymers with azobenzene moieties. Macromolecules 35:3739

46. Asaoka S, Uekusa T, Tokimori H, Komura M, Iyoda T, Yamada T, Yoshida H (2011) Normally oriented cylindrical nanostructures in amphiphilic PEO-LC diblock copolymers films. Macromolecules 44:7645–7658
47. Yu HF, Iyoda T, Ikeda T (2006) Photoinduced alignment of nanocylinders by supramolecular cooperative motions. J Am Chem Soc 128:11010–11011
48. Yu H, Kobayashi T, Hu G-H (2011) Photocontrolled microphase separation in a nematic liquid–crystalline diblock copolymer. Polymer 52:1554–1561
49. Morikawa Y, Nagano S, Watanabe K, Kamata K, Iyoda T, Seki T (2006) Optical alignment and patterning of nanoscale microdomains in a block copolymer thin film. Adv Mater 18:883–886
50. Morikawa Y, Kondo T, Nagano S, Seki T (2007) 3D photoalignment and patterning of microphase separated nanostructure in polystyrene-based block copolymer. Chem Mater 19:1540–1542
51. Nagano S, Koizuka Y, Murase T, Sano M, Shinohara Y, Amemiya Y, Seki T (2012) Synergy effect on morphology switching: real-time observation of photo-orientation of microphase separation in a block copolymer. Angew Chem Int Ed 51:5884
52. Schmidt K, Böker A, Zettl H, Schubert F, Hänsel H, Fisher F, Weiss TM, Abetz V, Zvelindovsky AV, Sevink GJA, Krausch G (2005) Influence of initial order on the microscopic mechanism of electric field induced alignment of block copolymer microdomains. Langmuir 21:11974–11980
53. Rochon P, Batalla E, Natansohn A (1995) Optically induced surface gratings on azoaromatic polymer films. Appl Phys Lett 66:136–138
54. Kim DY, Tripathy SK, Li L, Kumar J (1995) Laser-induced holographic surface relief gratings on nonlinear optical polymer films. Appl Phys Lett 66:1166–1168
55. Viswanathan NK, Kim DK, Bian S, Williams J, Liu W, Li L, Samuelson L, Kumar J, Tripathy SK (1999) Surface relief structures on azo polymer films. J Mater Chem 9:1941–1955
56. Yager KJ, Barrett CJ (2001) All-optical patterning of azo polymer films. Curr Opin Solid State Mater Sci 5:487–494
57. Ubukata T, Seki T, Ichimura K (2000) Surface relief gratings in host-guest supramolecular materials. Adv Mater 12:1675–1678
58. Zettsu N, Ubukata T, Seki T, Ichimura K (2001) Soft crosslinkable azo polymer for rapid surface relief formation and persistent fixation. Adv Mater 13:1693–1697
59. Zettsu N, Seki T (2004) Highly efficient photogeneration of surface relief structure and its immobilization in cross-linkable liquid crystalline azobenzene polymers. Macromolecules 37:8692–8698
60. Zettsu N, Ogasawara T, Arakawa R, Nagano S, Ubukata T, Seki T (2007) Highly photosensitive surface relief gratings formation in a liquid crystalline azobenzene polymer: new implications for the migration process. Macromolecules 40:4607–4613
61. Seki T (2006) Photoresponsive self-assembly motions in polymer thin films. Curr Opin Solid State Mater Sci 10:241–248
62. Isayama J, Nagano S, Seki T (2010) Phototriggered mass migrating motions in liquid crystalline azobenzene polymer films with systematically varied thermal properties. Macromolecules 43:4105–4112
63. Kato T, Frechet JMJ (1989) New approach to mesophase stabilization through hydrogen-bonding molecular interactions in binary mixtures. J Am Chem Soc 111:8533–8534
64. Kato T (2002) Self-assembly of phase-segregated liquid crystal structures. Science 295:2414–2418
65. Zettsu N, Ogasawara T, Mizoshita N, Nagano S, Ubukata T, Seki T (2008) Photo-triggered surface relief gratings in supramolecular liquid crystalline polymer system with detachable azobenzene unit. Adv Mater 20:516–521
66. Li W, Nagano S, Seki T (2009) Photo-crosslinkable liquid-crystalline azo-polymer for surface relief gratings and persistent fixation. New J Chem 33:1343–1348
67. Kulikovska O, Goldenberg LM, Stumpe J (2007) Supramolecular azobenzene-based materials for optical generation of microstructures. Chem Mater 19:3343–3348

68. Goldenberg LM, Kulikovsky L, Kulikovska O, Stumpe J (2009) New materials with detachable azobenzene: effective, colourless and extremely stable surface relief gratings. J Mater Chem 19:8068–8071
69. Darracq B, Chaput F, Lahlil K, Levy Y, Boilot JP (1998) Photoinscription of surface relief gratings on azo-hybrid gels. Adv Mater 10:1133–1136
70. Kulikovska O, Goldenberg LM, Kulikovsky L, Stumpe J (2008) Smart ionic sol-gel-based azobenzene materials for optical generation of microstructures. Chem Mater 20:3528–3534
71. Nishizawa K, Nagano S, Seki T (2009) Novel liquid crystalline organic–inorganic hybrid for highly sensitive photoinscriptions. Chem Mater 21:2624–2631
72. Nishizawa K, Nagano S, Seki T (2009) Micropatterning of titanium oxide via phototactic motions. J Mater Chem 19:7191–7194
73. Yonetake K, Masuko T, Morishita T, Suzuki K, Ueda M, Nagahata R (1999) Poly(propylene imine) dendrimers peripherally modified with mesogens. Macromolecules 32:6578–6586
74. Li W, Nagano S, Yonetake K, Seki T (2012) Photoinduced dewetting in thin films of liquid crystalline dendritic azobenzene derivatives. Mol Cryst Liq Cryst 563:112–120
75. Li W, Dohi T, Hara M, Nagano S, Haba O, Yonetake K, Seki T (2012) Phototriggered mass migration consorted with surface dewetting in thin films of a liquid crystalline azobenzene-containing dendrimer. Macromolecules 45:6618–6627

Chapter 6
Photochromism of Diarylethenes at Surfaces and Interfaces

Kenji Matsuda

Abstract Photochromic reactions of diarylethene derivatives are investigated at the surfaces of noble metal nanoparticles and at the interfaces between solution and highly ordered pyrolytic graphite (HOPG). The network prepared from diarylethene molecules and gold nanoparticles showed completely reversible 25-fold conductance photoswitching. Photo- and electrochromism of the diarylethene ligand overcame the quenching effect of the photoexcited state on metal nanoparticles. The switching behavior is attributed to the change in the π-conjugation in the molecules, that is, in the open-ring isomer the π-conjugation is discontinued, while in the closed-ring isomer the π-conjugation is delocalized throughout the molecule. By using scanning tunneling microscopy at a solution–HOPG interface, diarylethene derivatives that have a pyrene moiety showed reversible photoinduced molecular ordering change. The different photochromic isomers showed different orderings reflecting the differences in their molecular structures. For the diarylethene–pyrene–diarylethene triad, a new ordering appeared upon irradiation with UV light and returned to the original ordering upon subsequent irradiation with visible light. The new arrangement was assigned to the ordering of the closed–closed isomers based on the images of the isolated open and closed isomers. These results show that photochromic molecules are candidates for future switching units in molecular electronic devices.

Keywords Diarylethene • Metal nanoparticles • Molecular electronics • Photochromism • Scanning tunneling microscopy

---

K. Matsuda (✉)
Department of Synthetic Chemistry and Biological Chemistry,
Graduate School of Engineering, Kyoto University,
Katsura, Nishikyo-ku, Kyoto 615-8510, Japan
e-mail: kmatsuda@sbchem.kyoto-u.ac.jp

M. Irie et al. (eds.), *New Frontiers in Photochromism*,
DOI 10.1007/978-4-431-54291-9_6,

6.1 Introduction

Molecular electronics is a rapidly expanding field in nanoscience and nanotechnology. Molecular electronics can be defined as the study of electrical and electronic processes by accessing the individual molecules with electrodes and exploiting the molecular structure to control the flow of electrical signals through them [1].

Photochromic compounds change their geometric and electronic structures along with their color upon irradiation with appropriate wavelength of light. Not only the color but also various physical properties are different between the isomers because of the differences in the electronic and geometric structures. Diarylethenes undergo Woodward–Hoffmann-type 6π photocyclization/cycloreversion reactions. The colorless open-ring isomer has a hexatriene structure, while the structure changes to a cyclohexadiene one in the colored closed-ring isomer. The most important feature of the molecular systems is that while the π-systems of the two aryl rings are discontinued in the open-ring isomer, the closed-ring isomer has a planar structure and its π-electrons are delocalized throughout the molecule. This electronic structural change can be used to switch the interaction between the functional units that are placed at each end of the photochromic molecules.

The molecular conductance can be changed by controlling the molecular structure, and this is a key issue in molecular electronics [2, 3]. The relationship between the structure and the conductance has started to be unveiled. We have investigated the photoswitching behavior of the conductance in the network of photochromic diarylethene molecules and noble metal nanoparticles. Noble metal nanoparticles are currently attracting interest because they have unique physical properties [4]. As the size of the nanoparticles is several nanometers, which is comparable to the size of the molecules, the combined network structures fabricated from metal nanoparticles and organic molecules can be used to form multifunctional integrated materials. The relationship between the structural change and switching effect is discussed.

Two-dimensional (2D) self-assembly of organic compounds on solid surfaces has become a topic of considerable interest in materials science, organic electronics, and molecular electronics. Scanning tunneling microscopy (STM) is a powerful tool for investigating the 2D structures with atomic resolution [5]. STM measurements at a solution–solid interface provide information on an arrangement of self-assembled monolayer formed at the interface [6]. Using this technique, it is possible to determine the molecular orientation and packing in 2D crystalline states on surfaces and at interfaces [7]. We have investigated the reversible photoinduced 2D molecular ordering change of photochromic diarylethenes at the interface between a solution and highly oriented pyrolytic graphite (HOPG). The relationship between the molecular structure and the 2D assembled structure is discussed.

6.2 Photoreaction on Metal Nanoparticles

Photophysical properties of the metal nanoparticles differ from the bulk state. For example, plasmon absorption is clearly observed depending on the size of the particles. Gold nanoparticles have a surface plasmon resonance absorption band at around 520 nm [8], while silver nanoparticles have a surface plasmon resonance absorption at around 420 nm [9]. Photochemical properties of photochromic dyes attached to the metal nanoparticles are affected by the metal. Gold nanoparticles capped with azobenzene [10] and spiropyran [11] have been prepared and their photochromic reactivities were examined.

Gold and silver nanoparticles covered with diarylethene thiol **1a** were prepared and their photochromic properties were investigated [12, 13] (Fig. 6.1). Diarylethene **1a** has a thiol unit, which attaches to the metal nanoparticle core and forms self-assembled monolayer. Gold nanoparticles capped with **1a** (**Au-1a**) were prepared by the Brust and Schiffrin protocol [14]. Silver nanoparticles (**Ag-1a**) were synthesized by Kim's method which is a modification of the Brust's protocol [15]. Nanoparticles of different sizes were prepared by changing the molar ratio between diarylethene **1a** and the metal source.

Diarylethene attached to both gold and silver nanoparticles exhibits reversible photochromism. The plasmon absorption intensity decreases when the particle size decreases. **Ag-1a** (1:1) did not show any distinctive plasmon absorption because the size of this nanoparticle was very small. The absorption maxima of the closed-ring isomer were not strongly dependent on the metal element or the particle size. It is well known that metals readily quench the electronic excited states of molecule that are attached to their surfaces. Energy transfer and electron transfer to the metal surfaces are the main quenching processes [16, 17]. These quenching processes are faster than fluorescence lifetimes; energy transfer and electron transfer on metal surfaces occur on the picoseconds timescale, while fluorescence lifetimes are several nanoseconds. On the other hand, photocyclization reactions occur in the picoseconds timescale. Therefore, cyclization reactions compete with quenching processes.

The photochromic reactions of the gold and silver nanoparticles **Au-1** and **Ag-1** were investigated starting from the nanoparticles made of the closed-ring isomer. The conversion rate of the cyclization reaction at the photostationary state under irradiation with 313 nm light was 74 % and 64 % for gold and silver nanoparticles, respectively. The conversion rate gives information about the ratio of the cyclization and the cycloreversion quantum yields; a lower conversion compared to that of the free ligand indicates that the cyclization quantum yield is suppressed because of the metal. Although the conversion rate was suppressed by the metal, a photochromic reaction still takes place on the surface of noble metal nanoparticles. Provided that quenching occurs by an energy-transfer mechanism, the cyclization reaction should be quenched more effectively than the cycloreversion reaction. Gold nanoparticles have a plasmon band at a longer wavelength than silver nanoparticles; therefore, because of the weaker spectral overlap, the quenching ability of the gold nanoparticles is reduced.

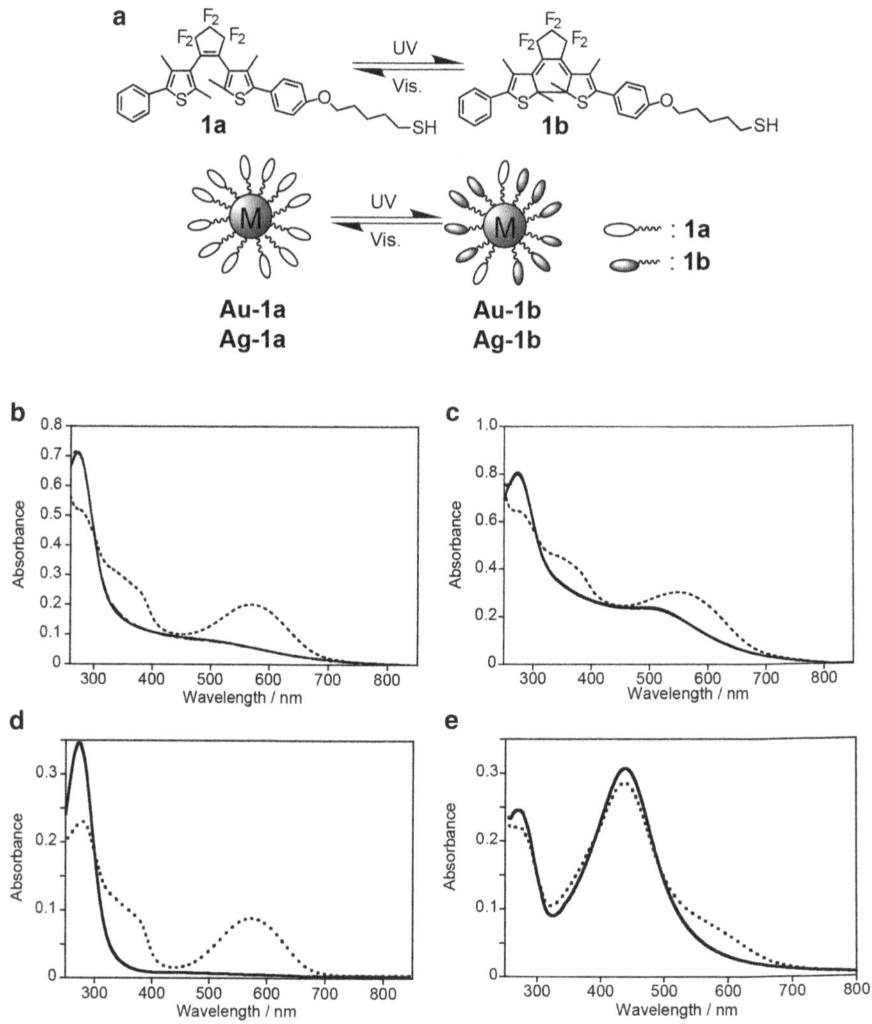

Fig. 6.1 (**a**) Photochromic reaction of diarylethene **1** and **Au-1** and **Ag-1**. Absorption spectra of (**b**) **Au-1a** (1:1), (**c**) **Au-1a** (3:1), (**d**) **Ag-1a** (1:1), and (**e**) **Ag-1a** (3:1) in ethyl acetate. *Solid lines* denote the open-ring isomer. *Dotted lines* denote the photostationary state under irradiation with 313-nm light. Reprinted with permission from Bull Chem Soc Jpn 79:1413–1419. Copyright (2006) Chemical Society of Japan

6.3 Conductance Photoswitching of Diarylethene–Gold Nanoparticle Network

Conductance photoswitching is achieved by using a diarylethene π-switching system sandwiched between electrodes. Although there are increasing numbers of reports dealing with the conductance of single molecules, studies on

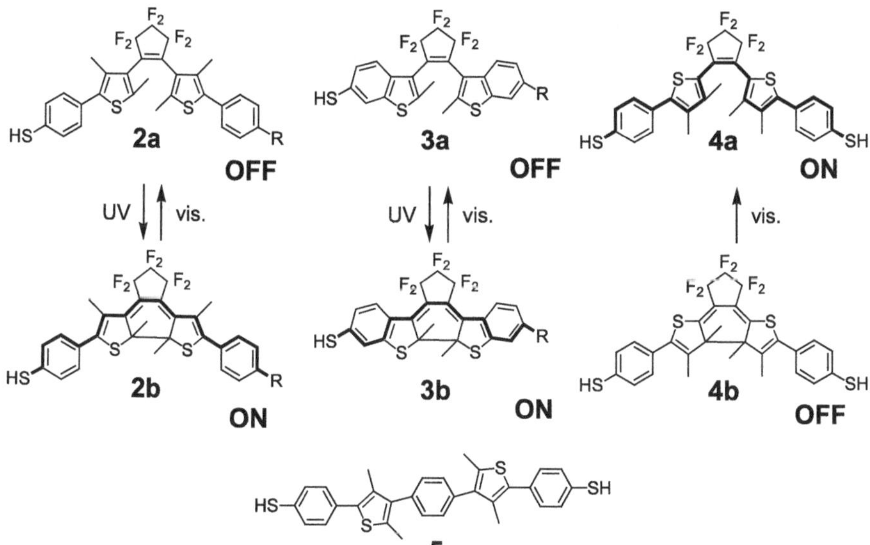

Fig. 6.2 Chemical structures of photochromic diarylethenes for making gold nanoparticles networks

photoswitchable molecules are rare [18–20]. Studies on the conductance of networks prepared with organic molecules and gold nanoparticles on interdigitated nanogap electrodes are attracting interest because of their facile preparation and the applicability to the small number of molecules [21, 22].

Diarylethene dithiols are designed to have a switching effect on molecular conductance (Fig. 6.2). Then the network is prepared from diarylethene dithiol and gold nanoparticles and the network makes a conducting path between the interdigitated nanogapped gold electrodes (Fig. 6.3a) [23, 24]. As the excited state of the organic molecule on noble metal surfaces is easily quenched by the surface plasmon resonance, the photoswitching unit should be placed a certain distance from the surface. However, considering that a conjugated molecule has much better conductance than a nonconjugated molecule, the sulfur atom should be attached directly to the π-conjugated system. The diarylethene–gold nanoparticle networks are characterized using TEM and SEM. TEM images of the nanoparticles network show the existence of an extended network. The SEM image of the network on the interdigitated nanogap gold electrode shows that the network bridges the electrodes.

Conductance is measured along with the alternate irradiation with UV and visible light (Fig. 6.3b–e). For **Au-2a** and **Au-3a**, upon UV irradiation, the conductance increased significantly, and then upon visible-light irradiation, the conductance decreased. This indicates that photoisomerization of the diarylethene unit induces switching of the π-conjugated system. The photocycloreversion reaction of **Au-2a** is very slow. Even after 56 h of visible-light irradiation, the conductance decreased

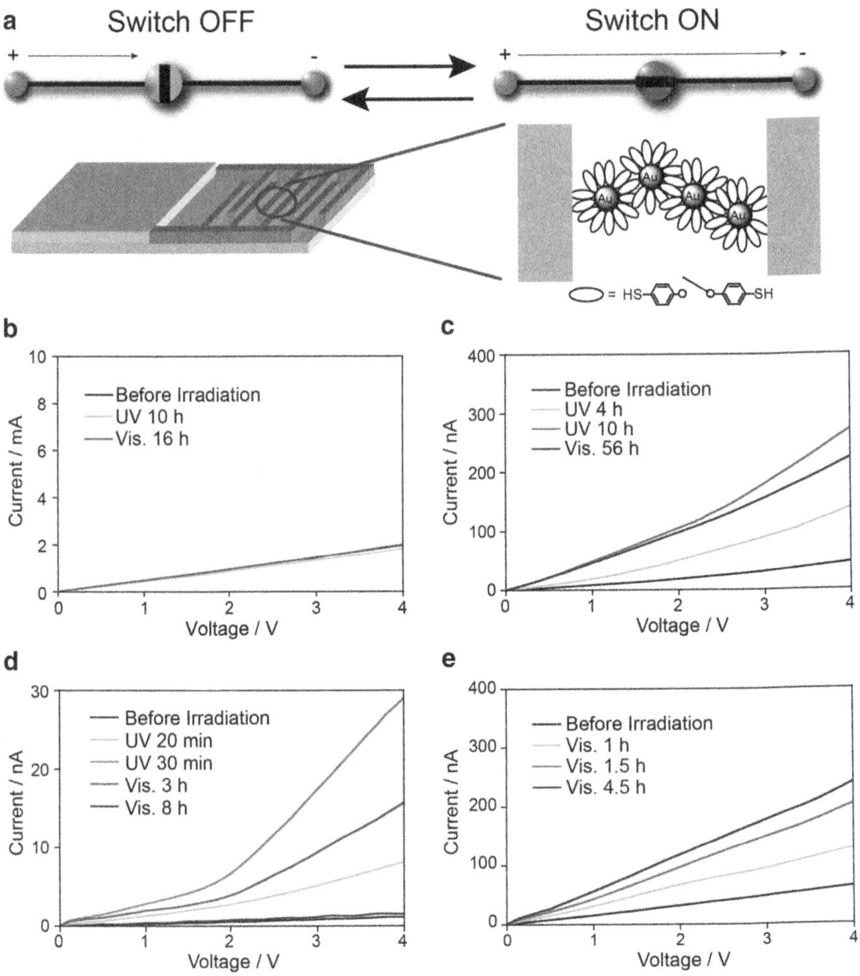

Fig. 6.3 (**a**) Schematic diagram of a diarylethene–gold nanoparticle network. Changes of the *I*–*V* curves of diarylethene–gold nanoparticle networks: (**b**) **Au-5**, (**c**) **Au-2a**, (**d**) **Au-3a**, and (**e**) **Au-4b**. Measurements were carried out at different stages during the photoreaction. For the open-ring isomer nanoparticle networks **Au-2a** and **Au-3a**, initially UV light was irradiated and then visible light was irradiated. For the closed-ring isomer nanoparticle network **Au-4b**, only visible-light irradiation was performed. Reprinted with permission from J Phys Chem C 112: 17005–17010. Copyright (2008) American Chemical Society

only 18 %. In contrast, the cycloreversion reaction of **Au-3a** was completed in 8 h and the system showed reversible photoswitching behavior. The maximum ON/OFF ratio of the conductance was 25-fold for **Au-3a**. Because **3a** has a high quantum yield for both cyclization and cycloreversion reactions, complete reversibility has been achieved. No changes in conductance were observed for the non-photoreactive **Au-5**.

To confirm that the origin of the switching is the photochromism of the diarylethene unit, 2-thienyl-type diarylethene **4**, in which switching occurs in the opposite direction, was investigated. Diarylethene **4** showed the opposite switching behavior to 3-thienyl-type diarylethenes **2** and **3**. **Au-4a** did not undergo cyclization under UV irradiation. However, **Au-4b**, prepared under dark conditions from the separated closed-ring isomer **4b**, undergoes cycloreversion reaction upon visible-light irradiation. Quenching by the gold surface suppressed the cycloreversion reaction of **Au-2b** and **Au-3b**. For **Au-4b**, upon visible-light irradiation (λ>470 nm), the conductance increased. The maximum ON/OFF ratio was 3.8-fold. This observed behavior of **Au-4** demonstrates that the switching of the diarylethene plays an essential role in controlling the conductance.

6.4 Conductance Switching of Diarylethene–Gold Nanoparticle Network by Oxidization

Switching should smoothly proceed in both directions for the realization of practical molecular switching devices. Excited-state quenching by metal nanoparticle surfaces should be avoided. As an alternative to photoisomerization, electrochromic reactions provide an alternative external stimulus in which the isomerizations are induced by electrochemical oxidation or chemical oxidation [25]. The diarylethene linker **6** shows both photo- and electrochromic reactivities [26] (Fig. 6.4). The cyclization reaction of the thiophene-substituted dithienylethene is strongly quenched when placed on gold nanoparticles and surfaces. Only one-way photoswitching properties have been reported for single-molecular devices. By using electrochromic oxidative cyclization, two-way photoswitching can be achieved [27].

The conductance of **Au-6a** network hardly changed under UV irradiation for 2 h. This result illustrates the strong quenching effect of gold nanoparticles on the cyclization reaction of the thienyl-substituted dithienylethene. The quenching effect of gold nanoparticles on phenyl-substituted dithienylethene has been reported to be less [28]. The fabricated electrode was soaked in $FeCl_3$ in acetone for 1 min to start the electrochromic reaction, and the conductance measurements were carried out in a vacuum chamber. The conductance of **Au-6a** network was increased about fivefold. The increase in the conductance is explained by the electrochromic reaction from “OFF” (the open-ring isomer) to “ON” (the closed-ring isomer). To confirm the increase in conductance, the cycloreversion reaction under visible light (λ>510 nm) was carried out. The conductance of the network slowly decreased under visible-light irradiation. This switching behavior suggests that the closed-ring isomer in the network underwent cycloreversion by photoexcitation.

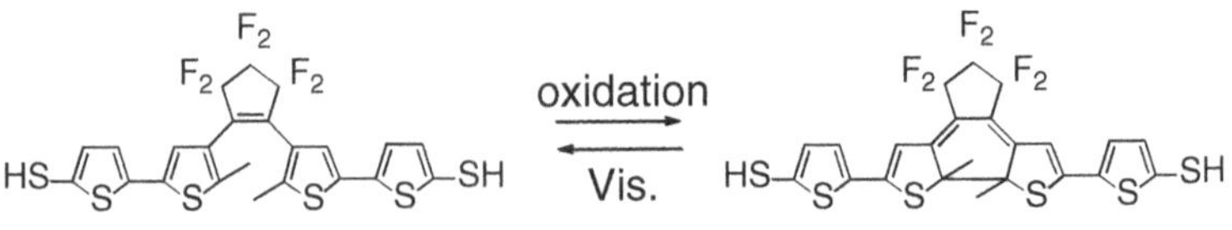

Fig. 6.4 Chemical structure of diarylethene which shows electrochromism

6.5 Photochromism of Diarylethenes Designed for STM Measurement

Molecular imaging of the 2D structure formed at a liquid–solid interface provides information on the molecular alignment and the molecular dynamics with distinguishing individual molecules. STM images at liquid–HOPG interfaces visualize self-assembled monolayers of solute molecules at the interface at very high resolution. At the liquid–solid interface, adsorbed molecules on the surface are in equilibrium with the molecules in solution and the formation of the 2D ordering is a very good model of crystallization. By measuring real-time STM observations at the liquid–solid interface, the dynamic behaviors of individual molecules can be accessed [29–34].

Here, changes in the STM images monitor the photochromism of diarylethene molecules [35, 36]. The photochromic reaction of diarylethene switches the π-conjugation and molecular planarity. Thermal effects are negligible as both isomers are thermally stable. The relationship between the molecular structure and the 2D assembled structure is discussed.

For the STM measurement at the solution–HOPG interface, diarylethene–pyrene diad **7** and diarylethene–pyrene–diarylethene triad **8** were designed and synthesized [37] (Fig. 6.5). The long alkyl chain and pyrene units were introduced into the diarylethene in order to stabilize the 2D structure. Pyrene interacts with HOPG due to extended planar π-conjugation [38]. The alkyl chains are also effective at interacting with the HOPG substrate. The presence of the short alkyl chain between the diarylethene and pyrene increases the flexibility. For the aromatic photochromic core, 1,2-bis(2-methyl-5-phenyl-3-thienyl)perfluorocyclopentene was chosen.

Compounds **7** and **8** underwent photochromic reactions upon irradiation with UV and visible light in solution. Upon irradiation with UV light ($\lambda = 313$ nm), the colorless solution turned blue and the solution returned to colorless by irradiation with visible light ($\lambda > 480$ nm). A comparison of the absorption spectra with the pure closed-ring isomer separated by HPLC showed that the conversion rate from the open- to the closed-ring isomer under irradiation with 313 nm light was 97 % for diarylethene **7**. For the diarylethene dimer **8**, the ratio of the open–open, the open–closed, and the closed–closed isomers in the photostationary state under irradiation with 313 nm light was 0.1:8.2:91.7, respectively, as determined by HPLC analysis.

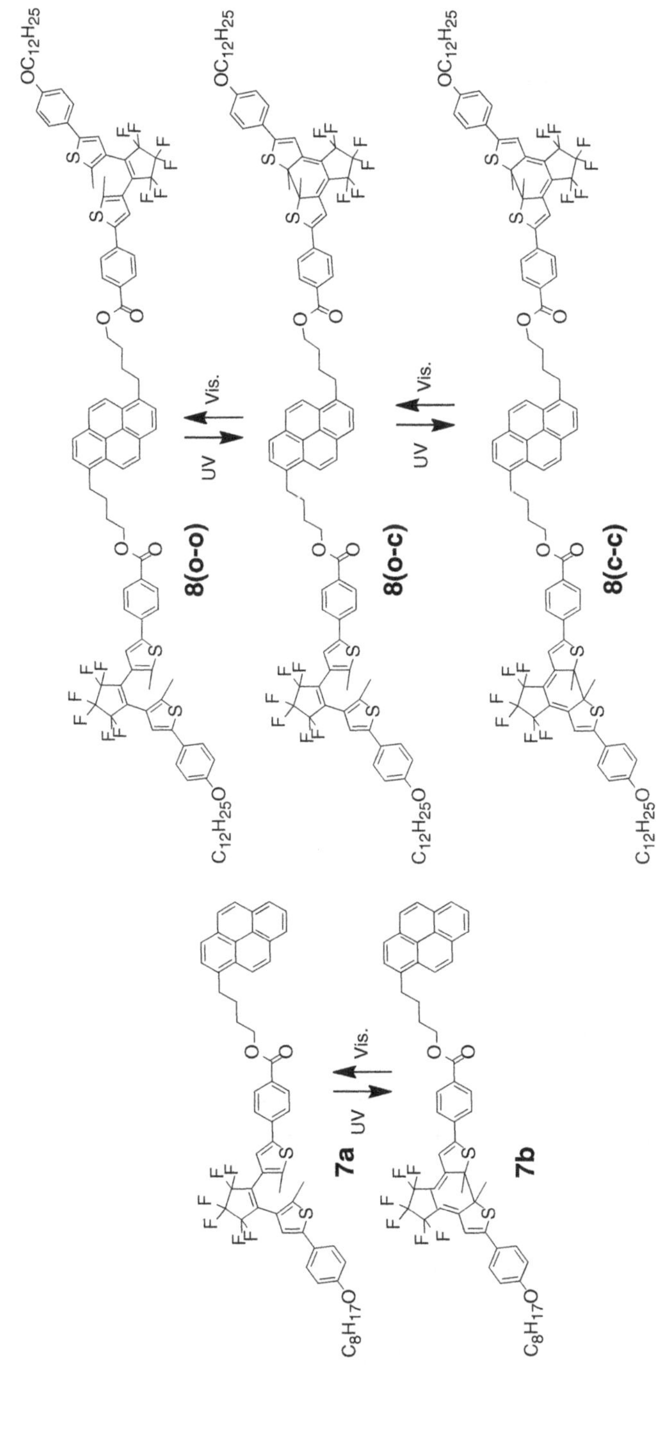

Fig. 6.5 Photochromism of diarylethenes with chemical structure of diarylethene which shows electrochromism

8(o–o) has two diarylethene moieties. Upon irradiation with UV light, both diarylethene units underwent efficient cyclization reactions and a very high conversion rate up to 92 % was attained. This result indicates that the two chromophores undergo cyclization independently and the interaction between the two closed-ring isomers is very small because of the broken π-conjugation and the large distance between the diarylethenes.

6.6 STM Measurements of Diarylethene–Pyrene Diad

STM images of the open-ring isomer **7a** were obtained at the 1-phenyloctane–HOPG interface. As shown in Fig. 6.6a, compound **7a** formed a well-ordered 2D molecular arrangement over a wide area (larger than the scanning area of 100×100 nm^2). According to a previous report on the 2D structure of a pyrene-containing molecule, the bright elliptic spots in Fig. 6.6b are assigned to the pyrene moiety of compound **7a**, and the small bright spots are assigned to the diarylethene moiety [38]. The octyloxy chains could not be observed at any bias voltage or set point current. The unit cell parameters of the 2D structure are $a=1.4\pm0.1$ nm, $b=2.6\pm0.1$ nm, and $\alpha=115\pm2°$. Since no stable 2D molecular structure was observed for those molecules synthesized without an octyloxy chain, we infer that the octyloxy chain is effectively physisorbed on HOPG and indispensable to stabilize 2D ordering.

This structure was explained on the basis of the molecular model constructed using molecular mechanics calculations (MacroModel, OPLS_2005) (Fig. 6.6c). In contrast to the bent structure observed in the bulk crystal, the molecule was calculated to have a planar structure in gas phase. The total energy of the twisted isomer, in which the pyrene ring and the diarylethene are perpendicular to each other, was ca. 0.5 kJ/mol higher than that of the planar isomer, in which the pyrene ring and the diarylethene are parallel. These calculation results indicate that the difference in the total energy between the rotational isomers is very small. The small energy difference suggests that the molecular conformation can be easily affected by intermolecular interaction between adjacent molecules or the interactions between the molecule and the HOPG surface.

By comparing the molecular packing in the bulk crystal with the molecular arrangement on the HOPG substrate, both sets of unit cell parameters are found to be similar, but the length of the 2D unit cell observed by STM is slightly longer than that observed in the bulk crystal. This suggests that the molecular packing on the HOPG surface is looser than that in the single crystal.

Figure 6.6d displays STM images of the closed-ring isomer **7b**. The unit cell parameters of the 2D structure are $a=0.5\pm0.04$ nm, $b=4.9\pm0.1$ nm, and $\alpha=109\pm6°$. This ordering was not always observed. In fact, most of the time, no ordering was observed. This result indicates that this ordering is less stable than that of **7a**. As shown in the small-area image, the 2D structure of compound **7b** differs from that of compound **7a** (Fig. 6.6e). The molecules pack more densely than those in the

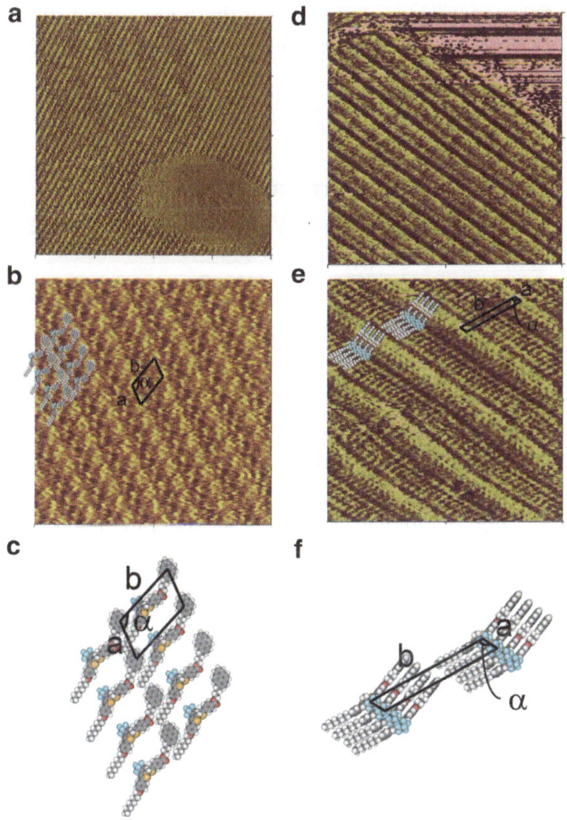

Fig. 6.6 STM images of **7 at** the interface of 1-phenyloctane–HOPG: (**a**) open-ring isomer **7a** (50×50 nm^2, $I_{set}=10$ pA, $V_{bias}=-1.5$ V). (**b**) Small-area image of **7a** (20×20 nm^2, $I_{set}=30$ pA, $V_{bias}=-1.35$ V). (**c**) Structural model of arrangement of **7a**. (**d**) Closed-ring isomer **7b** (50×50 nm^2, $I_{set}=20$ pA, $V_{bias}=-1.4$ V). (**e**) Small-area image of **7b** (20×20 nm^2, $I_{set}=40$ pA, $V_{bias}=-1.3$ V). (**f**) Structural model of arrangement of **7b**. Reprinted with permission from J Am Chem Soc 130: 9371–9379. Copyright (2008) American Chemical Society

structure of **7a**. The image of **7b** is explained by the calculated model shown in Fig. 6.6f. Unlike the open-ring isomer **7a**, in which the pyrene ring is lying parallel to the HOPG surface, the pyrene ring of **7b** is lying perpendicular to the HOPG surface. Compound **7b** is thought to have less interaction with HOPG in comparison with **7a**, and the 2D structure of **7b** is therefore less stable than the 2D structure of **7a**. The diarylethene open-ring isomer has a twisted structure but the closed-ring isomer has a flat thin structure. This difference significantly affects the stacking of the molecule, as suggested by studies on the molecule in solution and in the crystalline phase [39].

When the arrangement of the open-ring isomer **7a** was irradiated sufficiently with UV light in situ, the image disappeared. Because the adlayer is a monomolecular layer, most of the molecules exist in the solution rather than at the interface. Therefore, most of the photochromic reaction takes place in the solution not at the interface. When the content of the solution phase is thereby changed, adsorption and desorption processes exchange molecules between the interface and the solution [40, 41]. The photogenerated closed-ring isomer **7b** probably disturbs the adsorption of the open-ring isomer **7a**. In the STM image of a 6:4 mixture of **7a** and **7b**, no ordering was observed. This result indicates that the existence of **7b** disturbs the adsorption of **7a**.

6.7 STM Measurements of Diarylethene–Pyrene–Diarylethene Triad

Figure 6.7a shows STM images of the open–open isomer of the diarylethene dimer **8(o–o)** at the octanoic acid–HOPG interface. **8(o–o)** formed a well-ordered 2D molecular assembly, and the domain was larger than the scanning area of 100 nm^2. The arrangement consists of parallel lines with spacing of 5.8 nm. Each line has a lamellar structure of bright lines consisting of three bright spots. The brightest spot in the middle is assigned to the pyrene moiety, and the other bright spots are assigned to the diarylethene moieties in a similar manner as in the case of compound **7**. The unit cell parameters are $a = 2.4 \pm 0.2$ nm, $b = 6.3 \pm 0.2$ nm, and $\alpha = 87 \pm 2°$. The molecular packing was explained based on the results of molecular mechanics calculations of molecules without alkyl chains.

When the open–open isomer **8(o–o)** is irradiated with UV light for a few seconds in situ, a remarkable change in ordering was observed. After UV irradiation, a new arrangement with a different line spacing of 2.2 ± 0.2 nm was observed (Fig. 6.7b). This change is due to the photoisomerization of the diarylethene moiety. Furthermore, subsequent visible-light irradiation for about 10 min gave the original arrangement of **8(o–o)**.

As discussed for compound **7**, the surface exchange of adsorbed molecules with isomers in solution occurs after a photochromic reaction [40]. After UV irradiation, three isomers exist in the solution phase: the open–open, open–closed, and closed–closed isomers. In this system, not only the open–closed isomer but also the closed–closed isomer was easily formed because the two diarylethene moieties were separated from each other by a long distance. To investigate which isomer contributes to this new ordering, STM images of the pure open–closed isomer **8(o–c)** and the closed–closed isomer **8(c–c)**, which were separated by the HPLC, were obtained.

The arrangement of **8(o–c)** consists of parallel lines that are similar to those seen for **8(o–o)**. Each line consists of a lamellar structure and the line spacing of **8(o–c)** was 4.7 nm, which is slightly shorter than that of **8(o–o)**. Judging from this arrangement, the ordering of **8(o–c)** is not identical with either of the ordering of **8(o–o)** or the new ordering after UV irradiation. Although **8(o–c)** contains both open and closed isomers in one molecule, the two different isomers could not be distinguished in the STM image.

The ordering of the closed–closed isomer **8(c–c)** is completely different from that of **8(o–o)** (Fig. 6.8a). The ordering consists of parallel lines with a line spacing of 1.9 ± 0.2 nm which is narrower than that of **8(o–o)** and is very similar to the new ordering after UV irradiation. The new ordering in Fig. 6.7b is assigned to the arrangement of **8(c–c)**. Figure 6.8b shows an expanded image. One bright ellipse spot and the other spot make the unit cell. The unit cell parameters are $a = 2.6 \pm 0.2$ nm,

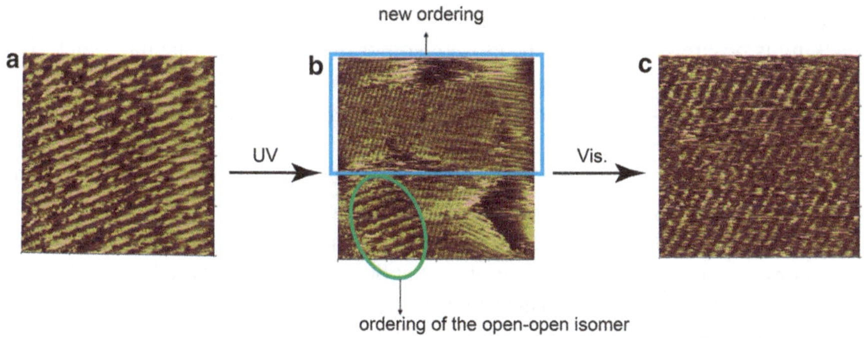

Fig. 6.7 The sequence of STM images of **8(o–o)** at the interface of octanoic acid–HOPG upon UV and visible-light irradiation (100×100 nm^2, the same areas of the sample): (**a**) before UV irradiation, (I_{set}=10 pA, V_{bias}=−2.0 V); (**b**) after UV irradiation for 5 s (I_{set}=10 pA, V_{bias}=−2.5 V); (**c**) after successive visible-light irradiation (I_{set}=10 pA, V_{bias}=−2.5 V). Reprinted with permission from J Am Chem Soc 130: 9371–9379. Copyright (2008) American Chemical Society

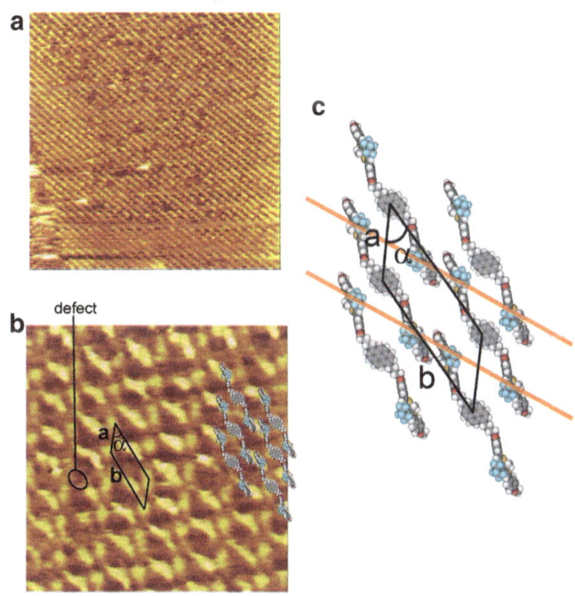

Fig. 6.8 (**a**) STM image of an arrangement of **8(c–c)** at the interface of octanoic acid–HOPG (100×100 nm^2, I_{set}=10 pA, V_{bias}=−1.5 V). (**b**) Small-area image of **8(c–c)** (20×20 nm^2, I_{set}=10 pA, V_{bias}=−1.5 V). (**c**) Structural model for **8(c–c)**. *Orange line* denotes the direction of the lines in the image. Reprinted with permission from J Am Chem Soc 130: 9371–9379. Copyright (2008) American Chemical Society

b=5.3±0.1 nm, and α=40±3°. Some of the brightest ellipses are only half the size because of defects, suggesting that the brightest ellipse consists of two spots, namely, the densely packed two diarylethene moieties. The calculated structural model of **8(c–c)** is displayed in Fig. 6.8c. The brightest ellipse represents two diarylethene moieties and the other spot represents the pyrene moiety and the void between them represents the alkyl chains. In order to reproduce the observed arrangement, the two diarylethene moieties need to be perpendicular to the HOPG surface. The closed-ring isomer of diarylethene has a flat structure, so that a dense packing is possible, as seen in the closed isomer of the diarylethene monomer **7b**.

Judging from these experiments, the new ordering in Fig. 6.7 is assigned to the arrangement of **8(c–c)**. As mentioned previously, the UV irradiation gives both the open–closed isomer **8(o–c)** and the closed–closed isomer **8(c–c)**, but the arrangement of **8(o–c)** was not observed. In order to examine the relative magnitudes of the adsorption process, STM images of the mixture with the ratio of **o–o**:**o–c**:**c–c** = 45:33:22 were obtained. The image predominantly consists of the closed–closed isomer **8(c–c)** arrangement. This implies that **8(o–o)** and **8(o–c)** are less likely to be physisorbed on the HOPG surface than **8(c–c)**. This result also suggests that the domain of the **8(o–o)** existing in Fig. 6.7b is a remnant of the initial **8(o–o)** arrangement domain.

For monomer **7**, the ordered structure disappeared upon light irradiation. A change in the ordering pattern is clearly observed for dimer **8**. This difference can be attributed to the higher stability of the 2D arrangement of the closed–closed isomer **8(c–c)** than that of the closed-ring isomer **7b**. The height of the pyrene moiety was around 0.17 nm in the STM images of **7a**, **8(o–o)**, **8(o–c)**, and **8(c–c)**, but in the STM images of **7b**, the pyrene moiety was 0.43 nm high. This confirmed that the pyrene adopts a perpendicular orientation to the HOPG surface in **7b**, and a parallel orientation in **7a**, **8(o–o)**, **8(o–c)**, and **8(c–c)**. The difference in the stability of the 2D structure is attributed to the different orientations of the pyrene ring.

One of the advantages of using a diarylethene photochromic unit is the thermal stability of the photochromic isomers. In the present study, the separate arrangement of each of the isomers could be determined by isolating each isomer. In our system, the intermolecular interaction should be weaker than in other systems having hydrogen bonds, metal–ligand interactions or charge–transfer interactions. Here, the molecular assemblies have only a molecule–substrate interaction and van der Waals interaction between the alkyl chains; therefore the 2D structure is not stabilized. However, this relatively weak adsorption energy on HOPG is preferable when studying ordering changes induced by external stimuli.

6.8 Conclusions

In this chapter, we have reviewed the photochromism and electrochromism of diarylethene derivatives as a molecular switch in the field of molecular electronics. Photochromism of diarylethene was studied on the surfaces of noble metal nanoparticles and at the interfaces between solution and HOPG. The conductance photoswitching of the network prepared from diarylethene molecules and noble metal nanoparticles was successfully demonstrated. Photo- and electrochromism of the diarylethene ligand are not quenched by the metal nanoparticles. STM images recorded the reversible photoinduced molecular ordering changes that occurred in the diarylethene–pyrene triad at the solution–HOPG interface. These molecules are candidates for future switching units in molecular electronic devices.

References

1. Metzger RM (2003) Unimolecular electrical rectifiers. Chem Rev 103:3803–3834
2. Joachim C, Gimzewski JK, Aviram A (2000) Electronics using hybrid-molecular and mono-molecular devices. Nature 408:541–548
3. Nitzan A, Ratner MA (2003) Electron transport in molecular wire junctions. Science 300:1384–1389
4. Daniel M-C, Austruc D (2004) Gold nanoparticles: assembly, supramolecular chemistry, quantum-size-related properties, and applications toward biology, catalysis, and nanotechnology. Chem Rev 104:293–346
5. Dri C, Peters MV, Schwarz J, Hecht S, Grill L (2007) Spatial periodicity in molecular switching. Nat Nanotech 2:687–691
6. Elemans JAAW, Lei S, De Feyter S (2009) Molecular and supramolecular networks on surfaces: from two-dimensional crystal engineering to reactivity. Angew Chem Int Ed 48:7298–7332
7. Florio GM, Klare JE, Pasamba MO, Werblowsky TL, Hyers M, Berne BJ, Hybertsen MS, Nuckolls C, Flynn GW (2006) Frustrated ostwald ripening in self-assembled monolayers of cruciform π-systems. Langmuir 22:10003–10008
8. Hostetler MJ, Wingate JE, Zhong C-J, Harris JE, Vachet RW, Clark MR, Londono JD, Green SJ, Stokes JJ, Wignall GD, Glish GL, Porter MD, Evans ND, Murray RW (1998) Alkanethiolate gold cluster molecules with core diameters from 1.5 to 5.2 nm: core and monolayer properties as a function of core size. Langmuir 14:17–30
9. Manna A, Imae T, Aoi K, Okada M, Yogo T (2001) Synthesis of dendrimer-passivated noble metal nanoparticles in a polar medium: comparison of size between silver and gold particles. Chem Mater 13:1674–1681
10. Manna A, Chen P-L, Akiyama H, Wei T-X, Tamada K, Knoll W (2003) Optimized photoisomerization on gold nanoparticles capped by unsymmetrical azobenzene disulfides. Chem Mater 15:20–28
11. Ipe I, Mahima S, Thomas KG (2003) Light-induced modulation of self-assembly on spiropyran-capped gold nanoparticles: a potential system for the controlled release of amino acid derivatives. J Am Chem Soc 125:7174–7175
12. Matsuda K, Ikeda M, Irie M (2004) Photochromism of diarylethene-capped gold nanoparticles. Chem Lett 33:456–457
13. Yamaguchi H, Ikeda M, Matsuda K, Irie M (2006) Photochromism of diarylethenes on gold and silver nanoparticles. Bull Chem Soc Jpn 79:1413–1419
14. Brust M, Walker, M, Bethell D, Schiffrin DJ, Whyman R (1994) Synthesis of thiol-derivatised gold nanoparticles in a two-phase liquid–liquid system. J Chem Soc Chem Commun 801–802
15. Kang SY, Kim K (1998) Comparative study of dodecanethiol-derivatized silver nanoparticles prepared in one-phase and two-phase systems. Langmuir 14:226–230
16. Dulkeith E, Morteani AC, Niedereichholz T, Klar TA, Feldmann J, Levi SA, van Veggel FCJM, Reinhoudt DN, Möller M, Gittins DI (2002) Fluorescence quenching of dye molecules near gold nanoparticles: radiative and nonradiative effects. Phys Rev Lett 89:203002
17. Barazzouk S, Kamat PV, Hotchaudani S (2005) Photoinduced electron transfer between chlorophyll a and gold nanoparticles. J Phys Chem B 109:716–723
18. Jong JJD, Bowden TN, van Esch J, Feringa BL, van Wees BJ (2003) One-way optoelectronic switching of photochromic molecules on gold. Phys Rev Lett 91:207402
19. Taniguchi M, Nojima Y, Yokota K, Terao J, Sato K, Kambe N, Kawai T (2006) Self-organized interconnect method for molecular devices. J Am Chem Soc 128:15062–15063
20. Whalley AC, Steigerwald ML, Guo X, Nuckolls C (2007) Reversible switching in molecular electronic devices. J Am Chem Soc 129:12590–12591
21. Ogawa T, Kobayashi K, Masuda G, Takase T, Maeda S (2001) Electronic conductive characteristics of devices fabricated with 1,10-decanedithiol and gold nanoparticles between 1-μm electrode gaps. Thin Solid Films 393:374–378

22. Bernard L, Kamdzhilov Y, Calame M, van der Molen SJ, Liao J, Schönenberger C (2007) Spectroscopy of molecular junction networks obtained by place exchange in 2D nanoparticle arrays. J Phys Chem C 111:18445–18450
23. Ikeda M, Tanifuji N, Yamaguchi H, Irie M, Matsuda K (2007) Photoswitching of conductance of diarylethene-Au nanoparticle network. Chem Commun 1355–1357
24. Matsuda K, Yamaguchi H, Sakano T, Ikeda M, Tanifuji N, Irie M (2008) Conductance photoswitching of diarylethene–gold nanoparticle network induced by photochromic reaction. J Phys Chem C 112:17005–17010
25. Koshido T, Kawai T, Yoshino K (1995) Optical and electrochemical properties of cis-1,2-dicyano-1,2-bis(2,4,5-trimethyl-3-thienyl)ethane. J Phys Chem 99:6110–6114
26. Peters A, Branda NR (2003) Electrochemically induced ring-closing of photochromic 12-dithienylcyclopentenes. Chem Commun 954–955
27. Yamaguchi H, Matsuda K (2009) Photo- and electrochromic switching of diarylethene–gold nanoparticle network on interdigitated electrodes. Chem Lett 38:946–947
28. Kudernac T, van der Molen SJ, van Wees BJ, Feringa BL (2006) Uni- and bi-directional light-induced switching of diarylethenes on gold nanoparticles. Chem Commun 3597–3599
29. Otsuki J, Komatsu Y, Kobayashi D, Asakawa M, Miyake K (2010) Rotational libration of a double-decker porphyrin visualized. J Am Chem Soc 132:6870–6871
30. Xu H, Minoia A, Tomovi Ž, Lazzaroni R, Meijer EW, Schenning APHJ, De Feyter S (2009) A multivalent hexapod: conformational dynamics of six-legged molecules in self-assembled monolayers at a solid–liquid interface. ACS Nano 3:1016–1024
31. Piot L, Marchenko A, Wu J, Müllen K, Fichou D (2005) Structural evolution of hexa-peri-hexabenzocoronene adlayers in heteroepitaxy on n-pentacontane template monolayers. J Am Chem Soc 127:16245–16250
32. Adisoejoso J, Tahara K, Okuhata S, Lei S, Tobe Y, De Feyter S (2009) Two-dimensional crystal engineering: a four-component architecture at a liquid–solid interface. Angew Chem Int Ed 48:7353–7357
33. Lackinger M, Griessl S, Kampschulte L, Jamitzky F, Heckl WM (2005) Dynamics of grain boundaries in two-dimensional hydrogen-bonded molecular networks. Small 1:532–539
34. Samorí P, Müllen K, Rabe JP (2004) Molecular-scale tracking of the self-healing of polycrystalline monolayers at the solid–liquid interface. Adv Mater 16:1761–1765
35. Katsonis N, Kudernac T, Walko M, van der Molen SJ, van Wees BJ, Feringa BL (2006) Reversible conductance switching of single diarylethenes on a gold surface. Adv Mater 18:1397–1400
36. Arramel, Pijper TC, Kudernac T, Katsonis N, van der Maas M, Feringa BL, van Wees BJ (2012) Electronic properties of individual diarylethene molecules studied using scanning tunneling spectroscopy. J Appl Phys 111:083716
37. Arai R, Uemura S, Irie M, Matsuda K (2008) Reversible photoinduced change in molecular ordering of diarylethene derivatives at a solution–HOPG interface. J Am Chem Soc 130:9371–9379
38. Uji-i H, Yoshidome M, Hobley J, Hatanaka K, Fukumura H (2003) Structural variations in self-assembled monolayers of 1-pyrenehexadecanoic acid and 4,4′-bipyridyl on graphite at the liquid–solid interface. Phys Chem Chem Phys 5:4231–4235
39. Hirose T, Matsuda K, Irie M (2006) Self-assembly of photochromic diarylethenes with amphiphilic side chains: reversible thermal and photochemical control. J Org Chem 71:7499–7508
40. Vanoppen P, Grim PCM, Rücker M, De Feyter S, Moessner G, Valiyaveettil S, Müllen K, De Schryver FC (1996) Solvent codeposition and cis–trans isomerization o isophthalic acid derivatives studied by STM. J Phys Chem 100:19636–19641
41. Xu L-P, Wan L-J (2006) STM investigation of the photoisomerization of an azobis-(benzo-15-crown-5) molecule and its self-assembly on Au(111). J Am Chem Soc 110:3185–3188

Chapter 7
Photo-Control of Magnetic Materials Utilizing Photochromic Compounds

Masayuki Suda and Yasuaki Einaga

Abstract Optically switchable magnetic materials are becoming increasingly important in the field of high-density information storage media because the photon mode allows us to access variety of different types of materials with high speed and superior resolution. In particular, the design of molecular compounds that exhibit photoinduced magnetization behavior has attracted great attention. Although reversible photo-control of magnetic properties or electronic states including spin states was realized in several compounds, most of the interesting phenomena were observed only at low temperature. Furthermore, the number of optically switchable molecular solids reported has not been so large. To realize practical optical switching at room temperature, we have proposed a novel strategy that involves the incorporation of organic photochromes into an inorganic magnetic system on a nanometer-scale order. In this chapter, several novel photo-functional magnetic compounds created, based on the above strategy, are introduced.

Keywords Ferromagnetic nanoparticles • Organic–inorganic interfaces • Photo-controllable materials • Superconductivity

7.1 Introduction

The design of molecular compounds that exhibit photoinduced magnetization and magnetic transitions is one of the main challenges in the field of materials science because of their possible application to future optical memory and switching devices [1].

M. Suda (✉)
Institute for Molecular Science, 38 Myodaiji, Okazaki 444-8585, Japan
e-mail: msuda@ims.ac.jp

Y. Einaga
Department of Chemistry, Keio University,
3-14-1 Hiyoshi, Yokohama 223-8522, Japan
e-mail: einaga@chem.keio.ac.jp

M. Irie et al. (eds.), *New Frontiers in Photochromism*,
DOI 10.1007/978-4-431-54291-9_7, © Springer Japan 2013

Photoinduced changes in magnetic order have been studied extensively in a variety of systems, including cyanometalate-based compounds [2–4], light-induced excited spin-state trapping (LIESST) compounds [5–7], diluted magnetic semiconductors [8, 9], and manganite films [10]. Although interesting photo-magnetic phenomena have been reported in the above systems, most of the observations of such phenomena were limited to very low-temperature operation. To realize practical optical switching at room temperature, we have proposed a novel strategy that involves the incorporation of organic photochromes into an inorganic magnetic system on a nanometer-scale order. The photochromes not only change their adsorption spectra, they also change their physical and chemical properties, meaning that the magnetic properties of such composite materials can be controlled by photoillumination. Indeed, on the basis of this idea, we have already reported some optically switchable magnetic materials [11]. Although the examples described above exhibit photo-switching even at room temperature, these systems still also show superparamagnetic behavior at room temperature. This is because decreases in particle size lead to thermal fluctuations in magnetization, resulting in the disappearance of ferromagnetism. From the viewpoint of their practical application to magnetic recording systems, the ability to fix their magnetic moments such that they still exhibit room temperature ferromagnetism is an absolute requirement. In this chapter, several novel photo-functional magnetic compounds that show room temperature operation are introduced. Furthermore, extended research results that show photo-control of superconductivity will be also introduced.

7.2 Reversible Photo-Control of Ferromagnetic Nanoparticles at Room Temperature

There has been a great interest in developing photo-switchable magnetic materials because of their possible applications for future high-density information storage media. In fact, however, the examples reported so far did not show ferromagnetic behavior at room temperature. From the viewpoint of their practical application to magnetic recording systems, the ability to fix their magnetic moments such that they still exhibit room temperature ferromagnetism is an absolute requirement. $L1_0$–FePt nanoparticles are promising materials for future high-density magnetic recording media due to their uniaxial magnetocrystalline anisotropy ($Ku = 7 \times 10^6$ J m^{-3}) [12] and excellent chemical stability. Here we have designed reversible photo-switchable ferromagnetic FePt nanoparticles whose surfaces were coated with azobenzene-derivatized ligands [13].

We have focused on the direct synthesis of the $L1_0$ phase, which is the so-called polyol process with the aim of achieving surface modification by the application of photo-functional ligands. The advantages of these direct syntheses techniques are that they prevent particle aggregation during the annealing process, and they enable the possibility for surface modification by using functional organic ligands.

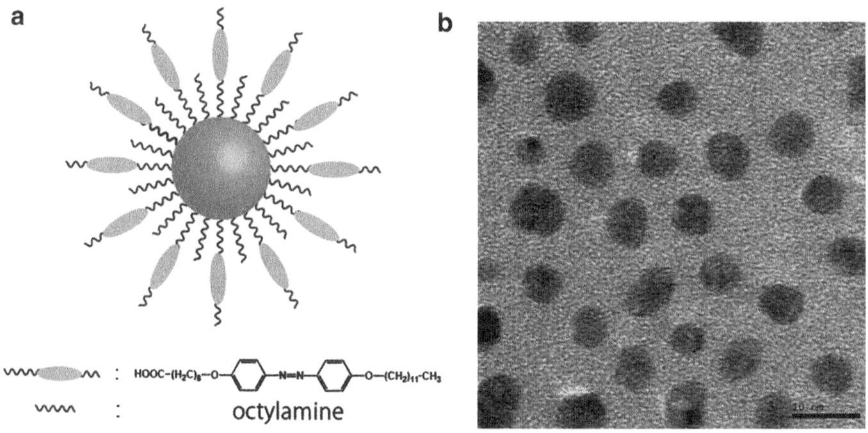

Fig. 7.1 Schematic illustrations (**a**) and TEM image (**b**) for azobenzene-passivated $L1_0$–FePt nanoparticles

Figure 7.1 shows schematic illustration and transmission electron microscope (TEM) image of FePt nanoparticles after surface modification. The ligand-coated FePt nanoparticles show good dispersibility and have a uniform nearest-neighbor spacing of 4.3 nm, while as-prepared "bare" particles tend to aggregate. This "neighbor spacing length" is in good agreement with the azobenzene ligand length of 4.1 nm calculated from the molecular model. This indicates the formation of a mixed monolayer of organic ligands on the surface of the FePt nanoparticles. The average diameter of the particles that appeared in the dispersion was 5.6 ± 1.4 nm.

On the surfaces of core particles, reversible photoisomerization of azobenzene in the solid state was realized by using spacer ligands which provide sufficient free volume. Normally, the photoisomerization of azobenzene-containing compounds does not occur in the solid state because photoisomerization is usually accompanied by a large change in volume. In this system, the presence of *n*-octylamine in the form of spacer ligands enables the reversible photoisomerization of the azo ligands (even in solid state) because it provides sufficient free volume between the azo ligands.

These photoisomerizations brought about changes in the electrostatic field around the core-FePt nanoparticles. During the UV illumination, the initial magnetization value at 10 G, 300 K increased from 3.13 to 3.44 emu/g. We then illuminated the composite FePt nanoparticles with visible light, and the magnetization value decreased from 3.44 to 3.21 emu/g. Even after the illumination was stopped, the increased magnetization value was maintained for at least 2 h. After this process, the UV light-induced increase and visible light-induced decrease in magnetization were repeated at least four times. The photoinduced changes in the magnetization values were estimated to be 9.9 % in the first cycle and 7.3 % after the second cycle. This decrease in the values after the second cycle is also consistent with the decrease in the efficiency of the photoisomerization after the second cycle. We also examined the changes in the magnetization curves caused by photoillumination (Fig. 7.2).

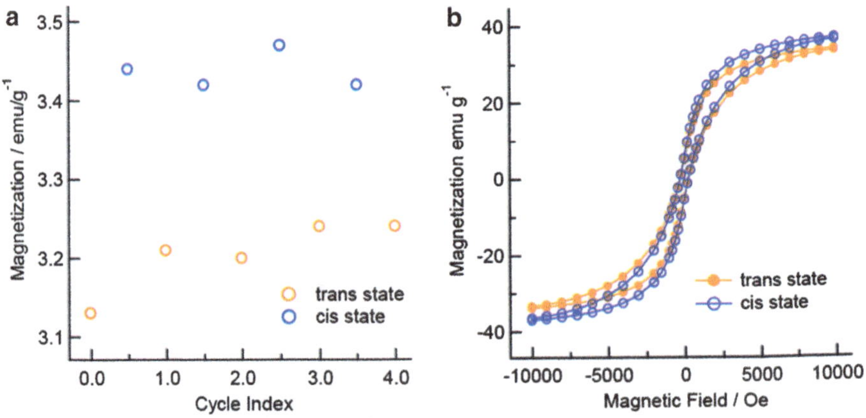

Fig. 7.2 Changes in magnetization by alternating photoillumination by UV and visible light at 300 K (**a**) and changes in magnetization curves at 300 K before and after the UV light illumination (**b**)

After UV illumination, the hysteresis loop became slightly larger compared to the initial state, with an increase in H_c from 250 to 300 G and in M_r from 4.38 to 5.87 emu/g. That is, UV illumination caused the composite FePt nanoparticles to become slightly harder magnets. As a result, we have succeeded in controlling the magnetic properties of these ferromagnetic composite nanoparticles by alternating the photoillumination in the solid state at room temperature.

In order to confirm the photoinduced changes in the electronic states of the iron atoms directly, the ^{57}Fe Mössbauer spectra of the composite nanoparticles were measured before and after the UV light illumination. It is known that organic-ligand-coated FePt nanoparticles exhibit two magnetic phases: one is a hyperfine interaction in the bulk-like core, while the other is a dipolar surface layer due to dipolar charges on the organic ligands. Actually, ^{57}Fe Mössbauer spectra for the current composite FePt nanoparticles show another sextet component with a hyperfine field of ca. 18 T which was not observed in bare-FePt nanoparticles. We attributed this component to the surface Fe atoms on the core particles, which were electronically polarized by the dipolar charge of the organic ligands. These differences in magnetic phase between the particle cores and the particle surfaces form electronic shell structures with significant gradients in their electronic charge densities. After UV illumination, the hyperfine fields of the surface-layer phase changed from ca. 18 to 21 T. These photoinduced changes in the hyperfine field mean that the photoisomerization of the azobenzene ligands had induced electronic polarization on the surfaces of the FePt nanoparticles. As a result, changes in the magnetic properties of the composite nanoparticles were caused by photoinduced changes in the electronic charge density on the surface iron atoms.

This system is the first report about photo-control of ferromagnetic properties at room temperature. By use of this system as a plot type, we developed the functions and expanded this strategy to other physicalities (see following sections).

7.3 Sequential Assembly of Photo-Controllable Ferromagnetic Nanoparticles with Perpendicular Magnetic Anisotropy

The controlled synthesis and assembly of magnetic nanoparticle-based compounds are the keys to important new technologies. In the course of developing such assemblies, it is also crucial to find a way of controlling the magnetic anisotropy of the nanoparticles, since it is expected that the application of perpendicular magnetization will enable the production of ultrahigh-density magnetic recording media. Although a number of papers about thin films that exhibit large perpendicular magnetic anisotropy have been reported, an alternative route to these conventional methods is to use a solution-based process utilizing chemically synthesized nanoparticles. Such processes have great potential to realize the storage of one bit per one particle, which is the ultimate goal in the development of ultrahigh-density recording media. Although various methods for fabricating well-ordered nanoparticle assemblies by wet processes have been proposed, no methods that enable the fabrication of nanoparticle assemblies featuring perpendicular magnetic anisotropy have been proposed as yet.

Herein we propose a novel method, named as the "magnetic field-assisted layer-by-layer method," for the fabrication of ferromagnetic FePt nanoparticle assemblies which exhibit large perpendicular magnetic anisotropy [14].

In the current system, we have modified the well-known layer-by-layer method by simply applying a strong magnetic field during the deposition of the FePt nanoparticles (Fig. 7.3).

Negatively charged hydrophobic $L1_0$–FePt nanoparticles were synthesized based on the "SiO_2 nanoreactor method [15]" and subsequent ligand-exchange reaction with 11-aminoundecanoic acid. Hybrid multilayer films were fabricated by means of the layer-by-layer deposition of cationic polyelectrolyte azobenzene (Fig. 7.4) and $L1_0$–FePt nanoparticles.

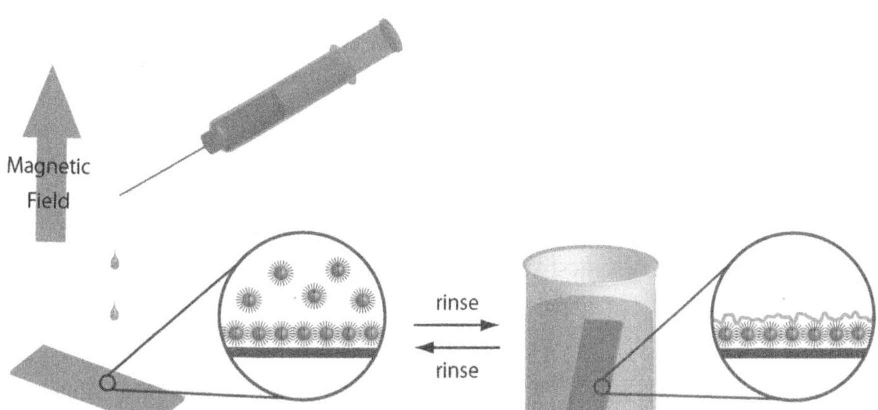

Fig. 7.3 Schematic illustration of film preparation procedure by "magnetic field-assisted layer-by-layer method"

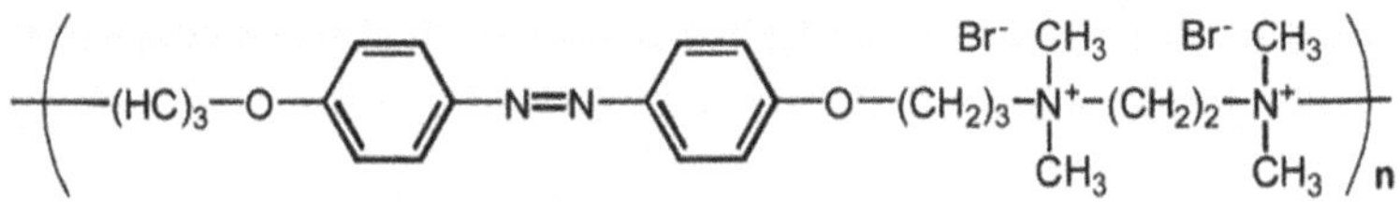

Fig. 7.4 Azobenzene–polyelectrolyte

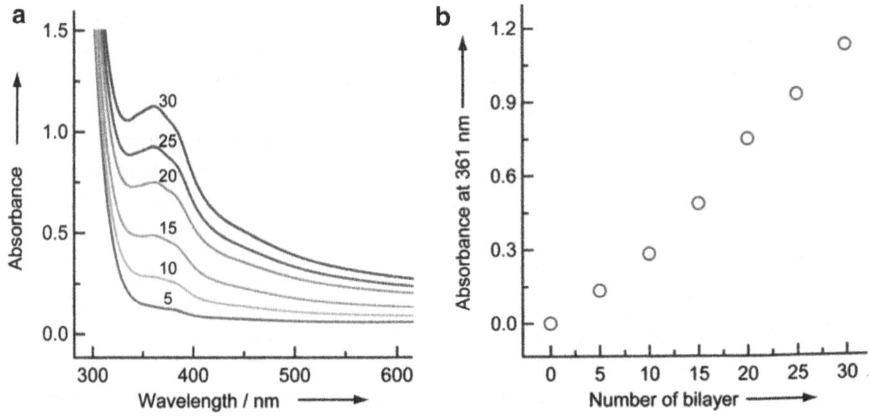

Fig. 7.5 Changes in UV–vis adsorption spectra for the hybrid films with increasing number of deposition cycles (**a**) and plot of maximum absorbance at 361 nm for the hybrid films versus the number of dipping cycles (**b**)

The deposition cycle was monitored by UV–vis spectroscopy. Adsorption spectra showing increasing numbers of deposition cycles are illustrated in Fig. 7.5. Each spectrum gave an intense absorption peak at 361 nm. This peak is ascribed to the π–π* transition of the *trans*-isomer of azobenzene–polyelectrolyte. The absorbance at 361 nm increased linearly as the number of deposition cycles increased. This linearity indicates that uniform deposition processes took place during each dipping cycle. Surface ζ potential measurements also indicate that uniform layer-by-layer deposition was taking place.

The magnetization curves for $L1_0$–FePt nanoparticles dispersed in water solution show superparamagnetic-like behavior with no hysteresis curve, while a large hysteresis loop with coercivity of 1.8 T was observed for powder $L1_0$–FePt nanoparticles. In this case, the magnetization curve could be fitted by the curve given by Langevin's function, which expresses the superparamagnetic magnetization theoretically. This means that each $L1_0$–FePt nanoparticle could be rotated by an applied magnetic field until its c-axis (magnetic easy axis) was parallel to the direction of the magnetic field when in the water solution because each of the

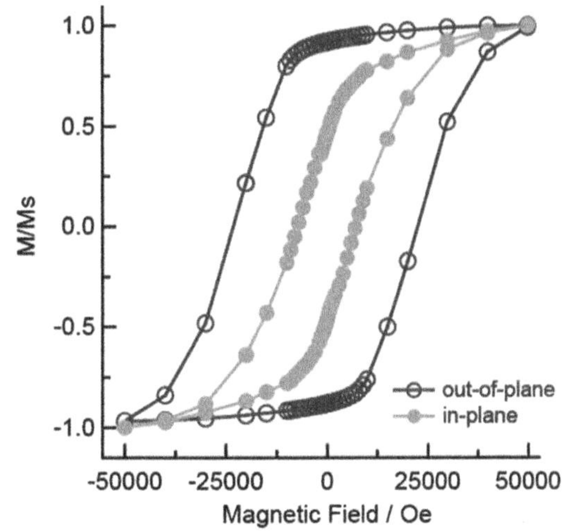

Fig. 7.6 Plot of normalized magnetization M/Ms versus applied magnetic field H at 300 K with the external magnetic field perpendicular (*filled circles*) and parallel (*open circles*) to the film surface

$L1_0$–FePt nanoparticles were magnetically isolated from each other. Hence, it is expected that the application of an external magnetic field during the layer-by-layer deposition process will result in the assembly of magnetic nanoparticles with oriented magnetic moments. Subsequently, we investigated the magnetic anisotropy of the hybrid films as a primary purpose. Figure 7.6 shows the *M*–*H* plots of the hybrid films. The external magnetic field was applied perpendicular and parallel to the film surface.

Importantly, the hybrid films exhibited an extremely high coercive field of 2.4 T and a higher degree of squareness (Mr/Ms = 92 %) in the out-of-plane direction than in the in-plane direction (H_c = 0.5 T, Mr/Ms = 40 %). That is, the hybrid films demonstrated a preferential orientation for their crystalline *c*-axes in the direction perpendicular to the film surface. To the best of our knowledge, this is the first report of the fabrication of ultrathin films consisting of magnetic nanoparticles with such large perpendicular magnetic anisotropy through a solution-based process. Furthermore, the magnetic anisotropy for the current system ($H_{c,easy}/H_{c,hard} = 0.20$) is comparable to the values obtained for the bulk systems realized previously ($H_{c,easy}/H_{c,hard} = 0.13$) [15].

We also investigated the photo-responsibility of the hybrid films. The photoisomerization of the hybrid films in the solid state at room temperature was monitored by UV–visible absorption spectroscopy. The reversible spectral changes indicate high-efficiency reversible photoisomerization, even in the solid state. Subsequently, we investigated the influence of photoillumination on the magnetic properties of the hybrid films at 300 K (Fig. 7.7) with an external magnetic field of 5 T applied perpendicular to the film surface. An increase of about 5 % for the initial saturation magnetization value was observed during the UV illumination, and recovery to the initial value was observed during the visible illumination. After this process, the UV light-induced increases and the visible light-induced decreases in

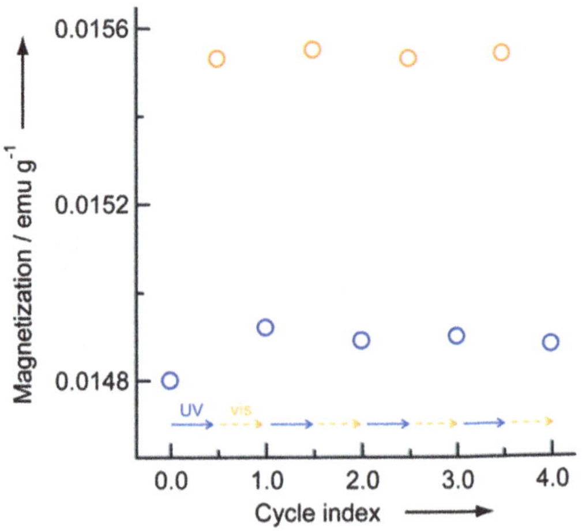

Fig. 7.7 Changes in magnetization by photoillumination at 300 K under an external magnetic field of 5 T perpendicular to the film surface

magnetization were repeated without any attenuation. Even when the external magnetic field was applied parallel to the film surface, very similar phenomena were observed, with almost the same rates of change in the magnetization.

7.4 Reversible Photo-Control of Ferromagnetism Appearing at Au–S Interfaces

Above systems that show photo-magnetic effects (Sects. 7.2 and 7.3) are the hybrid materials of organic photochromes and inorganic magnetic compounds whose magnetic properties are relatively superior to those of conventional compounds. Based on such a strategy, the systems have demonstrated room-temperature reversible photo-control of ferromagnetic order in photochrome-modified FePt nanoparticles. However, the occurrence of these photo-magnetic effects was limited to just the surface layers of the FePt nanoparticles. Hence, the design and synthesis of a new class of optically switchable magnetic compounds that exhibit both large magnetization changes and ferromagnetic order even at room temperature is still a challenging issue.

Here we propose a novel strategy that focuses on the two-dimensional ferromagnetism which appears at the interfaces between organic and inorganic hybrids such as self-assembled monolayer (SAM) films on gold [16, 17]. Recently, the emergences of paramagnetism or ferromagnetism in thiol-passivated gold films and gold nanoparticles despite their bulk diamagnetic component have been reported. Since the discovery by Naaman et al., the occurrence of magnetism at Au–S interfaces has been reported in a number of recent papers. Furthermore, Crespo et al. have demonstrated ferromagnetism in thiol-passivated gold nanoparticles even at room temperature [18].

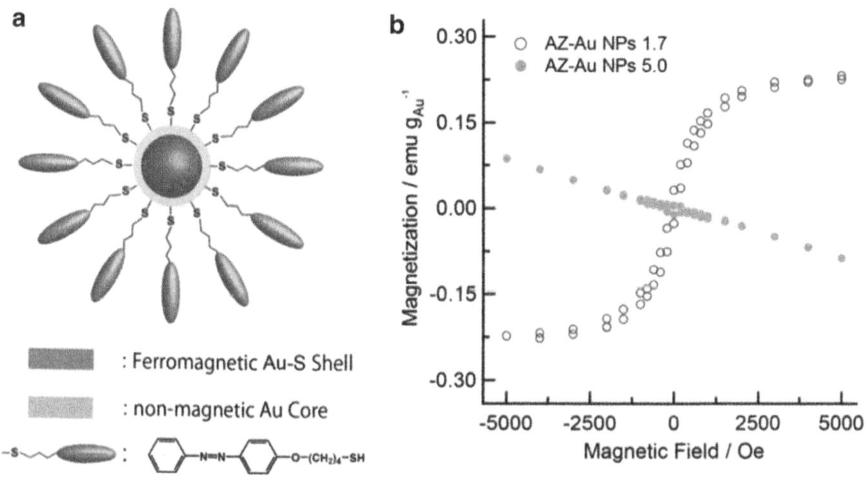

Fig. 7.8 Schematic illustrations for azobenzene-passivated gold nanoparticles (**a**) and magnetic properties of hybrid nanoparticles with different average particle size with 5.0 and 1.7 nm (**b**)

The ferromagnetism has been associated with Au 5d localized holes that are the result of charge transfer from the Au surface atoms to the S atoms of the organic ligands when forming the Au–S bonds. Since charge transfer from Au to S atoms acts as a "trigger" for the generation of ferromagnetism, the magnitudes of the magnetic moments will vary with the metal work functions. It is well known that the metal work function correlates with the surface dipole of organic layers, arising from the cooperative effect of intrinsic molecular dipole moments. Since *trans*-state and *cis*-state of azobenzene-derivatized thiol have opposite dipoles, they can be used to decrease and increase the work function of gold, respectively. As a consequence, the use of SAM-containing photochromes grafted onto gold surfaces has been shown to have great potential for the photo-control of magnetic properties. We now report reversible photoinduced magnetization changes that were observed in gold nanoparticles passivated with azobenzene-derivatized ligands.

The gold nanoparticles with tunable particle size have been prepared by the modified Brust method with different Au/S ratio.

The magnetization curves for the hybrid nanoparticles with average particle size with 5.0 and 1.7 nm are shown in Fig. 7.8. The large nanoparticles (5.0 nm) exhibit a diamagnetic behavior, similar to that of bulk gold. On the other hand, typical hysteresis loops with coercivity and remanent magnetization were observed for the small hybrid nanoparticles (1.7 nm).

The photoisomerization of the nanoparticles in the solid state at room temperature was monitored by UV–visible absorption spectroscopy (Fig. 7.9). Each spectrum of the initial state gave an intense absorption peak at 360 nm, which is ascribed to the π–π^* transition in the *trans*-isomer of the azobenzene ligands. After UV illumination of the Au nanoparticles (5.0 nm), the absorption at 360 nm decreased, suggesting the *trans*-to-*cis* photoisomerization of azobenzene occurred. Following

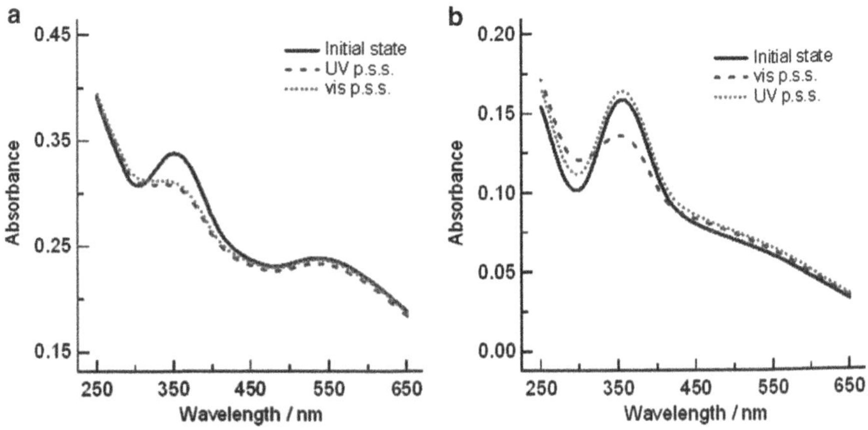

Fig. 7.9 Changes in the optical absorption spectra for the Au nanoparticles (5.0 nm) (**a**) and the Au nanoparticles (5.0 nm) (**b**) due to photoisomerization. The initial *trans*-state (*solid line*) was first illuminated with UV light for 1 min (*dashed line*). It was then illuminated with visible light for 1 min (*solid line*)

subsequent illumination with visible light, the absorption at 360 nm showed almost no spectral changes, suggesting the *cis*-to-*trans* photoisomerization did not occur. In the case of the Au nanoparticles (1.7 nm), the absorption at 360 nm also decreased after UV illumination of the *trans*-isomer. Following subsequent illumination with visible light, the absorption at 360 nm increased again. After the second cycle, the *cis–trans* photoisomerization was repeated without any attenuation of the area between the curves. These reversible spectral changes indicate high-efficiency reversible photoisomerization, even in the solid state.

From the above observations, it is proposed that the decrease in particle size provided free volume between each of the azobenzene ligands because the adjacent angles between two neighboring azobenzene ligands would be increased with decreasing particle size. The average distance between two neighboring nitrogen double bonds on the azobenzene ligands could be estimated as 7.7 Å for the Au nanoparticles (5.0 nm), while it was 13.0 Å for the Au nanoparticles (1.7 nm).

Another important remark concerning the optical absorption spectra is almost no surface plasmon resonance around 550 nm despite the observation of significant surface plasmon resonance for the AZ–Au NPs 5.0, indicating the strong Au–S interactions make the 5d electrons of the core gold nanoparticles behave as partially localized holes that have lost itinerancy because of the high-specific surface area for Au nanoparticles (1.7 nm).

Subsequently, we investigated the influence of photoillumination on the magnetic properties (Fig. 7.10). Reducing the particles size to 1.7 nm also gave rise to photo-magnetic properties because the decrease in particle size provided free volume between each of the azobenzene ligands. The photoinduced changes in the magnetization values were estimated to be ca. 27 %. As mentioned above, in Au nanoparticles, the apparent ferromagnetism is associated with 5d localized holes in

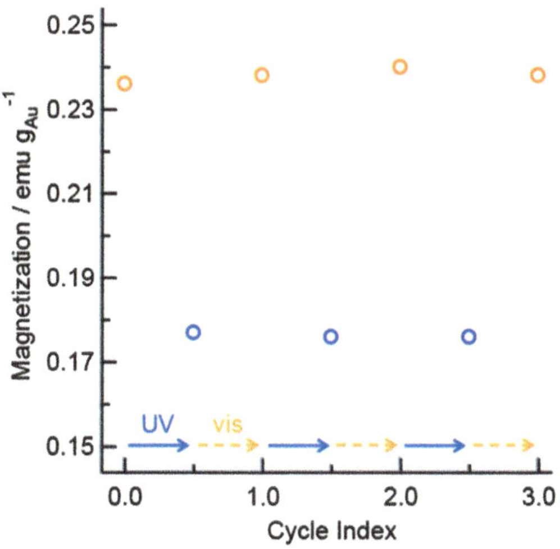

Fig. 7.10 Changes in magnetization by photoillumination at 300 K under an external magnetic field of 5 T at 300 K

the 5d shell generated through Au–S bonding. The magnetic ordering is not due to the large exchange interaction but rather is due to extremely high local magnetic anisotropy (estimated to be as high as 1×10^8 ergs cm^{-3}) which blocks the moments from switching, resulting in very high T_B in Au nanoparticles. From the saturated magnetization value of the Au nanoparticles under applied field of 5 T at 5 K, an estimation of the lower limit value of the magnetic moment of the gold atoms is straightforward. The values of the magnetic moment per Au atom bound to sulfur in the *trans*-state and the *cis*-state are estimated to be $0.033\mu_B$ and $0.024\mu_B$, respectively, where μ_B represents the Bohr magnetron. This result clearly suggests that the *d*-charge loss was decreased in the case of the *cis*-state. That is, the charge transfer from Au to S could be reversed with *trans*-to-*cis* photoisomerization.

These phenomena are also consistent with the reported cooperative effect of organic molecules in the electron transfer between a metal substrate and organic layers. In such organic–inorganic interfaces, charge transfer acts to reduce the dipole–dipole interaction between molecules but may either decrease or increase the molecules-to-surface dipole moment. The change in the work function (WF) is related to the dipole moment density perpendicular to the surface through the following relationship:

$$\Delta\phi = \frac{\mu N \cos\theta}{\varepsilon\varepsilon_0}.$$

In other words, the magnitude of the electron transfer to the organic layer becomes smaller for adsorbates with electronegative end groups, while it becomes larger for adsorbates with electropositive end groups.

In the current system, the values of $\mu \cos\theta$ for the azobenzene ligands were calculated as −0.24 D for the *trans*-state and 2.74 D for the *cis*-state, meaning that the sign of the work function change would be negative for the *trans*-state and positive

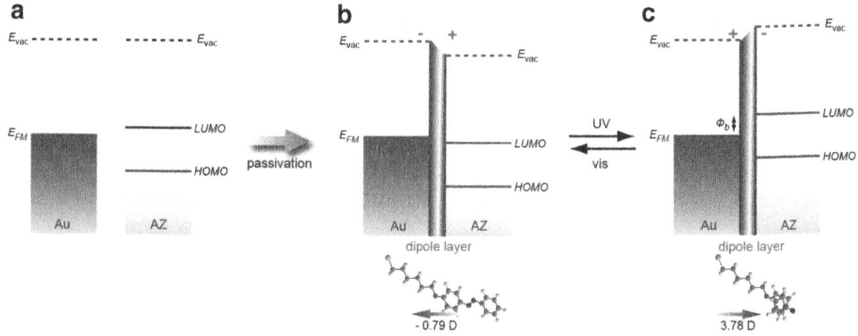

Fig. 7.11 Schematic illustrations of changes in the work function due to the photoisomerization of azobenzene ligands. (**a**) Schematic energy level diagrams for an untreated interface (without surface passivation). (**b**) Passivation of *trans*-azobenzene imposes an interface dipole that decreases local vacuum energy level (E_{vac}). (**c**) Photoisomerization to *cis*-azobenzene imposes an interface dipole that increases local vacuum energy level (E_{vac})

for the *cis*-state. Hence, the magnitude of electron transfer in the initial *trans*-state becomes smaller in the *cis*-state under UV light illumination, and it recovers to its initial value under visible light illumination. Schematic illustrations of the changes in the work function due to the photoisomerization are shown in Fig. 7.11. As a consequence, the magnetization value could be controlled by employing alternating photoillumination with UV and visible light and is caused by changes in the d-charge loss due to the photoisomerization of azobenzene ligands.

7.5 Reversible Photo-Control of Superconductivity

In the above systems, charge transfers at the organic–inorganic interfaces were the origin of the photo-magnetic effects. Such phenomena open new opportunities to control other properties, such as superconducting properties.

Since the disappearance of electrical resistivity was first observed in mercury by Onnes [19], much attention has been given to study superconducting materials. The main objectives of such studies are to discover new classes of superconducting materials, to improve their superconducting properties, or to clarify their mechanisms [20–24]. While manipulating the electronic states of condensed mater by external stimuli is a key topic in the field of modern electronics, most notably optical stimuli are of considerable importance because they present many possibilities for realizing future optical memory or switching devices. Until now, photoinduced changes in superconducting properties have been reported in several systems such as the YBCO films [25], YBCO/LCMO hetero films [26], and the alkali-metal fullerides [27]. However, such photoinduced effects were based on structural changes or on photoinduced excitation of the electrons, so no reversible photoinduced effects on superconductivity have been reported so far. In the field of superconducting materials, the

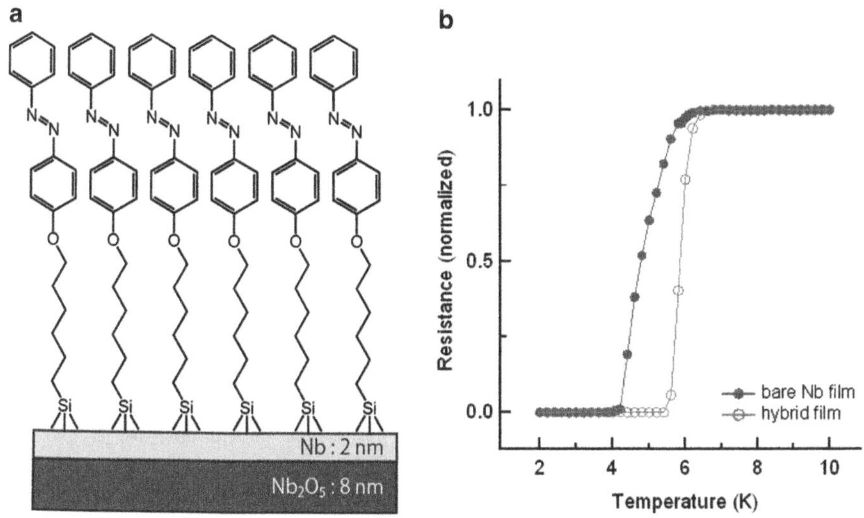

Fig. 7.12 Schematic illustration for azobenzene-monolayer-passivated Nb thin film (**a**) and temperature dependence of the Nb film resistance before and after the surface passivation (**b**)

electric-field tuning of surface carrier density using a metal–insulator–semiconductor field-effect-transistor (FET) structure is the most common method of manipulating electronic states [28–30]. As existing microelectronics approaches its technological and physical limits, alternative routes for realizing future nanoelectronics must be molecular-based or hybrid devices. Recent studies reveal that a cooperative effect involving charge transfer occurs at the interfaces between organic and inorganic hybrids such as SAM films on conductive substrates [31]. These charge transfer processes induce field-effect-like behavior, resulting in the alternation of conducting properties in metals [32], semiconductors [33], and even in superconductors [34].

Here, we have designed an azobenzene-containing SAM on an Nb thin film (10 nm thick including a ca. 2.7 nm native oxide layer) with which we can realize the reversible photo-tuning of superconducting properties [35]. It is known that well-organized SAMs on oxidized Nb surfaces can be obtained by using a silane functional group. In the present study, we used a silane-terminated azobenzene molecule. A schematic representation of such a system is illustrated in Fig. 7.12.

The superconducting properties of the bare Nb film and the hybrid film were determined from electrical measurements made using a four-point scheme. The bare Nb film was superconducting with a zero resistance temperature (T_c^{ZERO}) of 4.2 K, while an abrupt increase in T_c^{ZERO} to 5.2 K was found in the hybrid film (Fig. 7.12b). On the other hand, the initial I_c value was dramatically decreased after the formation of the AZ monolayer. Furthermore, the values of I_c, T_c, and resistance for the hybrid film could be reversibly controlled by employing alternating photoillumination with UV and visible light (Fig. 7.13).

Table 7.1 summarizes the values of T_c, I_c, and resistance for each state. Since our comparative experiments revealed that Nb films passivated with “non-azo” silane

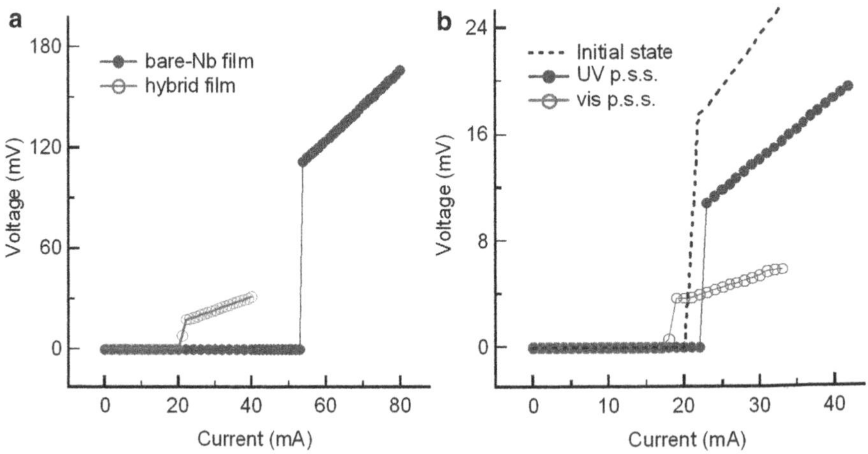

Fig. 7.13 *V–I* measurements of the Nb film at 2K before and after the surface passivation (**a**) and photoillumination (**b**)

Table 7.1 Superconducting transition temperature, critical current, and resistance value for the Nb films in each state

	Bare Nb	Hybrid film (initial state)	UV p.s.s.	Vis p.s.s.
T_c^{ZERO} (K)	4.2	5.2	Down	Up
I_c (mA)	53	20	23	17
Rn (Ω) on[a] T_c	2.2	0.71	0.50	0.21
Rn (Ω) on[b] I_c	2.1	0.79	0.47	0.18

[a]The resistance values on T_c were defined as the resistance value at 7 K
[b]The resistance values on I_c were extracted from the slope of the *V–I* plots

molecules (e.g., hexadecyltrichlorosilane) did not show any photoinduced changes in their superconducting properties, despite showing similar passivation-induced effects, the observed photoinduced effects must originate from photoisomerization of the azobenzene. Furthermore, the same measurements were also performed on thick Nb films (~100 nm). In this case, the hybrid system did not show any passivation-induced or photoinduced effects, confirming that the observed passivation induced and photoinduced effects are surface-dominated phenomena.

One possible explanation for the surface passivation-induced or photoinduced effects is the theory of critical current in type II superconductors. In a type II superconducting thin film, the surface pinning potential is very strong because of the nonuniformity of the electron density along the surface. When a SAM is formed on a type II superconducting thin film, charge is transferred at the surface between the superconductor and the organic layer in order to reduce repulsion within the organic layer, resulting in a uniform distribution of the charge density along the surface. As a result, variations in the electron density at the film surface that are induced by the

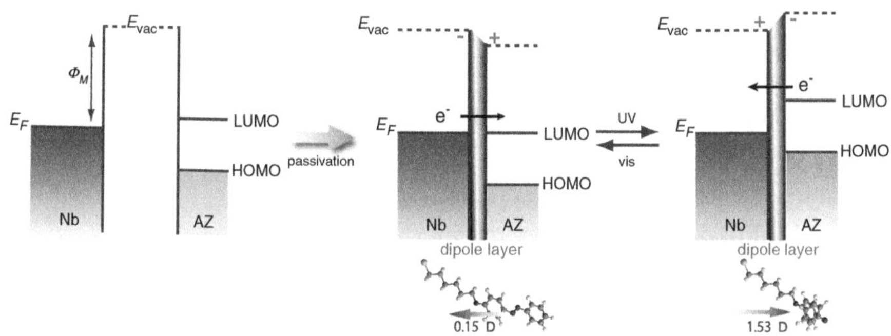

Fig. 7.14 Schematic illustrations of changes in the work function due to the photoisomerization of azobenzene. (**a**) Schematic energy level diagrams for an untreated interface (without surface passivation). (**b**) Passivation of *trans*-azobenzene imposes an interface dipole that decreases the local vacuum energy level (E_{vac}). (**c**) Photoisomerization to *cis*-azobenzene imposes an interface dipole that increases the local vacuum energy level (E_{vac})

absorbed organized molecular layer are responsible for the dramatic decrease in I_c. These explanations are also consistent with enhancement of T_c^{ZERO}. Type II superconductors exhibit superconducting mixed states around their superconducting transition temperature. Such effects make a superconducting mixed state extend over a wide temperature range, resulting to a lower T_c^{ZERO}. Hence, the presence of a more uniform distribution of charge density along the surface caused by the charge transfer process increases the value of T_c^{ZERO}.

Since charge transfer processes from the substrate to the organic monolayer act as a "trigger" for the alternation of superconducting properties, these effects must be controlled by using the metal work function, which is linearly related to the dipole moment density perpendicular to the surface. Since the *trans*-state and *cis-state* of the azobenzene monolayer have different dipoles, they can be used to change the work function of the substrate by photo-irradiation. In the current system, the values of $\mu \cos\theta$ for AZ were calculated as −0.15 D for the *trans*-state and 1.53 D for the *cis*-state. Hence, the magnitude of the electron transfer in the initial *trans*-state becomes smaller in the *cis*-state under UV light illumination, and it recovers to around its initial value under visible light illumination.

Consequently, the observed reversible photo-switching of superconductivity can be summarized as follows. The work function of bare Nb substrates was decreased by the negative surface dipole of *trans*-state azobenzene with surface passivation, and charge transfer from the Nb substrate to the organic monolayer was induced. Such charge transfer processes decrease the inhomogeneities of the electron distribution along the surface. After UV illumination, the charge transfer could be reversed with *trans*-to-*cis* photoisomerization because of the positive surface dipole of *cis*-state azobenzene. As a result, the values of I_c and T_c could be controlled by employing alternating photoillumination with UV and visible light, which caused changes in the electron density along the surface due to the photoisomerization of azobenzene. Schematic illustrations of the changes in the work function due to the photoisomerization of azobenzene monolayer are shown in Fig. 7.14.

Until now, photoinduced changes in superconducting properties have been reported in several systems. However, such photoinduced effects were based on structural changes or on photoinduced excitation of the electrons, so this system is a first report about reversible photoinduced effects on superconductivity.

7.6 Conclusion

In this chapter, several novel photo-functional magnetic compounds created based on the above strategy have been introduced. These results overcame the hard part of the designing condition for photo-functional magnetic compounds, which are (1) to determine how to develop such materials and (2) room temperature operation. Finally, these works have not only proposed a new strategy for developing photo-functional materials but have also opened up great possibilities for practical applications in future optical magnetic recording systems.

References

1. Sato O, Tao J, Zhang Y (2007) Control of magnetic properties through external stimuli. Angew Chem Int Ed 46:2152–2187
2. Sato O, Iyoda T, Fujishima A, Hashimoto K (1996) Photoinduced magnetization of a cobalt–iron cyanide. Science 272:704–705
3. Ohkoshi S, Yorozu S, Sato O, Iyoda T, Fujishima A, Hashimoto K (1997) Photoinduced magnetic pole inversion in a ferro-ferrimagnet. $(Fe^{II}{}_{0.40}Mn^{II}{}_{0.60})_{1.5}$Cr-III$(CN)_6$. Appl Phys Lett 70:1040–1042
4. Yamamoto T, Umemura Y, Sato O, Einaga Y (2005) Observation of the anisotropic photoinduced magnetization effect in Co–Fe Prussian blue thin films fabricated by using clay Langmuir–Blodgett films as a template. J Am Chem Soc 127:16065–16073
5. Decurtins S, Gütrich P, Kohler CP, Spiering H, Hauser A (1984) Light-induced exited spin state trapping in a transition-metal complex—the hexa-1-propyltetrazole-iron (II) tetrafluoroborate spin-crossover system. Chem Phys Lett 105:1–4
6. Renz F, Oshio H, Ksenofontov V, Waldeck M, Spiering H, Gütrich P (2000) Strong field iron(II) complex converted by light into a long-lived high-spin state. Angew Chem Int Ed 39:3699–3700
7. Hayami S, Gu ZZ, Shiro M, Einaga Y, Fujishima A, Hashimoto K (2000) First observation of light-induced excited spin state trapping for an iron(III) complex. J Am Chem Soc 122:7126–7127
8. Koshihara S, Oiwa A, Hirasawa M, Katsumoto S, Iye Y, Urano C, Takagi H, Munekata H (1997) Ferromagnetic order induced by photogenerated carriers in magnetic III–V semiconductor heterostructures of (In, Mn)As/GaSb. Phys Rev Lett 78:4617–4620
9. Haneda S, Munekata H, Koshihara S, Takatani Y (2000) Fe-based magnetic-semiconductor hybrid structures for photocarrier-induced magnetism. J Appl Phys 87:6445–6447
10. Matsuda K, Machida A, Moritomo Y, Nakamura A (1998) Photoinduced demagnetization and its dynamical behavior in a $(Nd0.5Sm0.5)_{0.6}Sr_{0.4}MnO_3$ thin film. Phys Rev B 58:R4203–R4206
11. Einaga Y (2006) Photo-control of magnetization by photochromic compounds. Bull Chem Soc Jpn 79:361–372

12. Ivanov OA, Solina LV, Demeshina VA, Magat LM (1973) Determination of the anisotropy constant and saturation magnetization, and magnetic properties of powders of an iron-platinum alloy. Phys Met Metallogr 35:81–85
13. Suda M, Nakagawa M, Iyoda T, Einaga Y (2007) Reversible photoswitching of ferromagnetic FePt nanoparticles at room temperature. J Am Chem Soc 129:5538–5543
14. Suda M, Einaga Y (2009) Sequential assembly of phototunable ferromagnetic ultrathin films with perpendicular magnetic anisotropy. Angew Chem Int Ed 48:1754–1757
15. Tamada Y, Yamamoto S, Takano M, Nasu S, Ono T (2007) Well-ordered L1(0)-FePt nanoparticles synthesized by improved SiO_2-nanoreactor method. Appl Phys Lett 90:162509
16. Suda M, Kameyama N, Suzuki M, Kawamura N, Einaga Y (2008) Reversible phototuning of ferromagnetism at Au–S interfaces at room temperature. Angew Chem Int Ed 48:160–163
17. Suda M, Kameyama N, Ikegami A, Einaga Y (2009) Reversible phototuning of the large anisotropic magnetization at the interface between a self-assembled photochromic monolayer and gold. J Am Chem Soc 131:865–870
18. Crespo P, Litran R, Rojas TC, Multigner M, de la Fuente JM, Sanchez-Lopez JC, Garcia MA, Hernando A, Penades S, Fernandez A (2004) Permanent magnetism, magnetic anisotropy, and hysteresis of thiol-capped gold nanoparticles. Phys Rev Lett 93:087204
19. Onnes KH (1911) The Superconductivity of Mercury. Comm Phys Lab Univ Leiden 122–124
20. Bednorz JG, Müller KA (1986) Possible high-Tc superconductivity in the BaLaCuO system. Z Phys B 64:189–193
21. Hebard AF, Rosseinsky MJ, Haddon RC, Murphy DW, Glarum SH, Palstra TT, Ramirez AP, Kortan AR (1991) Superconductivity at 18 K in potassium-doped C_{60}. Nature 350:600–601
22. Kamihara Y, Watanabe T, Hirano M, Hosono H (2008) Iron-based layered superconductor $La[O_{1-x}F_x]FeAs$ (x = 0.05–0.12) with Tc = 26 K. J Am Chem Soc 130:3296–3297
23. Bardeen J, Cooper L, Schrieffer JR (1957) Theory of superconductivity. Phys Rev 108:1175–1204
24. Emery VJ (1987) Theory of high-Tc superconductivity in oxides. Phys Rev Lett 58:2794–2797
25. Kudinov VI, Kirilyuk AI, Kreines NM, Laiho R, Lähderanta E (1990) Photoinduced superconductivity in YBaCuO films. Phys Lett A 151:358–364
26. Pena V, Gredig T, Santamaria J, Schuller IK (2006) Interfacially controlled transient photoinduced superconductivity. Phys Rev Lett 97:177005
27. On DT, Jiang L, Kitazawa K, Fujishima A, Hashimoto K (1999) Enhancement of superconductive shielding via photoinduced surface modification. J Phys Chem B 103:3511–3514
28. Takahashi KS, Gabay M, Jaccard D, Shibuya K, Ohnishi T, Lippmaa M, Triscone JM (2006) Local switching of two-dimensional superconductivity using the ferroelectric field effect. Nature 441:195–198
29. Ueno K, Nakamura S, Shimotani H, Ohtomo A, Kimura N, Nojima T, Aoki H, Iwasa Y, Kawasaki M (2008) Electric-field-induced superconductivity in an insulator. Nature Matter 7:855–858
30. Ahn CH, Triscone JM, Mannhart J (2003) Ferroelectricity at the nanoscale: local polarization in oxide thin films and heterostructures. Nature 424:1015–1018
31. Cahen D, Naaman R, Vager Z (2005) The cooperative molecular field effect. Adv Func Mater 15:1571–1578
32. Zhang YM, Terrill RH, Bohn PW (1999) Ozonolysis is the primary cause of UV photooxidation of alkanethiolate monolayers at low irradiance. Anal Chem 71:119–125
33. Cohen R, Kronik L, Shanzer A, Cahen D, Liu A, Rosenwaks R, Lorenz JK, Ellis AB (1999) Molecular control over semiconductor surface electronic properties: dicarboxylic acids on CdTe, CdSe, GaAs, and InP. J Am Chem Soc 121:10545–10553
34. Shvarts DM, Hazani M, Shapiro BY, Leitus G, Sidorov V, Naaman R (2005) Molecular induced field effect in superconducting Nb films. Europhys Lett 72:465–471
35. Ikegami A, Suda M, Watanabe T, Einaga Y (2010) Reversible optical manipulation of superconductivity. Angew Chem Int Ed 49:372–374

Part II
New Photochromic Molecules

Chapter 8
High Performance Thermally Irreversible Photochromism Based on 6π-Electrocyclization

Yasushi Yokoyama

Abstract In this chapter, we will first introduce a remarkably efficient switching of the fluorescence of a bisarylcoumarin with photochromism as well as pH change of the medium. Next, the development of a new thermally irreversible photochromic family featuring the indene group as the central ethene moiety is presented along with the realization of a highly photon-economic transformation of the open form to the closed form. Thirdly, the highly diastereoselective transformation (up to 100 %) for the chiral photochromic diarylethenes, from the open to closed form, is described. Finally, this research concept is extended to include the conformation control of photochromic bisarylindenols, achieving a high ring-closing quantum yield (up to 0.85) as well as high diastereoselectivity (up to 100 %) which were brought about by the conformational fixation of the hexatriene moiety by two intramolecular hydrogen bonds.

Keywords Diarylethene • Diastereoselectivity • Electrocyclization • Fluorescence switch • Quantum yield

8.1 Introduction

A number of thermally reversible photochromic compound families are known; however, few that are thermally irreversible have been reported [1]. In fact, until recently, only fulgides [2], diarylethenes [3], and phenoxynaphthacenequinones [4] had been recognized. Diarylethenes exhibit reversible transformation between hexatriene and cyclohexadiene structures by photochemical electrocyclization. And

Y. Yokoyama (✉)
Department of Advanced Materials Chemistry, Graduate School of Engineering, Yokohama National University, Tokiwadai, Hodogaya, Yokohama 240-8501, Japan
e-mail: yyokoyam@ynu.ac.jp

M. Irie et al. (eds.), *New Frontiers in Photochromism*,
DOI 10.1007/978-4-431-54291-9_8, © Springer Japan 2013

when the structure is adequately designed, they are extraordinarily fatigue resistant. Diarylethenes are, therefore, regarded as one of the best candidates for applications in light-driven switches and memory devices.

Most of the ethene moieties in present-day diarylethenes employ hexafluorocyclopentene. The hexafluorocyclopentene ring has two additional advantages to photochromic systems: (1) high durability towards fatigue by photoirradiation as well as thermal side reactions and (2) no occurrence of *cis–trans* isomerization. Despite such superb characteristics, hexafluorocyclopentene restricts molecular modification only on aromatic groups. If the central ethene moiety can be used as a part of a photochemically switchable functional group when durability is not an important issue, the capabilities of the molecular design can expand. We undertook to create a novel skeleton and found that the indene skeleton is an attractive alternative ethene moiety.

In order to use photochromic compounds as switches of functions and properties, the conversion ratios (CRs) to both stable open form (O-form) and closed form (C-form) should be high; otherwise, the properties of the O-form and C-form may compensate for each other. While the O-form tends to have an absorption maximum only in the UV region so that it is usually colorless, the C-form has absorption bands in the visible region as well as UV region. Thus, the CR to C-form is usually less than 100 % since irradiation of UV light causes not only O-form to C-form transformation but also C-form to O-form transformation. Since the conversion of the C-form to O-form by visible light irradiation results in the complete generation of the O-form, the problem is how to increase the CR to the C-form by UV irradiation. An increase in the quantum yield of the O-form to C-form is one of the possible ways to achieve such high CR. This chapter investigates these challenges in relation to novel compounds such as bisarylindenone derivatives.

We have realized from before that the hexatriene moiety of the O-forms of photochromic compounds does not take a planar conformation but a helical structure [5]. Upon photochemical cyclization, the hexatriene moiety composed of all-sp^2 carbon atoms changes to the cyclohexadiene ring with two newly formed sp^3 carbon atoms which may be stereogenic centers [6]. We have long been interested in the chirality in photochromism and have achieved the 94:6 diastereomer ratio of a bisbenzothienylethene which was governed by one point chirality [7]. Extended research results on this subject will be described here.

8.2 Fluorescence Switch

Diarylethenes [3] are one of the most promising thermally irreversible photochromic families to be applied to molecular switches [8]. However, since structurally less modifiable hexafluorocyclopentene is usually used as the central ethene moiety, replacement of the hexafluorocyclopentene ring with other functional groups has recently attracted much attention [9–15]. When a π-conjugating functional group is located as the central ethene moiety and its property is switched by changing the π-conjugation, a photoinduced molecular switch can be realized. Among the

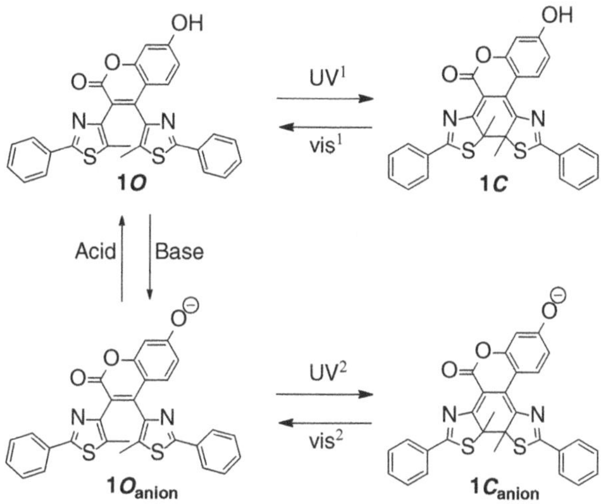

Fig. 8.1 Dual mode switching of fluorescence by pH change and photochromism

molecular properties that can be switched, fluorescence is particularly interesting and useful [16, 17] since fluorescent chemical species can be used as probes for intracellular processes [18–23] as well as for optical recording materials with fluorescence readout [24–26]. In addition, related to the former use, fluorescence imaging techniques have been developed to visualize biological materials. For example, ultrahigh resolution imaging that we could hardly have previously imagined has recently been achieved [27]. In order to follow the flow of materials in a living cell before and after an event, such as a stimulus applied to the cell, the activation of the fluorescent ability of the probe, which should coincide with stimulation at the same point where the stimulus is applied, is essential. It is also more useful if the fluorescence intensity is sensitive to a specific environment so that it works as a fluorescent probe, e.g., a light-switchable probe which works under specific pH conditions [28].

It is obvious that such a switchable fluorescence imaging system requires both a high fluorescence quantum yield and high fluorescence switching ability. Although photochromic compounds exhibiting good reversibility in fluorescence intensity have been reported, most of them have relatively low fluorescence quantum yields and fluorescence switching ability [18–23]. It is also known that coumarin-type compounds possess a high fluorescence quantum yield and that it increases when an electron-donating group is attached to C(7) of the coumarin skeleton [29–31]. Although a photochromic hexatriene structure possessing a coumarin as one of the terminal double bonds has been reported [32], its fluorescence switching ability was not excellent.

With these considerations in mind, we undertook the synthesis, photochromism, and fluorescence switching of a molecule ***1O*** having two thiazole rings and a 7-hydroxycoumarin skeleton as the central ethene moiety. Upon UV irradiation, we expected the fluorescence of ***1O*** to be quenched by photocyclization due to the deformation of the coumarin structure. In addition to photochromic structural changes, its fluorescence would become stronger by the generation of the phenolate anion upon base treatment [28, 31] so that it could be regarded as a dual-mode fluorescence switching molecule (Fig. 8.1).

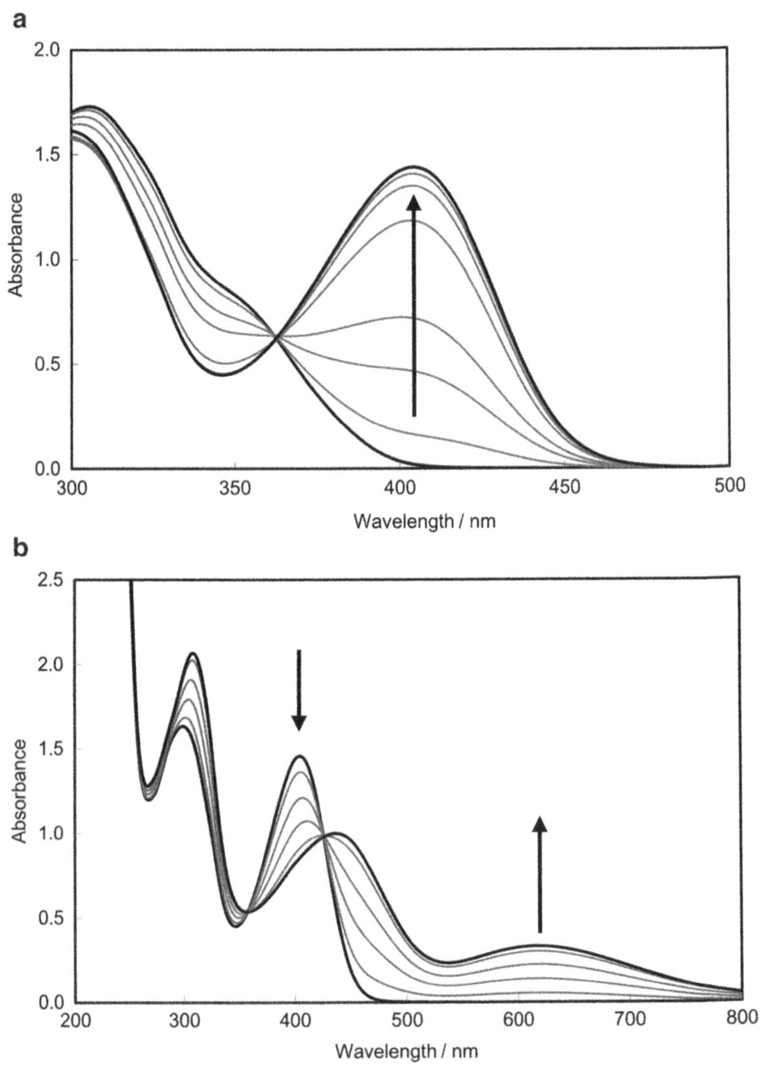

Fig. 8.2 (**a**) Change in absorption spectra of ***1O*** upon addition of DBU in CH_2Cl_2. Equivalence of DBU: 0–12; (**b**) change in absorption spectra of $\boldsymbol{1O}_{\mathbf{anion}}$ during irradiation of 405-nm light in the presence of 12.0 eq. DBU in CH_2Cl_2. Irradiation time: 0–40 min. Reproduced by permission of The Royal Society of Chemistry from [34]

Upon consecutive irradiation with 313- and 578-nm light in dichloromethane and acetonitrile, **1** showed good photochromism. High conversion ratios to ***1C*** at the photostationary state (PSS) of 313-nm light irradiation were as high as 98 % and 95 % for dichloromethane and acetonitrile, respectively. When 1,8-diazabicyclo[5.4.0]undec-7-ene (DBU) was added to a dichloromethane solution of ***1O*** to generate the phenolate, a new absorption band centered at 404 nm appeared, and its intensity was saturated when the 12.0 mol equivalent of DBU was added (Fig. 8.2a). The appearance of this

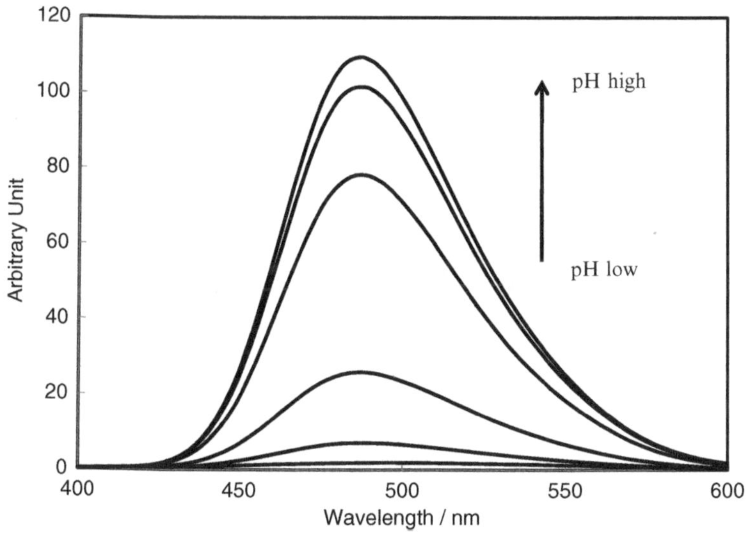

Fig. 8.3 Change in fluorescence spectra of **1O** with pH of the medium. Solvent: $CH_3OH/NaNO_3$ aqueous buffer solution = 1/2 (v/v), concentration: 5.1×10^{-6} mol dm^{-3}, excitation: 365 nm, pH of the solution: 2.64, 6.05, 6.38, 7.01, 7.58, 9.65. Reproduced by permission of The Royal Society of Chemistry from [34]

new band is ascribable to the generation of a phenolate anion ***1O***$_{anion}$ [33]. A reversible photochromic reaction between ***1O***$_{anion}$ and its PSS (***1C***$_{anion}$/***1O***$_{anion}$ > 100; calculated from the fluorescent intensity) in dichloromethane was observed by means of alternative irradiation with 405-nm (Fig. 8.2b) and 578-nm light.

The fluorescence of ***1O*** was then examined in water/methanol (2/1, v/v) with base treatment. Figure 8.3 illustrates the change in the fluorescence spectra upon acid–base titrations using aqueous sodium hydroxide solution. The pK_a value of the phenol moiety of ***1O*** was determined to be 6.84, which is consistent with that of 7-hydroxycoumarin [29]. This pK_a value is adequate in controlling the changes between ***1O*** and ***1O***$_{anion}$ by the slight change of the pH in biological systems. Indeed, a 14.2-fold increase was observed by changing the pH from 6.05 to 7.58. The fluorescence quantum yield of ***1O***$_{anion}$ with an excitation wavelength of 330 nm at pH 9.48 was determined to be 0.25, which is large enough to apply to fluorescence imaging, even under more neutral conditions (Fig. 8.3).

In the mixed solvent of a buffer solution of sodium tetraborate and methanol (2/1, v/v, pH 9.48), **1** showed excellent photochromism upon successive irradiation of 405-nm (Fig. 8.4a) and >500-nm light. When ***1O***$_{anion}$ at pH 9.48 was irradiated with 405-nm light and reached its PSS (***1C***$_{anion}$/***1O***$_{anion}$ = 98.5/1.5; calculated from the fluorescent intensity), the intensity of the fluorescence excited at 420 nm decreased to 1.5 % from the initial intensity before irradiation (Fig. 8.4b).

As the conclusion of this section [34], we have synthesized 7-hydroxy-3,4-bis(5-methyl-2-phenyl-4-thiazolyl)coumarin **1**. Its fluorescence intensity changed (quantum yield of 0.25 at pH 9.48 in buffered aqueous solution containing methanol) with photochromism as well as with changes in the pH of the media. The fluorescence

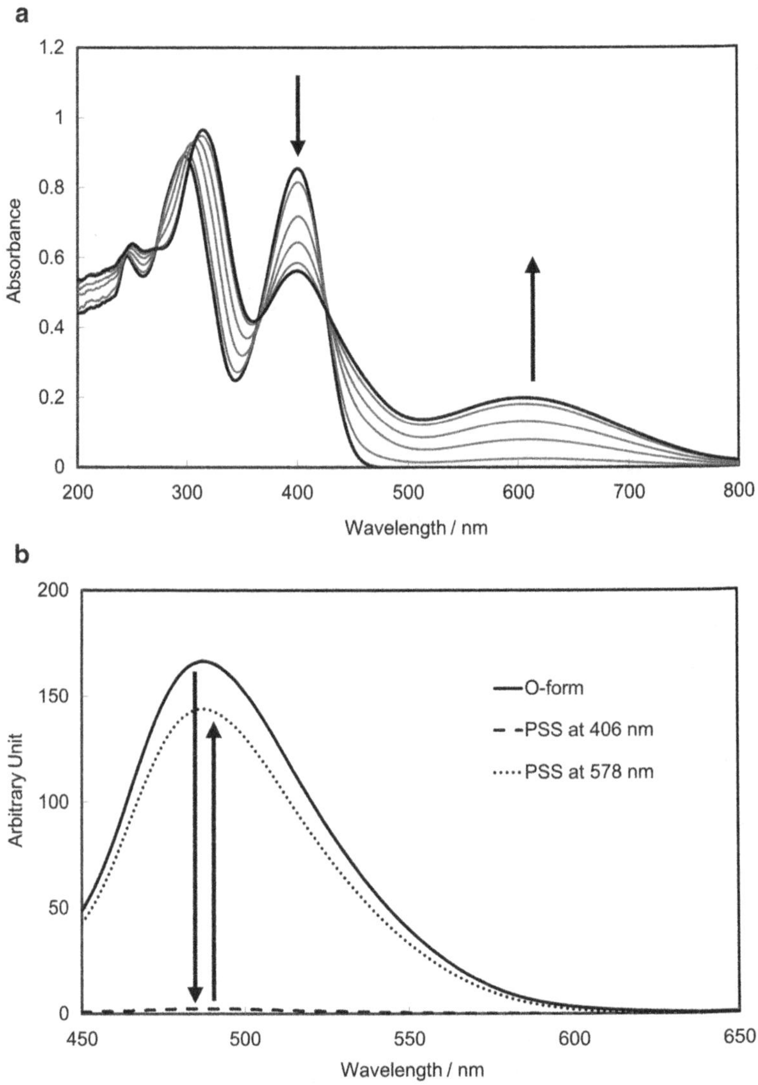

Fig. 8.4 (**a**) Change in absorption spectra of **1** during irradiation of 405-nm light in CH_3OH/10 mM sodium tetraborate buffer solution (pH 9.48). Concentration: 4.38×10^{-5} mol dm^{-3}, light intensity: 0.25 mW cm^{-2}, irradiation time: 0–7 min; (**b**) change in fluorescence spectra between $\mathbf{1}\boldsymbol{O}_{\mathbf{anion}}$ and $\mathbf{1}\boldsymbol{C}_{\mathbf{anion}}$ upon 405- and >500-nm light irradiation in CH_3OH/10 mM sodium tetraborate buffer solution (pH 9.51). Concentration: 2.75×10^{-6} mol dm^{-3}, excitation: 420 nm. Reproduced by permission of The Royal Society of Chemistry from [34]

intensity of the PSS generated from **1*O*** by 405-nm light irradiation at pH 9.48 was quenched 0.015-fold (1/60), while that of the open form at pH 7.58 was increased 14.2-fold that of pH 6.05. These results show that the fluorescence intensity of **1*O*** can be modulated efficiently by photochromic reactions as well as by the pH of the media.

8.3 Photochromic Compounds with Novel Skeletons: Indenone Derivatives

Diarylethenes [3] are a well-designed and superior class of thermally irreversible photochromic compounds; however, one drawback is that structural modifications are mainly performed on the aryl groups at both ends of the molecule and a modification of the central ethene moiety is sometimes difficult. Nowadays, most diarylethenes are hexafluorocyclopentene based and structural modification of this ethene moiety is almost impossible. When the central ethene is incorporated in an aryl group [14, 15, 35, 36], it usually becomes thermally reversible due to the large loss of aromatic stabilization energy upon photocyclization. Other modifications have proved that the central ethene moiety can act as a switchable functional group [11–13] or the monomer for polymerization [37].

We envision that if the central ethene moiety were a functional group that could be modified structurally as well as electronically in easily accessible ways, the diversity of the thermally irreversible photochromic molecules could be widely expanded. We decided to synthesize 2,3-bisarylindenones (Fig. 8.5).

For the aryl group, 5-methyl-2-phenylthiazole was employed, and 2,3-bis(5-methyl-2-phenyl-4-thiazolyl)indenone (**2*o***) was synthesized. When it cyclizes by photoirradiation, the closed form **2*c*** will be formed as the result of the photoinduced 6π-electrocyclization, with a large conjugated system showing absorption bands at longer wavelengths in the visible region.

In acetonitrile, **2*o*** showed an absorption maximum at 433 nm, in addition to the maxima in the UV region at 262 and 311 nm (Fig. 8.6a), while the color of the solution was yellow. Upon irradiation of 436 nm light, new absorption maxima appeared at 542, 429, 363, 307, and 253 nm, and the color of the solution turned brown. This is attributable to the formation of the cyclic form **2*c***. HPLC analysis revealed that the

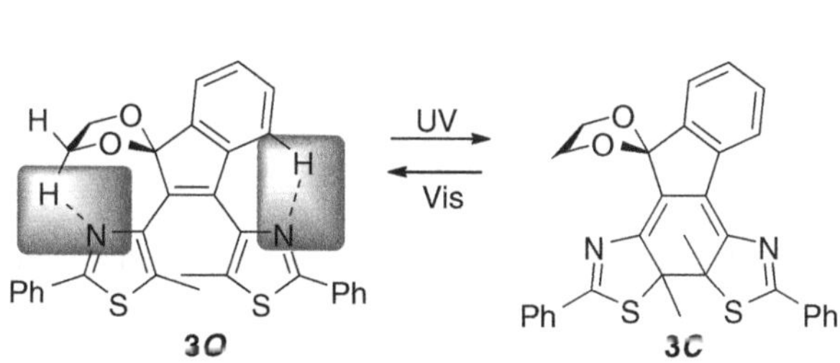

Fig. 8.5 Photochromism of 2,3-bisthiazolylindenone and its ethylene acetal

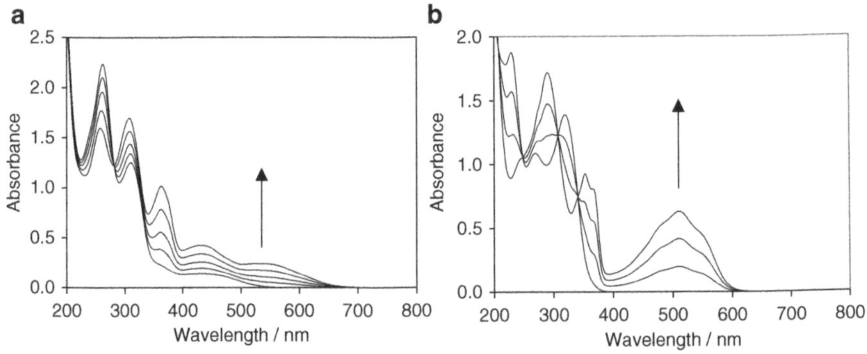

Fig. 8.6 (**a**) Absorption spectral change of **2** during irradiation of 436-nm light in acetonitrile. Concentration: 0.558×10^{-4} mol dm^{-3}, light intensity: 2.4 mW cm^{-2}, irradiation time: 0–12 min; (**b**) absorption spectral changes of **3** during irradiation of 313-nm light in hexane. Concentration: 0.628×10^{-4} mol dm^{-3}, light intensity: 0.24 mW cm^{-2}, irradiation time: 0–20 min. Reprinted with permission from [9]. Copyright 2009 American Chemical Society

conversion ratio to **2C** at the PSS of 436-nm light irradiation was 58 %. The cyclized form showed no change after 1.5 months storage in the dark at room temperature. When the 436-nm PSS solution was irradiated with 579 nm light, it turned yellow and HPLC analysis proved that **2o** was recovered again. However, due to the long absorption tail of **2o** in the visible region, the ratio of **2o/2C** at the PSS of 579-nm light irradiation was 99/1.

Thus, a novel photochromic system, in which back-and-forth reactions can be driven by visible light of two different wavelengths, was created.

The electronic influence of the carbonyl group was then changed by acetalization in a manner similar to that of Kühni and Belser for 3,4-bis(2-methyl-5-phenyl-3-thienyl)-3-cyclobutene-1,2-dione [38]. In their case, acetalization of both carbonyl groups could realize photochromic properties.

Photochromism of the acetal **3o** was investigated in several solvents with different polarities, and the photocyclization quantum yield in hexane proved to be as large as 0.81. The changes observed in the absorption spectra by photochromic reactions in hexane are shown in Fig. 8.6b. The conversion ratio to **3C** at a PSS of 313-nm irradiation was 99 %.

The reason for the high cyclization quantum yield of **3o** can be interpreted as two sets of intramolecular nitrogen–hydrogen interactions (highlighted with squares in Fig. 8.5; a thiazole nitrogen atom and the proximate hydrogen atom on the indenone phenyl group, and a thiazole nitrogen atom and a hydrogen atom on the carbon atom adjacent to an oxygen atom in the acetal group) that play an important role in constraining the conformation of **3o** in favor of cyclization. The solvent effect on the quantum yield for the ring closure of **3o** can be explained by polar CH–N interactions. The highest value was recorded in hexane, while the value decreased as the solvent polarity became larger (hexane: 0.81, toluene: 0.72, ethyl acetate: 0.70, acetonitrile: 0.63, ethanol: 0.68). As the polar solvent molecules prevent effective

CH–N interactions, the distribution of the conformers became less biased, thus decreasing the quantum yield. In addition, results obtained by DFT calculations suggest that an antiparallel (i.e., photocyclizable) conformation is particularly stable (more than 98 % population at 25 °C) in which the two sets of CH–N distances are around 260 pm, almost the same as the sum of the van der Waals radii of N and H. As the conclusion of this section [9], we have synthesized a novel thermally irreversible photochromic system based on 6π-electrocyclization. Bisarylindenone **2** showed photochromic back-and-forth reactions with visible light of two different wavelengths. The photochromic and absorption spectral properties can be switched by its acetalization. The photocyclization quantum yield of acetal ***3o*** became as large as 0.81 in hexane. Achievement of the large quantum yield was explained by the two sets of intramolecular interactions of nitrogen and hydrogen atoms to fix the conformation in favor of photochemical cyclization. This concept was extended to the study described in the final section of this chapter.

8.4 Diastereoselective Photochromism

8.4.1 Bisbenzothienylethenes: Stereocontrol with Point Chirality

The photochromic reaction of diarylethenes is, like another important class of thermally irreversible fulgides [2], based on the photochemically induced conrotatory 6π-electrocyclization of the 1,3,5-hexatriene moiety, which takes a C_2-symmetric helical structure because of steric congestion. As a result, the photocyclization reaction produces a racemic mixture of the colored closed form (C-form), possessing two new stereogenic sp^3 carbon atoms.

When some intermolecular or intramolecular stereoregulating force works on a diarylethene molecule to make one of the enantiomeric helical structures more stable than the other, then the ratio of the stereoisomers of the C-form molecules would be biased. If such stereoselective photochromic reactions are possible, the switch of functions can be quite useful for biological as well as materials sciences. We have reported several stereoselective photochromic reactions on fulgides [39–41]. As for diarylethenes, several studies had already been reported when we started our research. However, they were effective only at low temperatures [42], in crystals at low conversion [43–45], or in reactions that required some additives [46].

Our strategy of stereoregulation is to use "allylic 1,3-strain" [47, 48] as well as steric and electronic repulsion between the substituents. Allylic strain has been used in various aspects of organic synthesis [49–51]. When it is applied to the terminal part of the ring-closing hexatriene of a diarylethene, the smallest substituent R^S, such as hydrogen, faces the bulky hexafluorocyclopentene that locates *cis* to the stereogenic carbon atom. Accordingly, two different substituents (medium-sized R^M and largest R^L) hang down in the rather wide space, as shown in Fig. 8.7. The other benzothiophene will take a position close to R^M rather than R^L because of steric

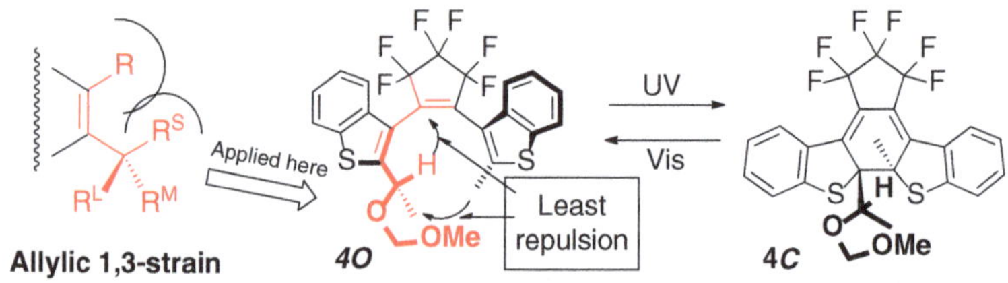

Fig. 8.7 Concept of allylic strain applied to bisbenzothienylethene

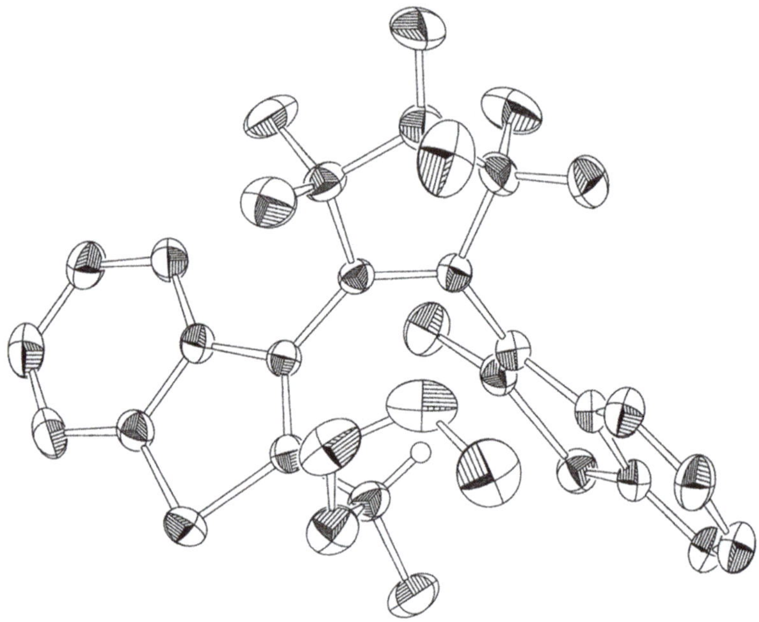

Fig. 8.8 ORTEP drawing of (*S*)-**4*O*** (30 % probability). Reprinted with permission from [7]. Copyright 2003 American Chemical Society

repulsion, resulting in the diastereomerically biased ratio of the helical conformers. If R^L has heteroatoms, their electronic repulsion with the sulfur atom on the second benzothiophene will make the conformer ratio more biased.

With this expectation in mind, we have designed and synthesized **4*o***, as depicted in Fig. 8.7, possessing a hydrogen atom, a methyl group, and a methoxymethoxy (MOMO) group on the stereogenic carbon atom on C2 of one of the benzothiophene rings.

When racemic **4*o*** was irradiated with 313-nm light in ethyl acetate at room temperature, two diastereomeric **4*c*** were formed. HPLC analysis proved that the diastereomer ratio (DR) was 94:6, with a conversion ratio to the C-form of 85 %. They then completely returned to **4*o*** by 519-nm light irradiation. In other solvents such as toluene (94:6 DR, 85 % conversion) and hexane (93.5:6.5 DR, 80 % conversion), the dr values were similarly high.

The crystal structure of **4*o*** (Fig. 8.8) revealed that (1) the allylic strain nicely works around the relevant double bond on the benzothiophene so that the hydrogen atom on the stereogenic center is facing the perfluorocyclopentene ring, (2) the second benzothiophene resides in close proximity to the methyl group, and (3) the hexatriene moiety takes the *s–cis–cis–s–cis* (i.e., *anti*parallel) conformation so that photochemical ring closure can occur easily.

In conclusion of this section [7], we have succeeded in obtaining a highly diastereoselective photochromic diarylethene system by introducing one asymmetric center and employing steric as well as electronic interactions.

8.4.2 Bisbenzothienylethenes: Stereocontrol with Point Chirality-Based Axial Chirality

When a photochromic compound is chiral and non-racemic [52], its chiroptical properties, such as circular dichroism spectra [53, 54] and optical rotation [39, 55, 56], also change reversibly. These chiral and chiroptical differences can be used to (1) detect the state of the system without inducing any property change, (2) induce and control the alignment of a chiral nematic liquid crystal in a sensitive manner [57–60], or (3) make a chiral complex [46] that may work as a chiral catalyst [61].

Diastereoselective ring closure of diarylethenes [3, 62, 63] through photochemical reactions, which can be achieved for molecules possessing at least one chirality unit, has been reported in solution [42, 64], crystals [43–45], amorphous solids [65], and gels [66]. We have also achieved diastereoselective ring closure of photochromic diarylethenes in solution by applying the concept of the allylic 1,3-strain [47–51, 67] to the hexatriene moiety, which gave a 94:6–97:3 diastereomer ratio (DR) by using only one chiral unit to control the stereochemistry [7, 68–70], as described in the previous section. We next expanded this approach to helical chirality by attaching two chiral substituents on both ends of the hexatriene moiety.

In our new strategy, electrostatic and steric repulsion between the two 1-methoxymethoxyethyl side chains are expected. This methodology is applicable to any diarylethene derivative as long as the chiral unit that controls the stereochemistry can be attached to both of the carbon atoms where ring closure occurs.

We synthesized three diarylethenes (**5**–**7**; Fig. 8.9) from two chiral benzothiophene components **8** and **9** obtained by way of the enantioselective reduction of the corresponding ketones with the Corey–Bakshi–Shibata oxazaborolidine [71, 72]. The final diarylethenes were purified by repeated recrystallization and/or HPLC on a chiral stationary phase.

Photochemical reactions were carried out in ethyl acetate with 366 nm (***5O***) or 313 nm (***6O***, ***7O***) light for ring closure and 437 nm (***5C***), 483 nm (***6C***), or 515 nm (***7C***) light for ring opening. These diarylethene derivatives showed excellent photochromism. The changes in the absorption spectra of **5** in ethyl acetate upon irradiation with UV light and visible light are shown in Fig. 8.10.

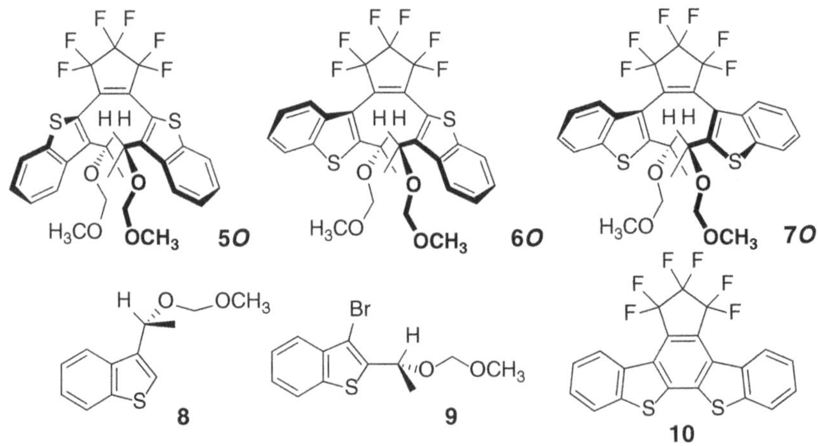

Fig. 8.9 Chiral diarylethenes with two side arms

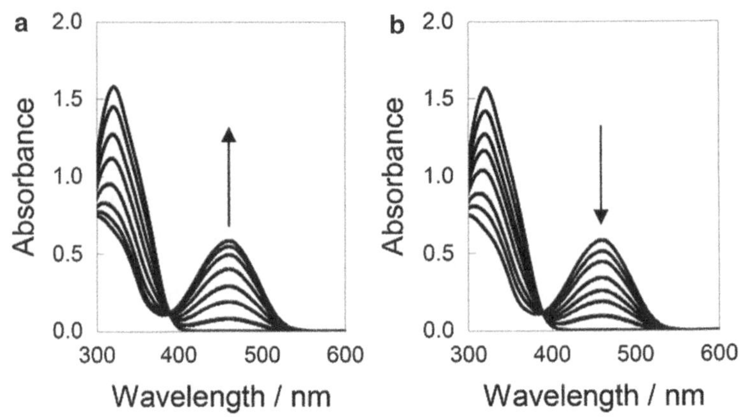

Fig. 8.10 Change in absorption spectra of **5** by photochromism in ethyl acetate. (**a**) Concentration: 1.01×10^{-4} mol dm^{-3}, light intensity: 0.49 mW cm^{-2} (366 nm), irradiation time: 0–35 min. (**b**) Concentration: 1.01×10^{-4} mol dm^{-3}, light intensity: 0.29 mW cm^{-2} (437 nm), irradiation time 0–16 min. Reproduced from [74]

The DR values for the ring-closing reaction at their PSS were then examined using HPLC on a chiral stationary phase. The results are summarized in Table 8.1. It is apparent that the dual-stereocontroller system worked very well.

We then examined the solvent effect on the diastereoselectivity of these compounds. In *n*-hexane, they are all slightly lower than those observed in ethyl acetate. This difference can be interpreted by the interaction of the solvent molecules with the polar side arms. As for the stable helical conformation, the polar side arms are expected to be sticking out of the molecule so that the polar solvent molecules can strongly interact with them. When the molecule takes the other unstable helical conformation, the polar side arms are located inside the molecule so that the solvent

Table 8.1 Diastereoselectivity of C_2-symmetrical diarylethenes [74]

	Diastereomer ratio (conversion ratio to *C*-form/%)			
	Ethyl acetate			Hexane
	Room temperature	−70 °C	67 °C	Room temp.
5	99:1 (81)	100:0 (50)	–	97:3 (82)
6	99:1 (83)	99.5:0.5 (89)	–	98:2 (81)
7	97.5:2.5 (94)	95.5:4.5 (97)	99:1 (56)	96:4 (94)

effect is less pronounced. The DR values of **5** in polar acetonitrile and ethanol were also 99:1, with conversion ratios of 77 % and 81 %, respectively.
Another way to change the ratio of the diastereomeric helical conformers is to change the temperature. We then carried out the photoreaction of **5 at** −70 °C. When the diastereomer ratio was examined by HPLC, the minor diastereomer of **5*C*** which was observed when the reaction was carried out at room temperature was not detected. Thus, 100 % diastereoselectivity was achieved and, similarly, **6** showed 99.5:0.5. However, **7** showed only 95.5:4.5 at −70 °C, which is lower than the ratio of 97.5:2.5 at room temperature. When the photoirradiation of **7** was carried out at 67 °C, the diastereomer ratio was raised to 99:1, although a small amount (1.6 %) of an aromatized compound **10** was detected, which was not observed during the photoirradiation at room temperature. This mysterious problem was solved by ^{1}H NMR measurements of the UV–PSS solution at periodic intervals over several months at room temperature. The less stable **7*C*** (**7*C***$_{\mathbf{minor}}$) slowly went back to **7*O*** thermally at room temperature, but the more stable **7*C*** (**7*C***$_{\mathbf{major}}$) did not. Therefore, at higher temperatures the thermal reaction rate of **7*C***$_{\mathbf{minor}}$-to-**7*O*** increased so that the diastereoselectivity increased, while at lower temperatures the rate of **7*C***$_{\mathbf{minor}}$-to-**7*O*** decreased so that the diastereoselectivity decreased [73].

In conclusion of this section [74], we have shown extremely high diastereoselectivity in polar solvents for photochemical 6π-electrocyclizations of diarylethenes, which have two side arms that control the stereochemistry. The cyclization of **5 at** a lower temperature afforded exceptional diastereoselectivity of 100 %. This structural modification can be applied to any diarylethene molecule possessing two alkyl groups on bond-forming carbon atoms at both ends of the hexatriene moiety by simply replacing these alkyl groups with 1-methoxymethoxyethyl groups having the same absolute stereochemistry.

8.4.3 Bisthienylethene: Stereocontrol with Facial Chirality

Loosely classified, there are three categories of asymmetry [75]: point chirality, axial chirality, and facial chirality. We have reported the use of point chirality [7, 67–70] and point chirality-based axial chirality [74, 76] to induce diastereoselectivity for the photochromic ring closure of diarylethenes which both exhibit thermally irreversible photochromic 6π electrocyclizations. Different from point chirality,

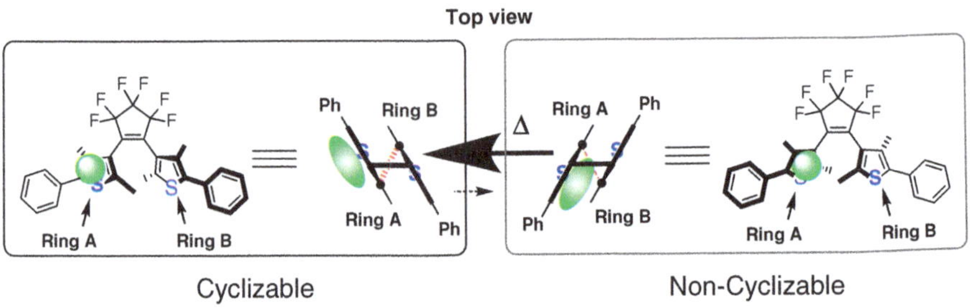

Fig. 8.11 Concept of facially chiral bisthienylethene

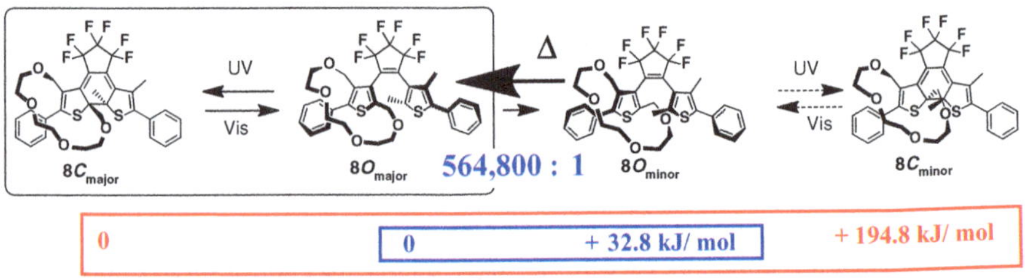

Fig. 8.12 Calculated stability of isomers of facially chiral diarylethene

facially and axially chiral compounds do not need to have asymmetric carbon atoms. In fact, facial chirality has been successfully introduced to thermally reversible photochromic azobenzenes [77, 78]. We then aimed for the use of facial chirality on a diarylethene to induce 100 % diastereoselectivity of photochromic ring closure in various solvents (Fig. 8.11). When one of the two surfaces of an aromatic ring (ring A) is occupied by a bulky attachment, ring closure is possible when the other ring (ring B) approaches from the back of ring A. Perfect control of the diastereoselectivity during ring closure could, thus, be realized.

Our molecular design and expected photochromic reaction pathways are shown in Fig. 8.12. The target open form **8*O*** has a bridge across the surface of a thiophene ring which prevents access of the other thiophene to the side of the bridge. Therefore, this molecule can take only one antiparallel (cyclizable) conformation. We chose the C2 and C4 carbon atoms to span the bridge since C3 is used to connect with hexafluorocyclopentene while C5 is useful in attaching to a functional substituent. As the bridge, we chose triethylene glycol (TEG) which is long enough to connect the methylene carbon atoms on C2 and C4, yet short enough to prohibit isomerization of the facial chirality with the rope-jumping action of the TEG bridge over a substituent such as the phenyl group on C5.

First, we carried out DFT calculations (Spartan008 with B3LYP/6-31G*) which proved that one of the two possible antiparallel conformers (**8*O***$_{\textbf{major}}$), in which the non-bridged thiophene is located at the back side of the bridge, is more stable by 32.8 kJ mol^{-1} than the other (**8*O***$_{\textbf{minor}}$). If the population at 25 °C is calculated from

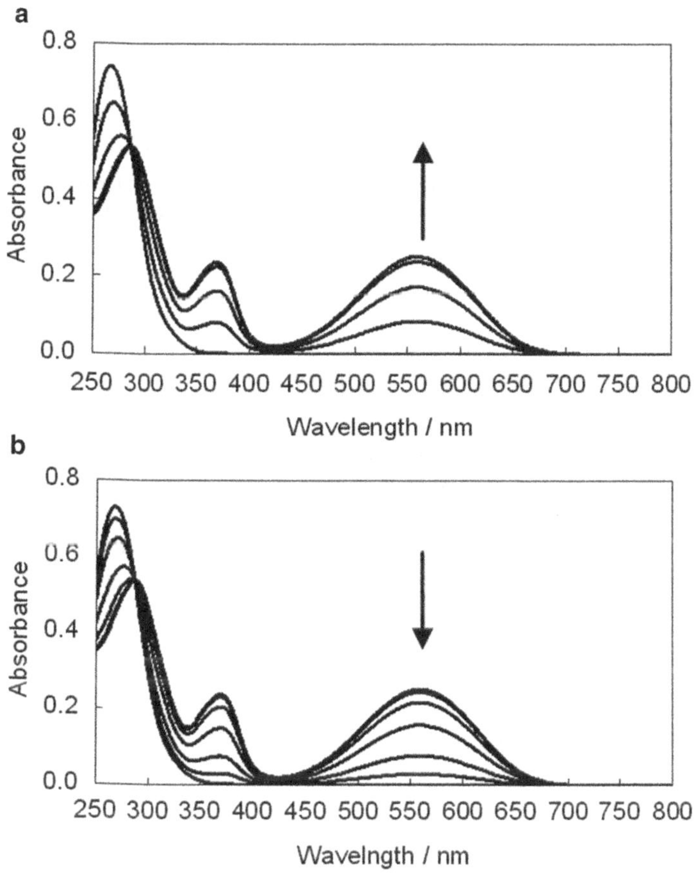

Fig. 8.13 Absorption spectral changes of **8** in hexane. (**a**) ***8O*** to PSS. 313 nm (0.71 mW cm^{-2}). 0–15 min. (**b**) PSS to ***8O***. 512 nm (1.24 mW cm^{-2}). 0–150 min. Reproduced by permission of The Royal Society of Chemistry from [79]

the difference in heat of formation (ΔH_f) by disregarding the entropy term, the ratio is 564,800/1. In addition, ΔH_f of the two C-forms (***8C***$_{\mathbf{major}}$ generated from the more stable ***8O***$_{\mathbf{major}}$ and ***8C***$_{\mathbf{minor}}$ generated from the less stable ***8O***$_{\mathbf{minor}}$) is 194.1 kJ mol^{-1}, indicating that the ground state of ***8C***$_{\mathbf{minor}}$ locates close to the excited state level of ***8C***$_{\mathbf{major}}$.

Photochromic reactions of ***8O*** were carried out in hexane, toluene, and ethyl acetate. A change in the absorption spectra during the ring-closing reaction of ***8O*** in hexane with 313-nm light irradiation until the PSS was achieved, and the ring-opening reaction upon 512-nm light irradiation to the PSS solution are shown in Fig. 8.13. A clear isosbestic point (287 nm) could be observed.

The components existing at the PSS were thoroughly examined by HPLC with three different columns and three different solvent systems. ^{1}H NMR analysis of the compounds was also carried out; however, not even slight evidence of the formation of the minor C-form was observed. From these results obtained along with the calculation results, it was concluded that **8*o*** showed ultimate diastereoselectivity within the detection limit of the minor diastereomer upon photochromic ring closure under any of the reaction conditions carried out.

In conclusion of this section [79], we have succeeded in synthesizing a bisthienylethene with hitherto unprecedented facial chirality imposed by a TEG bridge on a thiophene ring which showed photochromic ring closure with 100 % diastereoselectivity upon UV irradiation.

8.4.4 Bisarylindenols: Stereocontrol with Hydrogen Bonds

In Section 8.3, bisthiazolylindenone ethylene acetal showed a large quantum yield for photocyclization [9]. Sections 8.4.1–8.4.3 introduced several diarylethenes which cyclize with high diastereoselectivity [7, 67–70, 74, 76, 79]. If a compound has both characteristics of a large quantum yield and high diastereoselectivity, its application to biological systems [80–85] is favorable since a large quantum yield protects the living cells from prolonged exposure to light, and high diastereoselectivity creates a simple system with regard to the various interactions of the photo-generated isomers and biological chiral environment.

Compound **3** [9], an ethylene acetal of 2,3-bisthiazolylindenone **2**, showed a thermally irreversible photochromic reaction to give its closed form **3*c*** with a cyclization quantum yield of 0.81. In the most stable conformation of **3** obtained by DFT calculations, we found that two sets of CH–N interactions between two hydrogen atoms on the indenone acetal and two nitrogen atoms on the thiazole rings fix the molecule in antiparallel conformation, which favors photocyclization. As these interactions are obviously weak, introduction of a stronger interaction in this photochromic system was undertaken to enhance the controlling ability of the conformation. As **2** has a carbonyl group, nucleophilic alkylation will generate tertiary alcohol which can form a strong hydrogen bond with the nitrogen atom on the thiazole ring (**9*o*–12*o***). In addition, as the carbon atom bearing the hydroxy group in **9*o***, for example, is now stereogenic, the photochemical ring closure of **9*o*** to give **9*c*** will generate a pair of diastereomers. Judging from the close proximity of the hydroxy group and nitrogen atom, the most favorable conformation will predominantly generate one of the diastereomers (**9*c***$_{\mathbf{major}}$) (Fig. 8.14).

With these expectations in mind, we have synthesized four bisthiazolylindenols, **9*o*–12*o***, and examined their diastereoselectivity and quantum yields for photocyclization. X-ray crystallographic analysis of **10*o*** revealed its structure in the crystalline state (Fig. 8.15). Notable characteristics are the existence of a hydrogen bond between the hydroxy group and nitrogen atom (OH–N distance: 215.7 pm), and a hydrogen bond-like interaction between the phenyl hydrogen on the indenol moiety and the other nitrogen atom (C(indenol Ph)H–N distance: 274.3 pm). These two interactions fix the conformation of **10*o*** in favor of cyclization.

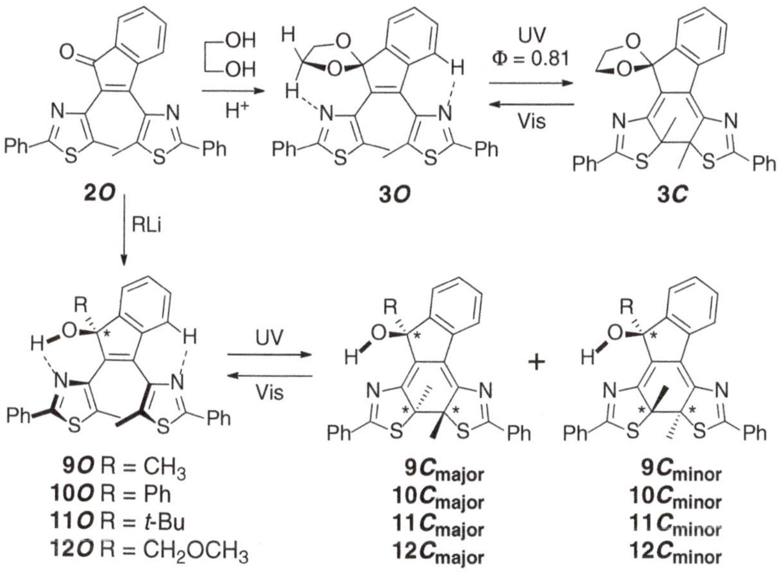

Fig. 8.14 Indenone, its acetal, and indenols

Fig. 8.15 X-ray crystallographic structure of bisarylindenol **10O**

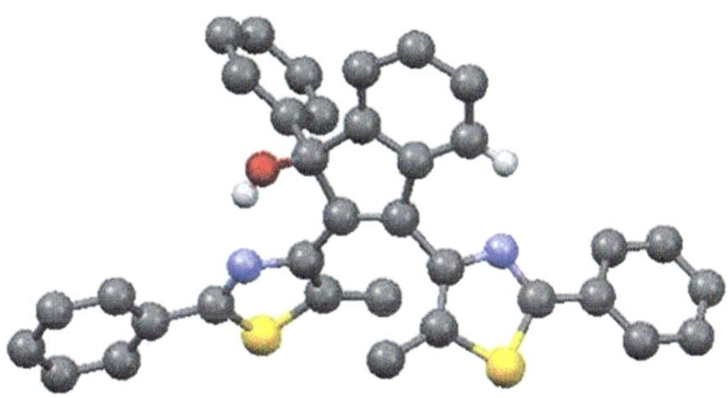

The indenols were then subjected to photochromic reactions with 313-nm and 514-nm light irradiation in hexane, ether, and acetonitrile, with different polarities to see the effect of the intramolecular hydrogen bonds. Typical absorption spectral changes of **11** in hexane upon irradiation of 313-nm light are depicted in Fig. 8.16, and the diastereoselectivity and quantum yields of the photoreactions of **9–12** are summarized in Table 8.2.

The diastereoselectivity is apparently higher when the alkyl/aryl substituents are larger. Steric repulsion is enhanced for the less stable conformation when the substituent is large. This effect is the largest for **11*O*** with a *t*-butyl group. When the photoreaction was carried out in hexane, the minor diastereomer was not observed, although tiny amounts were detected in ether and acetonitrile. This is because the solvent molecules did not interfere with the intramolecular hydrogen bonds in hexane. For **12*O*** possessing a methoxymethyl (MOM) group, lower diastereoselectivity was

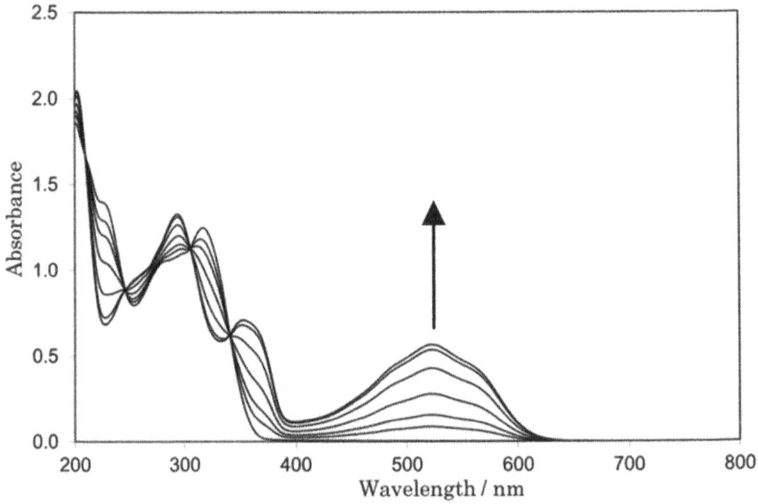

Fig. 8.16 Change in absorption spectra of **11** by 313-nm light irradiation in hexane. Concentration: 0.450×10^{-4} mol dm^{-3}, light intensity: 0.153 mW cm^{-2}, irradiation time: 0, 0.25, 0.5, 1, 2, 4, 8, 16 min. Reproduced by permission of The Royal Society of Chemistry from [90]

Table 8.2 Diastereomer ratio, conversion ratio to *C*-form, and quantum yield of photocyclization of indenols in solution[a]

	9 (Me)	**10** (Ph)	**11** (*t*-Bu)	**12** (MOM)
Hexane	94.9/5.1	99.1/0.9	100/0	90.8/9.2
	97.6 %	96.6 %	97.9 %	98.4 %
	0.74	0.78	0.85	0.56
Ether	75.8/24.2	92.0/8.0	99.8/0.2	74.6/25.4
	97.6 %	97.3 %	97.0 %	97.2 %
	0.65	0.70	0.75	0.43
CH_3CN	71.6/28.4	86.7/13.3	99.7/0.3	75.6/24.4
	93.0 %	92.9 %	94.9 %	96.2 %
	0.49	0.60	0.64	0.44

[a]*Upper numbers*: ratio of diastereomers. *Middle numbers*: conversion ratio to the *C*-form at the photostationary state of 313-nm light irradiation. *Lower numbers*: quantum yield for major *C*-form generation by 313-nm light irradiation

exhibited than the other compounds. This can be interpreted as a partial replacement of the hydrogen bond (OH–N) by the other hydrogen bond (OH–OMe) which forms a five-membered ring. The latter hydrogen bond enables the rotation of one of the single bonds between the thiazole rings and the indenol so that diastereoselectivity is significantly lowered.

Among the three solvents employed, hexane always exhibited the highest diastereoselectivity. In ether or acetonitrile, heteroatoms with the lone pair on the solvent molecules interfered with the formation of intramolecular OH–N hydrogen bonds so that fixation of the conformation was not sufficient.

As the alkyl/aryl substituent becomes larger, the cyclization quantum yield to form the major C-form becomes larger. **11** possessing a *t*-butyl group showed the largest quantum yield of $\mathbf{11}C_{\mathrm{major}}$ generation. In contrast, **12**, possessing a functional group capable of forming another hydrogen bond, showed rather low cyclization quantum yields. Partial replacement of the OH–N hydrogen bond with the OH–OMe hydrogen bond increases the amount of O-form molecules with parallel conformation which cannot cyclize in a conrotatory manner upon light irradiation.

Judging from the three properties, i.e., diastereoselectivity, quantum yield of photocyclization, and conversion ratio to the C-form, **11** having a *t*-butyl group can be considered the best performing among the compounds examined. With the collaboration between the intramolecular OH–N and CH–N hydrogen bonds and the sterically demanding *t*-butyl group, generation of the less stable conformation could almost entirely be prevented, and the molecules were fixed in the more stable antiparallel conformation, leading to high diastereoselectivity. Fixation of the conformation also resulted in an increase in the cyclization quantum yield to the highest level (≥0.8) for 6π-electrocyclization of the thermally irreversible photochromic compounds in solution [9, 86–89].

In conclusion of this section [90], we have synthesized bisarylindenols, a fascinating photochromic family which show both excellent diastereoselectivity and an extremely high ring-closing quantum yield with a high conversion ratio to the closed form in hexane. Collaborative interactions between two sets of intramolecular hydrogen bonds and the steric restriction worked very efficiently to fix the conformation in favor of cyclization in a diastereoselective manner. These properties are highly beneficial when photochromic compounds are applied to biological systems for the switching control of organic functions within cells.

8.5 Conclusions

We are engaged in the construction of photochromic systems exhibiting 6π-electrocyclization with high diastereoselectivity and a high cyclization quantum yield. Our interests also include photochromic systems with coumarin or indene as the central ethene moiety. The former involves an efficient two-way fluorescence switching molecule upon pH change and photochromism. The latter forms a system of conformations governed by hydrogen bonds which achieved high diastereoselectivity as well as a high cyclization quantum yield, a long-sought goal.

Acknowledgments I am greatly indebted to my coauthors, i.e., the colleagues and students who have been engaged in the research work described here. I would like to gratefully acknowledge support from the Grants-in-Aid for Scientific Research on Priority Area "New Frontiers in Photochromism" (471) of the Ministry of Education, Culture, Sports, Science and Technology (MEXT) and the Grant-in-Aid for Scientific Research (B) (23350096) of the Japan Society for the Promotion of Sciences (JSPS).

References

1. Durr H, Bouas-Laurent H (eds) (2003) Photochromism: molecules and systems, revised edn. Elsevier, Amsterdam
2. Yokoyama Y (2000) Fulgides for memories and switches. Chem Rev 100:1717–1739
3. Irie M (2000) Diarylethenes for memories and switches. Chem Rev 100:1685–1716
4. Barachevsky VA (1999) Photochromic quinones. In: Crano JC, Guglielmetti RJ (eds) Organic photochromic and thermochromic compounds, vol 1. Plenum, New York, pp 267–314
5. Yokoyama Y, Iwai T, Yokoyama Y, Kurita Y (1994) Helical chirality and enantiotopomerization of a photochromic furylfulgide. Chem Lett 1994:224–226
6. Yokoyama Y (2009) Chiral photochromism based on 6π-electrocyclization. New J Chem 33:1314–1319
7. Yokoyama Y, Shiraishi H, Tani Y, Yokoyama Y, Yamaguchi Y (2003) Diastereoselective photochromism of a bisbenzothienylethene governed by steric as well as electronic Interactions. J Am Chem Soc 125:7194–7195
8. Feringa BL, Browne WR (eds) (2011) Molecular switches, 2nd edn. Wiley, Weinheim
9. Morinaka K, Ubukata T, Yokoyama Y (2009) Structurally versatile novel photochromic bisarylindenone and its acetal: achievement of large cyclization quantum yield. Org Lett 11:3890–3893
10. Kose M, Şekerci ÇY, Suzuki K, Yokoyama Y (2011) Synthesis of photochromic 2,3-bis(5-methyl-2-phenyl-4-thiazolyl)-1,4-naphthoquinone derivatives. J Photochem Photobiol A Chem 219:58–61
11. Lemieux V, Gauthier S, Branda NR (2006) Selective and sequential photorelease using molecular switches. Angew Chem Int Ed 45:6820–6824
12. Sud D, Wigglesworth TJ, Branda NR (2007) Creating a reactive enediyne by using visible light: photocontrol of the Bergman cyclization. Angew Chem Int Ed 46:8017–8019
13. Lemieux V, Spantulescu MD, Baldridge KK, Branda NR (2008) Modulating the Lewis acidity of boron using a photoswitch. Angew Chem Int Ed 47:5034–5037
14. Nakashima T, Atsumi K, Kawai S, Nakagawa T, Hasegawa Y, Kawai T (2007) Photochromism of thiazole-containing triangle terarylenes. Eur J Org Chem 2007:3212–3218
15. Kawai S, Nakashima T, Atsumi K, Sakai T, Harigai M, Imamoto Y, Kamikubo H, Kataoka M, Kawai T (2007) Novel photochromic molecules based on 4,5-dithienyl thiazole with fast thermal bleaching rate. Chem Mater 19:3479–3483
16. Fukaminato T, Doi T, Tamaoki N, Okuno K, Ishibashi Y, Miyasaka H, Irie M (2011) Single-molecule fluorescence photoswitching of a diarylethene-perylenebisimide dyad: non-destructive fluorescence readout. J Am Chem Soc 133:4984–4990
17. Uno K, Niikura H, Morimoto M, Ishibashi Y, Miyasaka H, Irie M (2011) In situ preparation of highly fluorescent dyes upon photoirradiation. J Am Chem Soc 133:13558–13564
18. Ando R, Mizuno H, Miyawaki A (2004) Regulated fast nucleocytoplasmic shuttling observed by reversible protein highlighting. Science 306:1370–1373
19. Giordano L, Jovin TM, Irie M, Jares-Erijman EA (2002) Diheteroarylethenes as thermally stable photoswitchable acceptors in photochromic fluorescence resonance energy transfer (pcFRET). J Am Chem Soc 124:7481–7489
20. Zhu L, Wu W, Zhu MQ, Han JJ, Hurst JK, Li ADQ (2007) Reversibly photoswitchable dual-color fluorescent nanoparticles as new tools for live-cell imaging. J Am Chem Soc 129:3524–3526
21. Soh N, Yoshida K, Nakajima H, Nakano K, Imato T, Fukaminato T, Irie M (2007) A fluorescent photochromic compound for labeling biomolecules. Chem Commun 2007:5206–5208
22. Piao X, Zou Y, Wu J, Li C, Yi T (2009) Multiresponsive switchable diarylethene and its application in bioimaging. Org Lett 11:3818–3821
23. Fölling J, Belov V, Kunetsky R, Medda R, Schönle A, Egner A, Eggeling C, Bossi M, Hell SW (2007) Photochromic rhodamines provide nanoscopy with optical sectioning. Angew Chem Int Ed 46:6266–6270

24. Irie M, Fukaminato T, Sasaki T, Tamai N, Kawai T (2002) A digital fluorescent molecular photoswitch. Nature 420:759–760
25. Golovkova TA, Kozlov DV, Neckers DC (2005) Synthesis and properties of novel fluorescent switches. J Org Chem 70:5545–5549
26. Jiang G, Wang S, Yuan W, Jiang L, Song Y, Tian H, Zhu D (2006) Highly fluorescent contrast for rewritable optical storage based on photochromic bisthienylethene-bridged naphthalimide dimer. Chem Mater 18:235–237
27. Fernández-Suárez M, Ting AY (2008) Fluorescent probes for super-resolution imaging in living cells. Nat Rev Mol Cell Biol 9:929–943
28. Lee M, Gubernator NG, Sulzer D, Sames D (2010) Development of pH-responsive fluorescent false neurotransmitters. J Am Chem Soc 132:8828–8830
29. Sun WC, Gee KR, Haugland RP (1998) Synthesis of novel fluorinated coumarins: excellent UV-light excitable fluorescent dyes. Bioorg Med Chem Lett 8:3107–3110
30. Schiedel MS, Briehn CA, Bäuerle P (2001) Single-compound libraries of organic materials: parallel synthesis and screening of fluorescent dyes. Angew Chem Int Ed 40:4677–4680
31. Lin W, Yuan L, Cao Z, Feng Y, Song J (2010) Through-bond energy transfer cassettes with minimal spectral overlap between the donor emission and acceptor absorption: coumarin-rhodamine dyads with large pseudo-Stokes shifts and emission shifts. Angew Chem Int Ed 49:375–379
32. Traven VF, Bochkov AY, Krayushkin MM, Yarovenko VN, Nabatov BV, Dolotov SM, Barachevsky VA, Beletskaya IP (2008) Coumarinyl(thienyl)thiazoles: novel photochromes with modulated fluorescence. Org Lett 10:1319–1322
33. Mizoguchi H, Kubo K, Sakurai T, Inoue H (1997) Solvent effects on the proton-transfer reactions of 7-hydroxycoumarin and its 4-methyl derivative with tertiary amines in the ground and excited singlet states [1]. Ber Bunsenges Phys Chem 101:1914–1920
34. Suzuki K, Ubukata T, Yokoyama Y (2012) Dual-mode fluorescence switching of photochromic bisthiazolylcoumarin. Chem Commun 48:765–767
35. Krayushkin MM, Ivanov SN, Martynkin AY, Lichitsky BV, Dudinov AA, Uzhinov BM (2001) Photochromic dihetarylethenes 7. Synthesis of bis(thienylazoles), photochromic analogs of diarylethenes. Russ Chem Bull 50:116–121
36. Krayushkin MM, Lichitsky BV, Mikhalev AP, Nabatov BV, Dudinov AA, Ivanov SN (2006) Behavior of benzoins and hydroxy ketones in acid medium: part II. Reactions of 1,2-bis(2,5-dimethyl-3-thienyl)-2-hydroxyethan-1-one with N, S-binucleophiles in trifluoroacetic acid. Russ J Org Chem 42:860–864
37. Wigglesworth TJ, Branda NR (2004) Ultra-high-density photochromic main-chain 1,2-dithienylcyclopentene polymers prepared using ring-opening metathesis polymerization. Adv Mater 16:123–125
38. Kühni J, Belser P (2007) Gated photochromism of 1,2-diarylethenes. Org Lett 9:1915–1918
39. Yokoyama Y, Uchida S, Yokoyama Y, Sugawara Y, Kurita Y (1996) Diastereoselective photochromism of (R)-binaphthol-condensed indolylfulgide. J Am Chem Soc 118:3100–3107
40. Yokoyama Y, Sagisaka T, Yamaguchi Y, Yokoyama Y (2000) Highly diastereoselective photochromic cyclization of a bisthienylfulgide. Chem Lett 2000:220–221
41. Yokoyama Y, Okuyama T, Yokoyama Y, Asami M (2001) Highly diastereoselective photochromic cyclization of an indolylfulgide derivative possessing *C2*-symmetric chiral diol as an auxiliary. Chem Lett 2001:1112–1113
42. Yamaguchi T, Uchida K, Irie M (1997) Asymmetric photocyclization of diarylethene derivatives. J Am Chem Soc 119:6066–6071
43. Kodani T, Matsuda K, Yamada T, Kobatake S, Irie M (2000) Reversible diastereoselective photocyclization of a diarylethene in a single-crystalline phase. J Am Chem Soc 122:9631–9637
44. Matsuda K, Yamamoto S, Irie M (2001) Diastereoselective cyclization of a diarylethene having a chiral N-phenylethylamide substituent in crystals. Chem Lett 42:7291–7293
45. Yamamoto S, Matsuda K, Irie M (2003) Diastereoselective cyclization in chiral diarylethene crystals: polymorphism and selectivity. Org Lett 5:1769–1772

46. Murguly E, Norsten TB, Branda NR (2001) Nondestructive data processing based on chiroptical 1,2-dithienylethene photochromes. Angew Chem Int Ed 40:1752–1755
47. Johnson F (1968) Allylic strain in six-membered rings. Chem Rev 68:375–413
48. Hoffmann RW (1989) Allylic 1,3-strain as a controlling factor in stereoselective transformations. Chem Rev 89:1841–1860
49. Yokoyama Y, Kawashima H, Masaki H (1989) A(1,3) strain – controlled cyclic hydroboration of 1,4- and 1,5-dienes. Chem Lett 1989:453–456
50. Yokoyama Y, Kawashima H, Kohno M, Ogawa Y, Uchida S (1991) Stereospecific construction of three contiguous asymmetric centers via cyclic hydroboration. Tetrahedron Lett 32:1479–1482
51. Yokoyama Y, Tsuchikura K (1992) Doubly allylic strain-controlled diastereoselective intramolecular Michael addition and a synthesis of (±)-iridomyrmecin. Tetrahedron Lett 33:2823–2824
52. Yokoyama Y, Saito M (2004) Chirality in photochromism. In: Inoue Y, Ramamurthy V (eds) Chiral photochemistry. Dekker, New York, pp 235–259
53. Feringa BL, Jager WF, de Lange B, Meijer EW (1991) Chiroptical molecular switch. J Am Chem Soc 113:5468–5470
54. Yokoyama Y, Hosoda N, Osano YT, Sasaki C (1998) Absolute stereochemistry and CD spectra of resolved enantiomers of the colored form of a photochromic diarylethene. Chem Lett 1998:1093–1094
55. Wigglesworth TJ, Sud D, Norsten TB, Lekhi VS, Branda NR (2005) Chiral discrimination in photochromic helicenes. J Am Chem Soc 127:7272–7273
56. Takeshita M, Yamato T (2004) Reversible optical rotation change according to the enantiospecific photochromic reaction of [2.2]metacyclophan-1-ene. Chem Lett 33:844–845
57. Huck NPM, Jager WF, de Lange B, Feringa BL (1996) Dynamic control and amplification of molecular chirality by circular polarized light. Science 273:1686–1688
58. Yokoyama Y, Sagisaka T (1997) Reversible control of pitch of induced cholesteric liquid crystal by optically active photochromic fulgide derivatives. Chem Lett 1997:687–688
59. Sagisaka T, Yokoyama Y (2000) Bull Chem Soc Jpn 73:191–196
60. Mallia VA, Tamaoki N (2004) Design of chiral dimesogens containing cholesteryl groups; formation of new molecular organizations and their application to molecular photonics. Chem Soc Rev 33:76–84
61. Sud D, Norsten TB, Branda NR (2005) Photoswitching of stereoselectivity in catalysis using a copper dithienylethene complex. Angew Chem Int Ed 44:2019–2021
62. Tian H, Yang S (2004) Recent progresses on diarylethene based photochromic switches. Chem Soc Rev 33:85–97
63. Tian H, Wang S (2007) Photochromic bisthienylethene as multi-function switches. Chem Commun 2007:781–792
64. Takeshita M, Jin-nouchi H (2010) Fully diastereospecific photochromic reaction of a thiophenophan-1-ene. Chem Commun 46:3994–3995
65. Yamaguchi T, Nomiyama K, Isayama M, Irie M (2004) Reversible diastereoselective photocyclization of diarylethenes in a bulk amorphous state. Adv Mater 16:643–645
66. de Jong JJD, Lucas LN, Kellog RM, van Esch JH, Feringa BL (2004) Reversible optical transcription of supramolecular chirality into molecular chirality. Science 304:278–281
67. Yokoyama Y (2004) Seasoning materials chemistry by using a well-matured organic concept. Chem Eur J 10:4389–4394
68. Kose M, Shinoura M, Yokoyama Y, Yokoyama Y (2004) Diastereoselective photochromism of bisbenzothienylethenes with an oxycarbonyl-related functional group on the side chain. J Org Chem 69:8403–8406
69. Okuyama T, Tani Y, Miyake K, Yokoyama Y (2007) Chiral helicenoid diarylethene with large change in specific optical rotation by photochromism. J Org Chem 72:1634–1638
70. Tani Y, Ubukata T, Yokoyama Y, Yokoyama Y (2007) Chiral helicenoid diarylethene with highly diastereoselective photocyclization. J Org Chem 72:1639–1644

71. Corey EJ, Bakshi RK, Shibata S (1987) Highly enantioselective borane reduction of ketones catalyzed by chiral oxazaborolidines. Mechanism and synthetic implications. J Am Chem Soc 109:5551–5553
72. Corey EJ, Helal CJ (1998) Reduction of carbonyl compounds with chiral oxazaborolidine catalysts: a new paradigm for enantioselective catalysis and a powerful new synthetic method. Angew Chem Int Ed 37:1986–2012
73. Delbaere S, Berthet J, Shiozawa T, Yokoyama Y (2012) Photochromic C2-symmetric chiral diarylethene: from the initial state to the final state. J Org Chem 77:1853–1859
74. Yokoyama Y, Shiozawa T, Tani Y, Ubukata T (2009) A unified strategy for exceptionally high diastereoselectivity of photochromism of chiral diarylethenes. Angew Chem Int Ed 48:4521–4523
75. Robinson MJT (2000) Organic stereochemistry. Oxford University Press, Oxford
76. Yokoyama Y, Hasegawa T, Ubukata T (2011) Highly diastereoselective photochromic ring closure of bisbenzothienylethenes possessing dual fluorinated stereocontrollers. Dyes Pigm 89:223–229
77. Mathews M, Tamaoki N (2008) Planar chiral azobenzenophanes as chiroptic switches for photon mode reversible reflection color control in induced chiral nematic liquid crystals. J Am Chem Soc 130:11409–11416
78. Basheer MC, Oka Y, Mathews M, Tamaoki N (2010) A light-controlled molecular brake with complete ON-OFF rotation. Chem Eur J 16:3489–3496
79. Shiozawa T, Hossain MK, Ubukata T, Yokoyama Y (2010) Ultimate diastereoselectivity in the ring closure of photochromic diarylethene possessing facial chirality. Chem Commun 46:4785–4787
80. Mayer G, Heckel A (2006) Biologically active molecules with a "light switch". Angew Chem Int Ed 45:4900–4921
81. Willner I (1997) Photoswitchable biomaterials: En route to optobioelectronic systems. Acc Chem Res 30:347–356
82. Zhu MQ, Zhang GF, Li C, Aldred MP, Chang E, Drezek RA, Li ADQ (2011) Reversible two-photon photoswitching and two-photon imaging of immunofunctionalized nanoparticles targeted to cancer cells. J Am Chem Soc 133:365–372
83. Beharry AA, Wong L, Tropepe V, Wooley GA (2011) Fluorescence imaging of azobenzene photoswitching in vivo. Angew Chem Int Ed 50:1325–1327
84. Stafforst T, Hilvert D (2010) Modulating PNA/DNA hybridization by light. Angew Chem Int Ed 49:9998–10001
85. Vomasta D, Högner C, Branda NR, König B (2008) Regulation of human carbonic anhydrase I (hCAI) activity by using a photochromic inhibitor. Angew Chem Int Ed 47:7644–7647
86. Kiji J, Okano T, Kitamura H, Yokoyama Y, Kubota S, Kurita Y (1995) Synthesis and photochromic properties of fulgides with a t-butyl substituent on the furyl- or thienylmethylidene moiety. Bull Chem Soc Jpn 68:616–619
87. Fukumoto S, Nakashima T, Kawai T (2011) Photon-quantitative reaction of a dithiazolylarylene in solution. Angew Chem Int Ed 50:1565–1568
88. Aloïse S, Sliwa M, Pawlowska Z, Réhault J, Dubois J, Poizat O, Buntinx G, Perrier A, Maurel F, Yamaguchi S, Takeshita M (2010) Bridged photochromic diarylethenes investigated by ultrafast absorption spectroscopy: evidence for two distinct photocyclization pathways. J Am Chem Soc 132:7379–7390
89. Massaad J, Micheau JC, Coudret C, Sanchez R, Guirado G, Delbaere S (2012) Gated photochromism and acidity photomodulation of a diacid dithienylethene dye. Chem Eur J 18:6568–6575
90. Ogawa H, Takagi K, Ubukata T, Okamoto A, Yonezawa N, Delbaere S, Yokoyama Y (2012) Bisarylindenols: fixation of conformation leads to exceptional properties of photochromism based on 6π-electrocyclization. Chem Commun 48:11838–11840

Chapter 9
Fast Photochromism of Bridged Imidazole Dimers

Jiro Abe

Abstract Switching of the physical and chemical properties of materials by photochromic compounds has been the subject of considerable research. Increasing switching rates, particularly the thermal bleaching rate, is required for certain applications, such as optical data processing and light modulators. We recently developed the photochromic bridged imidazole dimers that show instantaneous coloration upon exposure to UV light and rapid fading in the dark. We designed and synthesized several bridged imidazole dimers, having a naphthalene or a [2.2]paracyclophane moiety that bridge two triphenylimidazole or diphenylimidazole units, and succeeded in accelerating the thermal bleaching rate. These photochromic compounds show photoinduced homolytic bond cleavage of the C–N bond between the imidazole rings and successive fast C–N bond formation. Rapid thermal bleaching kinetics enables a solution color change only where it is irradiated with UV light because the thermal bleaching rate is much faster than the diffusion rate of the colored species at room temperature. Photochromic materials showing such intense photocoloration and rapid thermal bleaching performance are promising materials for prospective fast light modulator applications.

Keywords Fast photochromism • Hexaarylbiimidazole • Imidazole dimer • [2.2]Paracyclophane • T-type

J. Abe (✉)
Department of Chemistry, School of Science and Engineering, Aoyama Gakuin University, 5-10-1 Fuchinobe, Chuo-ku, Sagamihara, Kanagawa 252-5258, Japan

CREST, Japan Science and Technology Agency (JST), 7 Gobancho, Chiyoda-ku, Tokyo 102-0076, Japan
e-mail: jiro_abe@chem.aoyama.ac.jp

M. Irie et al. (eds.), *New Frontiers in Photochromism*,
DOI 10.1007/978-4-431-54291-9_9, © Springer Japan 2013

9.1 Introduction

Photochromic materials are a well-known class of molecules that change their color upon irradiation with light. The photogenerated species can be reversed to the initial species either thermally or by subsequent irradiation with a specific wavelength of light [1, 2]. Switching of the physical and chemical properties of materials using photochromic compounds has been the subject of considerable research. There is an increasing interest in the use of organic photochromic compounds to modulate conductivity, fluorescence, magnetism, and shape at the bulk level [3–9]. One of their most common applications is their use in photochromic lenses that darken in sunlight. The ideal conditions for photochromic lenses are satisfied with molecules that can be stimulated by sunlight to rapidly develop an intense coloration in a wide range of visible light and can be returned to the initial state with fast fading kinetics in addition to fatigue resistivity for many coloring–decoloring cycles. Increasing switching rates, particularly the thermal bleaching rate, is also indispensable for applications such as optical data processing and light modulators. For the application to real-time image processing at video frame rates [10, 11], the return to the initial state should be accomplished within tens of milliseconds. However, the thermal back reaction of colored species toward their colorless form is generally on the timescale of tens of seconds to hours, which precludes their practical use in certain applications.

We recently developed two types of novel fast photochromic molecules, naphthalene-bridged imidazole dimers [12] and [2.2]paracyclophane-bridged imidazole dimers [13] (Fig. 9.1b, c), which show instant coloration upon exposure to UV light and rapid fading in the dark. Both of the photochromic molecules show photoinduced homolytic bond cleavage of the C–N bond between the imidazole rings and successive fast C–N bond formation. The half-life of colored species of 1-NDPI-8-TPI-naphthalene and *pseudogem*-bisDPI[2.2]PC in solution are 170 and 33 ms at room temperature, respectively. The diversity of the molecular design and the ease of synthesis make this class of photochromic compounds attractive for high-performance ophthalmic plastic lenses and revolutionary optical switching devices.

9.2 Photochromism of Hexaarylbiimidazole

Hexaarylbiimidazole (HABI) was discovered in the early 1960s by Hayashi and Maeda, and it has since attracted significant interest because of its unusual physical properties (Fig. 9.1a) [14]. Various stimuli, such as heat, light, and pressure, readily cleave HABI into a pair of 2,4,5-triphenylimidazolyl radicals (TPIRs), which thermally recombine to reproduce their original imidazole dimer. The solution of HABI changes from colorless to purple upon UV irradiation. This photochromic behavior is attributed to the photoinduced homolytic cleavage of the C–N bond between the imidazole rings. The photochromism of diarylethene is a closed-ring/open-ring-type switching, while that of HABI is a closed-shell/open-shell-type switching, which can be regarded as a diamagnetic/paramagnetic photoswitching.

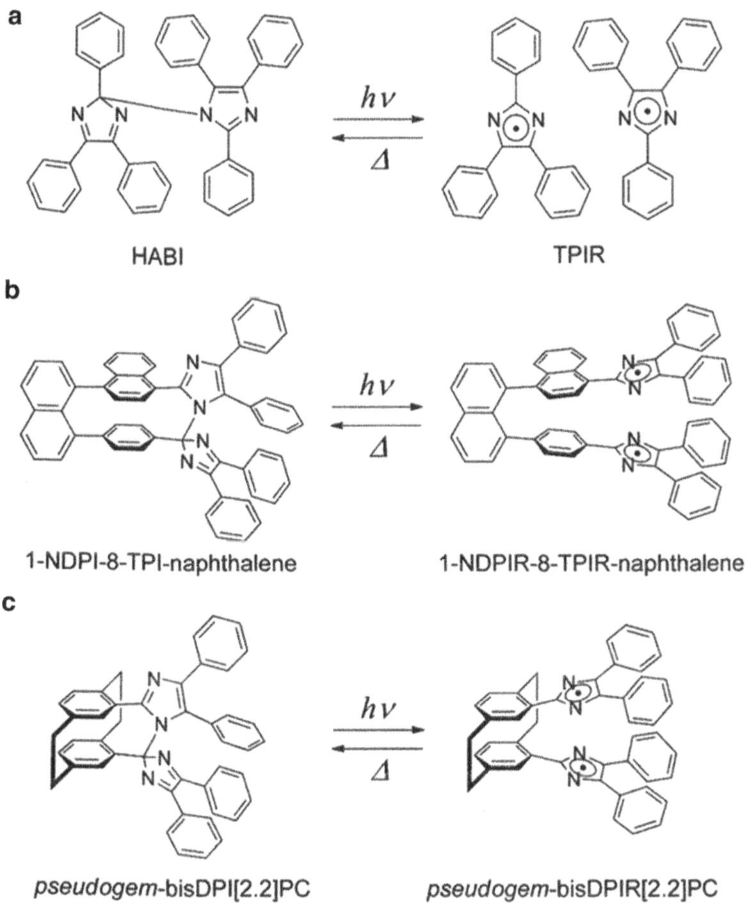

Fig. 9.1 Photochromism of (**a**) hexaarylbiimidazole, (**b**) 1-NDPI-8-TPI-naphthalene, and (**c**) *pseudogem*-bisDPI[2.2]PC

The dissociation of HABI into TPIRs occurs along the repulsive potential energy surface of the first excited singlet state with a time constant of 80 fs [15, 16]. The high yield of TPIRs in solid matrices and their low sensitivity to the presence of oxygen have stimulated industrial interest in the use of HABIs as a photoinitiator for a variety of imaging materials.

There are a number of spectroscopic studies for the photochemical reaction of HABI and their derivatives [17–25]. The thermal transformation of TPIR into HABI in solution requires several minutes at room temperature since the radical–radical reactions generally obey second-order kinetics. The recombination of the radical pairs (RPs) formed by photodissociation provides fundamental information on the solvent cage effect and the dynamic behavior of the encounter RPs in bimolecular chemistry. The nonreactive random encounter RP, which subsequently separates, is formed by diffusion, while the reactive random encounter RP is formed only when the RP approach with their reactive sites facing each other. Thus, the diffusion process and spatial arrangement of the encounter RPs made it difficult to study of the reaction kinetics of the radical–radical reactions in solution. A number of investigations on the photodissociation processes yielding two of the same radicals via photoexcitation have been published for tetraphenylhydrazine [26–28] and diphenyl disulfide derivatives [29–32]. Ernsting et al. reported that bis(*p*-aminophenyl) disulfide undergoes the photodissociation in the time region of 40–100 fs in various solvents. Following excitation, Ernsting et al. observed that the geminate recombination between the nascent radicals in viscous solutions occurs in the first 2 ps [31]. Similar results were confirmed for the other diphenyl disulfide derivative [32]. Although the time constant of the dissociation in HABI is comparable to those for the diphenyl disulfide derivatives, the radical recombination process of TPIRs is very slow as described above. The recombination process in the time frame of seconds to hours required large activation entropy, indicating that the geometrical restriction for the recombination is severe. This restriction might contribute to the inhibition of the rapid geminate recombination in the nascent RP.

Although UV irradiation of a single crystal of *o*-Cl-HABI at room temperature has no photochemical reaction, the crystal turns from pale yellow to reddish brown upon irradiation with UV light at low temperatures ($T < 160$ K). We reported the in situ direct observation of the photoinduced RP in a single crystal of *o*-Cl-HABI by X-ray diffraction analysis using cryotrapping [33] and described the first measurements of intermolecular exchange coupling for the photoinduced RP by ESR spectroscopy [34]. Our crystallographic study for a single crystal of *o*-Cl-HABI measured after UV irradiation at 103 K revealed the molecular geometry of the photoinduced RP. The resulting spatial arrangement of the RP in a crystal offers a structural suggestion for the transition state of the radical dimerization reaction. There are two types of RPs, pair A and pair B, differing only in the spatial arrangement of the two lophyl radicals. The temperature dependence of the ESR spectral pattern shows that pair A converts irreversibly into pair B on increasing the temperature. The Curie analyses on the ESR signal intensities show that the spin triplet state of pair A is the ground state or nearly degenerates with the spin singlet state. Conversely, the triplet state of pair B is thermally populated and the singlet state is the ground state. The exchange coupling between the unpaired electrons in pair B in the crystal was determined as $-77\ \mathrm{cm}^{-1}$ by assuming thermal equilibrium between the ground singlet state and excited triplet state. The experimental value is in fair agreement with the theoretical value of $-45\ \mathrm{cm}^{-1}$ evaluated by the DFT calculations [34]. Thus, the RPs created by homolysis in a single crystal of *o*-Cl-HABI can present an opportunity to observe electron–electron through-space coupling between the well-characterized radical species.

9.3 Molecular Design of the Bridged Imidazole Dimers

The thermal recombination behavior of the TPIR derivative, tF-BDPI-2Y (1,4-bis-(4,5-diphenylimidazol-2-ylidene)-2,3,5,6-tetrafluorocyclohexa-2,5-diene), was reported in 2004 [35]. tF-BDPI-2Y dimerizes to form photochromic

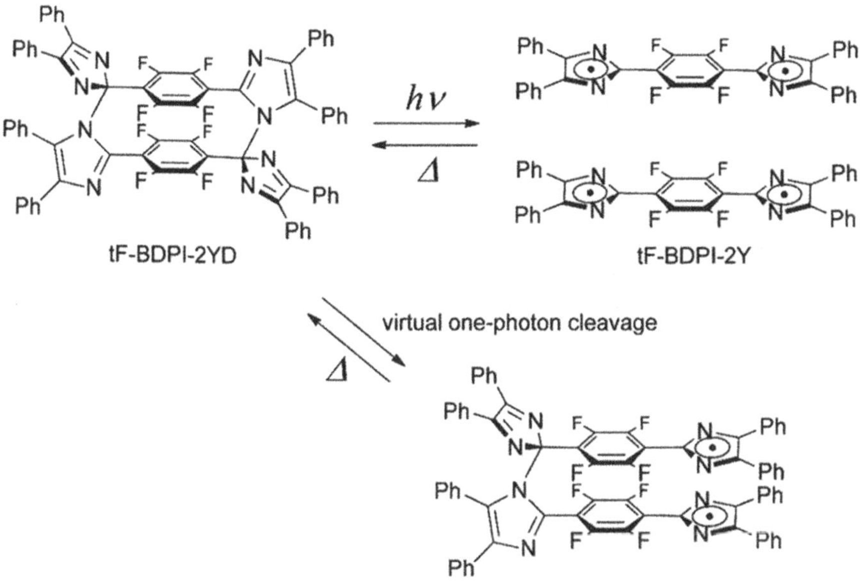

Fig. 9.2 Photochromism and virtual one-photon cleavage of tF-BDPI-2YD

tF-BDPI-2YD at room temperature, while it crystallizes at 200 K because of the activation energy barrier toward the recombination reaction. On irradiation with 360 nm light, the colorless solution of tF-BDPI-2YD quickly turns to blue–purple and gives rise to a broad absorption band centered at 609 nm. On standing in the dark, the blue–purple color gradually fades with the decrease in the visible light absorbance. The thermal recombination reaction of tF-BDPI-2Y to form tF-BDPI-2YD is observed over a period of two days at 293 K in the dark, indicating that the photoinduced radical species are stable in solution at room temperature. We conceptualized that a fast thermal bleaching photochromic molecule could be developed on the basis of a virtual one-photon reaction of tF-BDPI-2YD (Fig. 9.2) [36]. The photoinduced fluorinated TPIR pair will immediately recombine because the RP is restricted in their diffusion and the reaction centers are held in close proximity even in solution. Hence, this class of structure is assumed to be optimum for achieving the fast back reaction of the imidazole dimers. In restricted conditions such as a crystalline phase, it is known that the thermal back reaction of TPIR pairs instantly proceeds. For example, the recombination reaction of chlorine substituted TPIR pairs is completed within several milliseconds in the crystal at room temperature. We directed towards fabricating a fast thermal bleaching photochromic imidazole dimer with the assistance of naphthalene linker as a diffusion inhibition unit.

9.4 Naphthalene-Bridged Imidazole Dimers

9.4.1 Thermal Back Reaction of Colored Species

1,8-TPID-naphthalene cleaves photochemically into 1,8-bisTPIR-naphthalene, and the solution changes from colorless to green [36]. Unlike the conventional HABI derivatives, the photoinduced RP of 1,8-bisTPIR-naphthalene cannot diffuse into the medium to yield free radicals. A kinetic study of the thermal back reaction from the biradical to the dimerized product in solution showed that the reaction obeys first-order kinetics with a 730 ms half-lifetime at 298 K. A new type of a naphthalene-bridged imidazole dimer, 1-NDPI-8-TPI-naphthalene (Fig. 9.1b), which consists of two different TPIR units, i.e., 2-(1-naphthyl)-4,5-diphenylimidazolyl radical (NDPIR) and TPIR units, was also developed [12]. The colored species (1-NDPIR-8-TPIR-naphthalene) of 1-NDPI-8-TPI-naphthalene can be expected to give a superposed absorption spectrum in the visible region for the corresponding NDPIR and TPIR. Therefore, a wide range of visible light would be absorbed by 1-NDPIR-8-TPIR-naphthalene because TPIRs absorb light between 500 and 600 nm, while NDPIRs absorb light between 550 and 900 nm in addition to a sharp band at approximately 460 nm. 1-NDPI-8-TPI-naphthalene has the chirality resulting from the 1,1′-binaphthyl unit, and the two enantiomeric forms of the molecule are found to make up the racemic crystal. 1-NDPI-8-TPI-naphthalene shows a photochromic reaction changing its color from colorless to green under UV irradiation in both solid and solution. Under continuous irradiation, the solution of 1-NDPI-8-TPI-naphthalene reaches the photostationary equilibrium quickly, and the absorption band decreases soon after ceasing the irradiation. The complete bleaching is achieved within 1 s at 295 K. The colored species absorbs the whole range of visible light, which is attributed to NDPIR and TPIR chromophores of 1-NDPIR-8-TPIR-naphthalene. The thermal back reaction obeys first-order kinetics and its half-life at 295 K is 260 ms. The fast fading kinetics makes it possible to change the color of solution where UV light is irradiated because the diffusion rate of colored species is slower than the decoloration rate at room temperature. The enthalpies and entropies of activation ($\Delta H^{\ddagger}$ and $\Delta S^{\ddagger}$, respectively) for the thermal back reaction were estimated from the Eyring plots to be 53.0 kJ mol^{-1} and −57.1 J K^{-1} mol^{-1}, respectively, over a temperature ranging from 298 to 325 K. Thus, the naphthalene-bridged imidazole dimers were the first photochromic compounds that show both intense coloration upon UV light irradiation and successive fast decoloration in the dark.

9.4.2 Chiroptical Switching

Besides the fast bleaching rate, the intrinsic chirality of the binaphthyl moiety in 1-NDPI-8-TPI-naphthalene should guarantee the construction of chiral molecular switches [37]. Axially chiral binaphthalenes show strong circular dichroism (CD) signals and large optical rotation values, which are dependent on the dihedral angle of the two naphthyl moieties. Chiral molecular switches refer to a class of chiral materials whose chiroptical properties can be reversibly modulated under external stimuli such as light irradiation, heat, pH, chemicals, and electric fields [38–47]. Reversible photoregulation of chiral structures along with a photochromic reaction will lead to the switching of the chiroptical properties of molecules [48]. Moreover, a series of the imidazole dimers is the photochromic molecule that results in the formation of paramagnetic imidazolyl radicals with an unpaired electron by the homolytic C–N bond cleavage upon UV light irradiation. Therefore, the rapid optical switching of magneto-optical properties can be expected for the chiral fast photochromic molecules forming the chiral radicals. The X-ray crystallographic analysis of a single crystal of 1-NDPI-8-TPI-naphthalene affirmed the presence of a couple of enantiomers in a unit cell [12]. The resolution of the racemic mixture by chiral HPLC gave each enantiomer up to 96 % ee. The CD and UV–vis absorption spectra of each enantiomer in toluene (3.5×10^{-5} M) are shown in Fig. 9.3b. As described above, the half-life at room temperature of the colored species (1-NDPIR-8-TPIR-naphthalene) is too short to measure a CD spectrum by using a conventional CD spectrometer. Therefore, the CD spectra of the colored species were measured at 200 K immediately after the irradiation with 365 nm of UV light to slow down the rate of the thermal recombination reaction. The CD spectra of each enantiomer of 1-NDPI-8-TPI-naphthalene exhibit the strong Cotton effect at wavelengths shorter than 400 nm (Fig. 9.3b, left). This is consistent with the spectra for the axially chiral binaphthyl derivatives [49, 50], whereas the CD spectra of each enantiomer of the colored species are different from those of the parent 1-NDPI-8-TPI-naphthalene as shown in Fig. 9.3b (right). The broad absorption band in the visible light region can be attributed to the electronic transitions of NDPIR and TPIR chromophores. Each enantiomer of the colored species shows the mirror-image CD bands in the visible light region where the absorption band of the colored species is located. This indicates that the electronic transitions attributable to the radical chromophores have become optically active. It is known that the intensity of CD signals depends on the dissymmetric spatial arrangement of interacting chromophores, in particular on the distance and relative twisting angles between chromophores. Therefore, the Cotton effect in the visible light region of the colored species is tentatively ascribable to the chiral exciton interaction between NDPIR and TPIR chromophores [51, 52].

Several studies have derived optically active products from cage reactions. Stowell et al. reported the stereoselectivity in the cage effect of the homolytic decomposition of an optically active azo compound [53]. The RPs generated closer together recombine with higher stereoselectivity because the radical has a reactive partner initially positioned in a stereoselective manner within the solvent cage. Though the solvent cage could suppress the diffusion of the radicals generated by homolytic decomposition of the molecule, the change in orientation of one radical relative to the radical of the caged RP would decrease the stereoselectivity in the product of the radical–radical coupling [54–61]. We investigated the possibility of the photoracemization of 1-NDPI-8-TPI-naphthalene through the transient colored

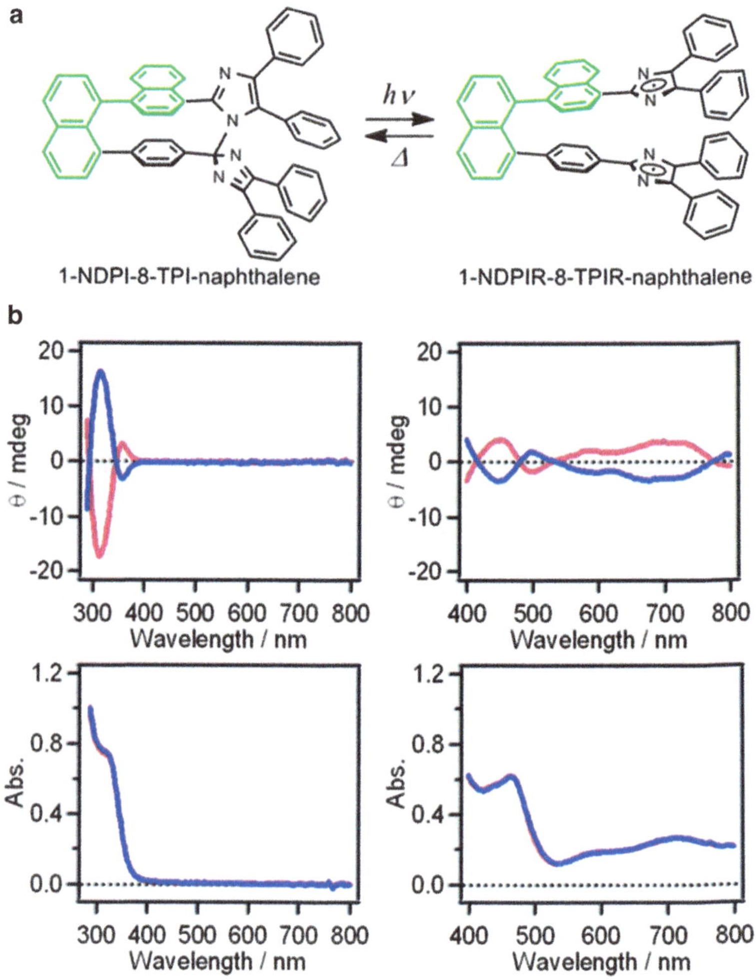

Fig. 9.3 (**a**) Photochromism of 1-NDPI-8-TPI-naphthalene. (**b**) CD and UV–vis absorption spectra of the enantiomer A (*red*) and the enantiomer B (*blue*) of 1-NDPI-8-TPI-naphthalene in toluene at 293 K (*left*) and those of 1-NDPIR-8-TPIR-naphthalene in toluene immediately after irradiation with 365 nm of UV light at 200 K (*right*)

biradical at room temperature. It should be noted that the chiral HPLC analysis confirmed that no photoracemization takes place along with the photochromic reaction, suggesting the suppression of the rotation of NDPIR moiety relative to the naphthyl unit. In darkness, 1-NDPI-8-TPI-naphthalene keeps the chiral structure because of the presence of C–N bond between two imidazole rings, which inhibit the rotation of NDPI moiety.

We achieved the first demonstration of reversible photogeneration of the stable chiral RP from the fast photochromic molecule. The photogenerated radical species of 1-NDPI-8-TPI-naphthalene maintains the chiral conformation, and colored radical exhibits the Cotton effect in the visible light region. This result suggests that the chirality resulting from binaphthyl moiety induces the exciton coupling between the two radical chromophores by a through-space interaction. In addition, the rapid magneto-optical switching can be expected because of the fast switching ability between the diamagnetic colorless state and paramagnetic colored state.

9.5 [2.2]Paracylophane-Bridged Imidazole Dimers

9.5.1 *Photochromic Properties*

As described in the previous section, the thermal bleaching rates of the naphthalene-bridged imidazole dimers are fast enough to be utilized in ophthalmic lenses. Although this development was noteworthy in which the thermal back reaction was drastically accelerated while maintaining the optical density in the colored state, it was not satisfactory for application to real-time image processing at video frame rates. A green-colored spot following the movement of UV light irradiation could be displayed in solution, but an afterimage can be recognized for 1 s or less with the naked eye. To increase the thermal bleaching rate, the photogenerated RPs should be more closely spaced together. Thus, we designed a novel-bridged imidazole dimer, *pseudogem*-bisDPI[2.2]PC, with a [2.2]paracyclophane (PC) moiety that tightly couples two TPIR units (Fig. 9.1c) [13]. *Pseudogem*-bisDPI[2.2]PC undergoes photochromic reaction, changing its color from colorless to blue, on UV irradiation in both solid form and solution at room temperature. Under continuous irradiation, the solution of *pseudogem*-bisDPI[2.2]PC reaches photostationary equilibrium quickly, and the absorption decreases rapidly following monoexponential thermal bleaching kinetics after ceasing the irradiation. The complete bleaching of the light-irradiated solution is achieved within 200 ms in benzene at 25 °C. Figure 9.4a shows the transient absorption spectra of *pseudogem*-bisDPI[2.2]PC in benzene (2.1×10^{-4} M) at 25 °C, measured by a nanosecond laser flash photolysis experiment (excitation wavelength, 355 nm; pulse width, 5 ns; power, 8 mJ/pulse). A sharp absorption band at 400 nm and a broad absorption band ranging from 500 to 900 nm can be ascribed to the colored species, *pseudogem*-bisDPIR[2.2]PC. The half-life of the colored species is 33 ms at 25 °C. Figure 9.4b shows the time profile of the transient absorbance at 400 nm, measured in the temperature range from 5 °C to 40 °C. The thermal bleaching process obeys first-order kinetics, and the half-life of the colored species varies from 198 ms at 5 °C to 10 ms at 40 °C as shown in Fig. 9.4c. The $\Delta H^{\ddagger}$ and $\Delta S^{\ddagger}$ values estimated from standard least-squares analysis of the Eyring plots are 59.8 kJ mol^{-1} and −19.1 J K^{-1} mol^{-1}, respectively. The free energy barrier ($\Delta G^{\ddagger} = \Delta H^{\ddagger} - T\Delta S^{\ddagger}$) is 65.5 kJ mol^{-1} at 25 °C. Fast thermal bleaching kinetics enable a solution color change where it is irradiated with UV light at room temperature. Note that the afterimage seen in the photochromic reaction of the naphthalene-bridged imidazole dimer is invisible to the naked eye for *pseudogem-bisDPI*[2.2]PC system.

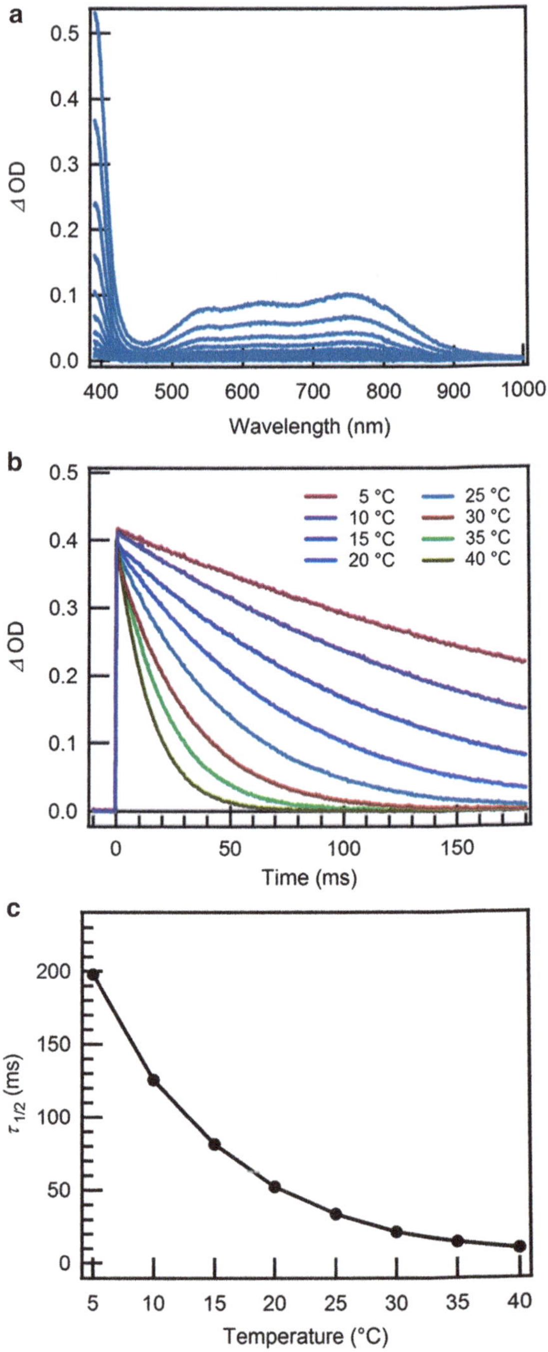

Fig. 9.4 (**a**) Transient absorption spectra of *pseudogem*-bisDPI[2.2]PC. Each of the spectra was recorded at 20 ms intervals after excitation with a nanosecond laser pulse. (**b**) Decay profiles of the colored species monitored at 400 nm. (**c**) Temperature dependence of the half-life of the colored species

In general, the faster the thermal bleaching rate is, the lighter the color of photostationary equilibrium because of the fundamental difficulty in increasing the stationary concentration of colored species [62, 63]. In contrast to any other currently available photochromic system, the high quantum yield of the bond cleavage reaction of the photochromic imidazole dimers can enable visual inspection of the coloration upon UV light irradiation. A moderate thermal bleaching rate is important for recognition by the human eye. It is difficult to detect a phenomenon that is faster than 10 ms by the human eye, but the colored species of *pseudogem*-bisDPI[2.2]PC has a half-life of tens of milliseconds, which is favorable for detection by the human eye. Thus, *pseudogem*-bisDPI[2.2]PC can potentially be applied to real-time image processing at video frame rates.

9.5.2 Remarkable Acceleration for Back Reaction

The stability of the colored species generated from the photochromic reaction of *pseudogem*-bisDPI[2.2]PC can be attributed to the inhibition of the diffusion of reactive radicals and rapid geminate recombination in the nascent RP. However, a molecular design that significantly accelerates the thermal back reaction is necessary for practical use in fast light modulators. A significant feature of the synthetic character of the [2.2]PC-bridged imidazole dimer is a stepwise formation of two imidazole rings by the condensation reaction between aromatic 1,2-diketones and two aldehyde groups of *pseudogem*-bisformyl[2.2]paracyclophane. This type of stepwise reaction permits rational design for achieving desired performances such as thermal bleaching rate, coloring, and photosensitivity.

Although Marcus theory was originally developed for electron-transfer reactions [64–66], it has been successfully applied to a variety of organic reactions [67–70]. Following standard Marcus theory, an increase in the change in Gibbs energy, ΔG^0, between the reactant and product would lead to a decrease in the free energy of activation, $\Delta G^\ddagger$, and consequently, the rate constant for the reaction would be accelerated. Thus, Marcus theory gave us a crucial insight into controlling the rate of the radical recombination reaction. We have considered that the ΔG^0 of the thermal back reaction could be enlarged by destabilizing the colored species, and designed *pseudogem*-DPI-PI[2.2]PC (Fig. 9.5), with a PC moiety that couples diphenylimidazole and phenanthroimidazole groups [71]. The steric repulsion between the rigid phenanthroimidazole group and the phenyl rings facing each other should destabilize the biradical state. Conversely, the rotational motion along the C–C single bond between the imidazole group and phenyl ring in *pseudogem*-bisDPIR[2.2]PC could relax the steric hindrance between the phenyl rings facing each other.

On irradiation with UV light in the solution of *pseudogem*-DPI-PI[2.2]PC, no color change is discernible at room temperature but the solution changes from colorless to blue at liquid nitrogen temperature. That is, the thermal back reaction is too rapid for detection by a human eye at room temperature. Laser flash photolysis measurements confirm the photochromic properties of *pseudogem*-DPI-PI[2.2]PC

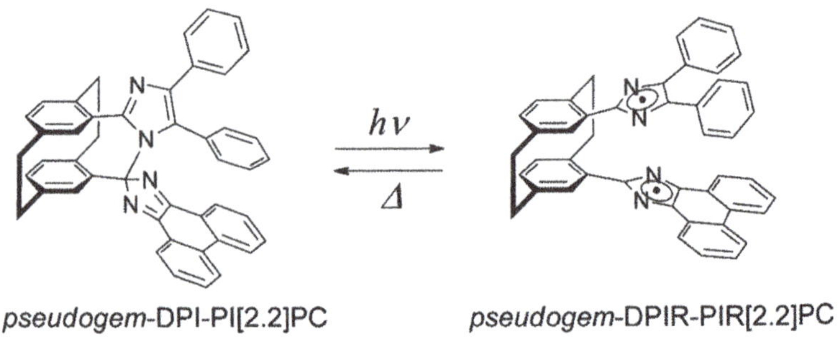

Fig. 9.5 Photochromism of *pseudogem*-DPI-PI[2.2]PC

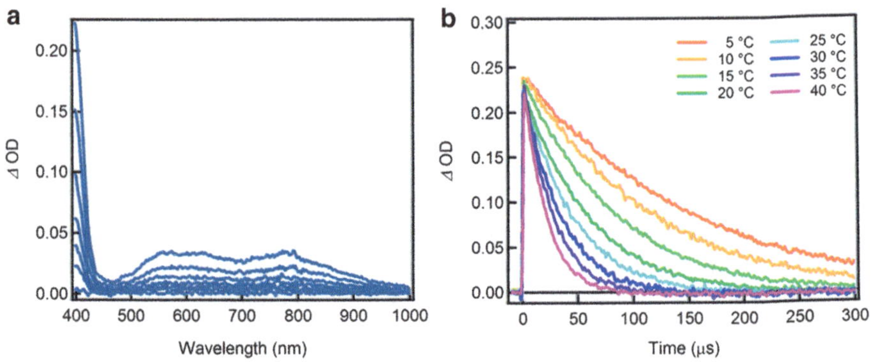

Fig. 9.6 (**a**) Transient absorption spectra of *pseudogem*-DPI-PI[2.2]PC in benzene at 25 °C. Each of the spectra was recorded at 20 μs intervals after excitation with a nanosecond laser pulse. (**b**) Time profiles of transient absorbance of the colored species monitored at 400 nm

with the formation of a transient colored species. Figure 9.6a shows the transient absorption spectra of *pseudogem*-DPI-PI[2.2]PC in benzene (3.4×10^{-5} M) at 25 °C (excitation wavelength, 355 nm; pulse width, 5 ns; power, 8 mJ/pulse). A sharp absorption band at 400 nm and broad absorption band ranging from 450 to 1,000 nm can be ascribed to the colored biradical, *pseudogem*-DPIR-PIR[2.2]PC. Indeed, the transient absorption spectra are almost identical to those for *pseudogem*-bisDPI[2.2]PC. The half-life of the colored species is 35 μs at 25 °C. Figure 9.6b shows the time profiles of the transient absorbance at 400 nm, measured at temperatures ranging from 5 °C to 40 °C. The thermal bleaching process obeys first-order kinetics, and the half-life of the colored species varies from 110 μs at 5 °C to 17 μs at 40 °C. The original state is fully restored within 200 μs in benzene at 25 °C. This fast coloration–decoloration cycle would make it possible to operate light modulator applications at a repetition frequency of 5 kHz. The $\Delta H^{\ddagger}$ and $\Delta S^{\ddagger}$ values estimated from standard least-squares analysis of the Eyring plots are 35.4 kJ mol^{-1} and −44.1 J K^{-1} mol^{-1}, respectively. The $\Delta G^{\ddagger}$ is 48.6 kJ mol^{-1} at 25 °C.

Note that the thermal back reaction of *pseudogem*-DPIR-PIR[2.2]PC is accelerated approximately 1,000 times compared with that of *pseudogem*-bisDPIR[2.2]PC while maintaining its optical density in the colored state. Moreover, ΔOD values immediately after laser excitation are not influenced by a temperature change, as can be seen in Fig. 9.6b. These photochromic behaviors demonstrate the distinguishing feature of the [2.2]PC-bridged imidazole dimer. As described above, controlling the stability of the biradical state is effective in controlling the thermal bleaching rate for the photochromic [2.2]PC-bridged imidazole dimer. The method for controlling the stability of the biradical state is not restricted to utilizing the steric effect since the electronic effect induced by substituent groups would also be effective. As compared with another photochromic system, the [2.2]PC-bridged imidazole dimers are characterized by their diverse molecular design. A variety of [2.2]PC-bridged imidazole dimers can be prepared by selecting appropriate aromatic 1,2-diketones that meet the requirements.

9.5.3 Efficient Strategy for Enhancing the Photosensitivity

Though the large optical density in the visible light region is necessary when the [2.2]PC-bridged imidazole dimers are applied to fast color-changing ophthalmic lenses, a fast bleaching rate decreases in the optical density in the photostationary state. One possible approach to increase the optical density for the colored state is to enhance the photosensitivity to solar radiation. Unfortunately the photosensitivity of the earliest [2.2]PC-bridged imidazole dimer is poor because it does not have an efficient absorption band in the UVA radiation region. To improve the photosensitivity to sunlight, we developed *pseudogem*-bisTMDPI[2.2]PC, which shows excellent photochromic color change even under sunlight [72]. The [2.2]PC-bridged imidazole dimer has two types of imidazole rings—Im1 and Im2—as shown in Fig. 9.7. The two imidazole rings are not equivalent in their electronic environment. Im1 is a resonant planar structure that has a typical bond distance for a 6π electron system with an electron-donating characteristic. Im2 has two localized C=N double bonds and one sp^3 carbon connecting Im1, consistent with a 4π electron system with an electron-withdrawing characteristic. The absorption band in the UVA radiation region of *pseudogem*-bisTMDPI[2.2]PC was assigned to the intramolecular charge-transfer (CT) transition from the electron-donating dimethoxy-substituted phenyl rings to the electron-withdrawing Im2 on the basis of the TDDFT calculation. Moreover, the relationships of the CT characteristics and photochromic properties of two types of newly designed [2.2]PC-bridged imidazole dimers (*pseudogem-DPI*-TMDPI[2.2]PC (**1**), *pseudogem*-TMDPI-DPI[2.2]PC (**2**)) were investigated to determine the accuracy of the molecular design for enhancing the photosensitivity [73].

The molecular structures revealed by X-ray crystallographic analysis are shown in Fig. 9.8. Each C–N bond lengths connecting two imidazole rings (**1**; 1.4851(18) Å,

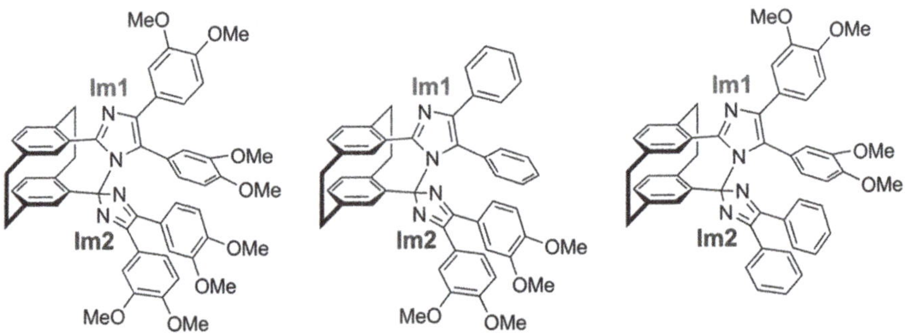

pseudogem-bisTMDPI[2.2]PC *pseudogem*-DPI-TMDPI[2.2]PC *pseudogem*-TMDPI-DPI[2.2]PC

Fig. 9.7 Chemical structures of methoxy-substituted [2.2]PC-bridged imidazole dimers

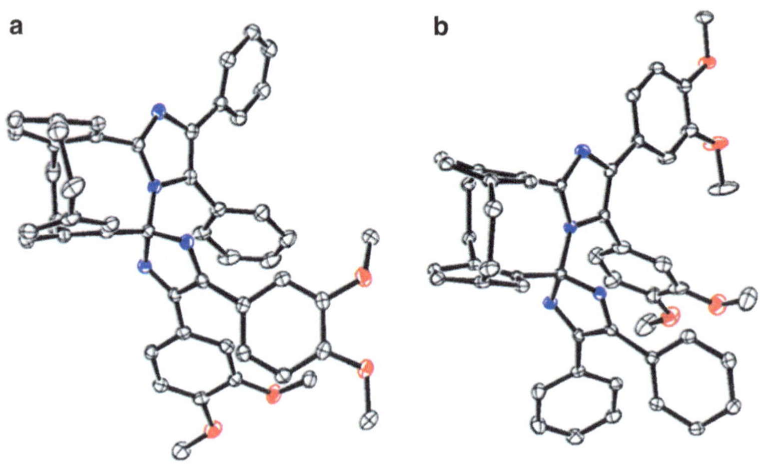

Fig. 9.8 ORTEP representation of the molecular structures of (**a**) **1** and (**b**) **2** with thermal ellipsoids (50 % probability), where nitrogen and oxygen atoms are highlighted in *blue* and *red*, respectively

2; 1.480(3) Å) are approximately equal to that of *pseudogem*-bisDPI[2.2]PC (1.4876(15) Å). The imidazole ring of **1** attached with the dimethoxy-substituted phenyl rings is an electron-withdrawing Im2, whereas the imidazole ring of **2** attached with the dimethoxy-substituted phenyl rings is an electron-donating Im1. Though the significant difference in the bond lengths of **2** compared with those of *pseudogem*-bisDPI[2.2]PC was not recognized, the bond lengths of the C–C bond between the imidazole ring and the dimethoxy-substituted phenyl rings of **1** were considerably different from those in *pseudogem*-bisDPI[2.2]PC. These differences in the molecular geometries can be plausibly explained by the contribution from the intramolecular CT resonance between the electron-donating dimethoxy-substituted phenyl rings and the electron-withdrawing imidazole ring.

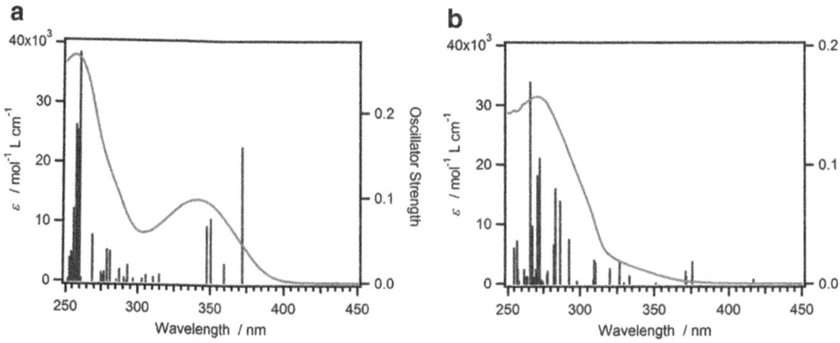

Fig. 9.9 UV–vis absorption spectra and TDDFT calculations for (**a**) **1** and (**b**) **2**. The calculated spectra (MPW1PW91/6-31+G(d)//MPW1PW91/6-31G(d)) are shown by the *blue perpendicular lines*

Figure 9.9 shows the UV–vis absorption spectra of **1** and **2** in the acetonitrile solution, along with those calculated by the TDDFT method (MPW1PW91/6-31+G(d)//MPW1PW91/6-31G(d)) for the optimized molecular geometries. The green curves are the experimental spectra and the calculated oscillator strengths are shown by the blue perpendicular lines. It should be noted that a clearly distinct absorption band is found in the UVA radiation region for the UV–vis absorption spectra of **1**, whereas **2** does not show a similar absorption band in the same region. The $S_0 \rightarrow S_2$ transition at 372 nm ($f = 0.159$) of **1** demonstrated by the TDDFT calculation is described by HOMO−1→LUMO transition. As can be found in Fig. 9.10, this transition can be attributed to the intramolecular CT transition from the dimethoxy-substituted phenyl rings to the electron-withdrawing Im2. Moreover, both of the $S_0 \rightarrow S_4$ transition at 351 nm ($f = 0.075$) and $S_0 \rightarrow S_5$ transition at 348 nm ($f = 0.066$) are associated with HOMO−3→LUMO transitions. These excited states are also characterized by the same type of the intramolecular CT transition. Conversely, as shown in Fig. 9.8, the dimethoxy-substituted phenyl rings are connected to the electron-donating Im1 for **2**. Thus, the intramolecular CT transition from the electron-donating dimethoxy-substituted phenyl rings to the electron-withdrawing Im2 would have small oscillator strength. This is because of the small orbital overlap between the molecular orbitals (MOs) delocalized over the dimethoxy-substituted phenyl rings and the MOs delocalized over Im2. As shown in Fig. 9.9, significant absorption bands are not found in the experimental or the theoretical absorption spectra of **2** in the UVA radiation region. The solvent effects on the absorption spectrum of **1** are also investigated. The maximum wavelength of the absorption in the UVA radiation region is slightly red shifted, depending on the increase in the permittivity of the solvent. This result also supports the CT nature of the absorption in the UVA radiation region. These structure–spectra relationships are the definitive evidence for the presence of the intramolecular CT band from the electron-donating substituents to the electron-withdrawing Im2.

The photosensitivity of the [2.2]PC-bridged imidazole dimers can be readily controlled by appropriately designing of the phenyl rings attached to the imidazole

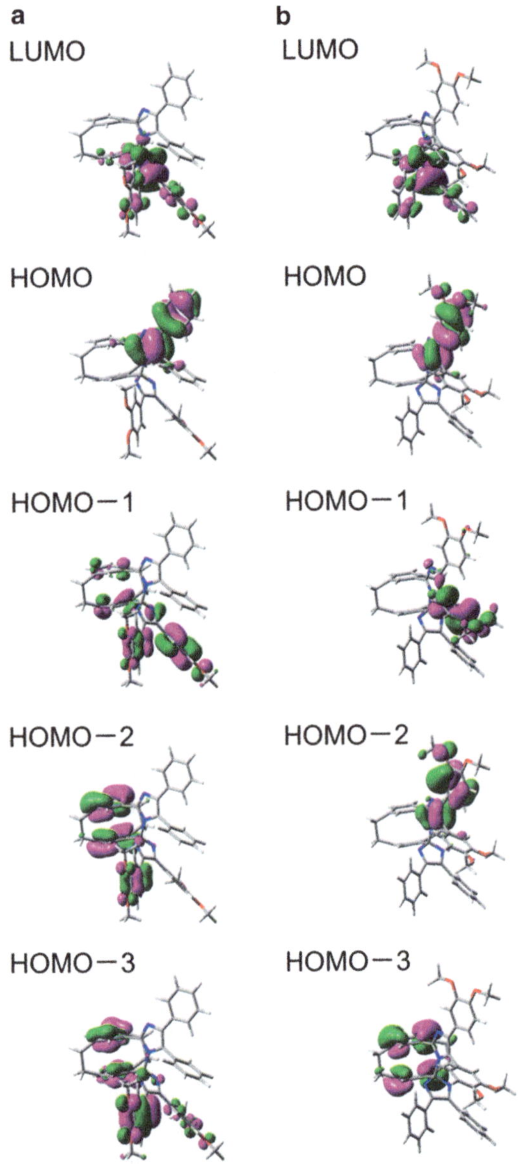

Fig. 9.10 The relevant molecular orbitals of (**a**) **1** and (**b**) **2** obtained at the MPW1PW91/6-31+G(d) level

rings. The introduction of electron-donating substituents into the phenyl rings attached to the electron-withdrawing Im2 would enhance the photosensitivity with the aid of the intramolecular CT transitions. Both of the solutions of **1** and **2** undergo photochromic reaction involving a color change from colorless to blue upon UV light irradiation and generate the identical colored biradical species. The solutions reach the photostationary equilibrium very rapidly under continuous irradiation, and after irradiation ceases, the absorption decreases quickly according to monoexponential thermal bleaching kinetics. The oxidation of the precursor gives the crude mixture of **1** and **2**, and consequently, the photochromic reaction of **1** leads to the formation of the mixture of **1** and **2** via the colored biradical species. The formation of **1** by the photochromic reaction of **2** is also true. Thus, the repeated irradiation of UV light to the solution of **1** would gradually change to the mixture of thermal equilibrium of **1** and **2**.

9.6 Conclusion

We have developed a wide variety of the photochromic [2.2]PC-bridged imidazole dimers, which show instantaneous coloration on exposure to UV light and rapid fading in the dark [74–79]. The diversity of the molecular design and the ease of synthesis make this class of photochromic compounds highly attractive for high-performance ophthalmic plastic lenses and revolutionary optical switching devices. In view of the thermal bleaching rate, the fast photochromism of the [2.2]PC-bridged imidazole dimers can be applied to real-time image processing at video frame rates. Thus, molecular design based on the photochromism of HABI can lead to the development of a new family of photochromic compounds with unprecedented switching speeds and remarkable stability. These could eventually evolve into solid-state photonic materials with unique photoresponsive characters.

Acknowledgments I would like to gratefully acknowledge support from the Grants-in-Aid for Scientific Research on Priority Area "New Frontiers in Photochromism" (No. 471) of the Ministry of Education, Culture, Sports, Science and Technology (MEXT) and the Grant-in-Aid for Scientific Research (A) (22245025) of the Japan Society for the Promotion of Sciences (JSPS).

References

1. Crano JC, Guglielmetti RJ (1999) Organic photochromic and thermochromic compounds. Plenum, New York
2. Duerr H, Bouas-Laurent H (2003) Photochromism: molecules and systems. Elsevier, Amsterdam
3. Kawai T, Nakashima Y, Irie M (2005) A novel photoresponsive π-conjugated polymer based on diarylethene and its photoswitching effect in electrical conductivity. Adv Mater 17:309–314
4. Irie M, Fukaminato T, Sasaki T, Tamai N, Kawai T (2002) A digital fluorescent molecular photoswitch. Nature 420:759–760
5. Matsuda K, Irie M (2000) A diarylethene with two nitronyl nitroxides: photoswitching of intramolecular magnetic interaction. J Am Chem Soc 122:7195–7201
6. Yu Y, Nakano M, Ikeda T (2003) Directed bending of a polymer film by light. Nature 425:145
7. Yamada M, Kondo M, Mamiya J, Yu Y, Kinoshita M, Barrett C, Ikeda T (2008) Photomobile polymer materials: towards light-driven plastic motors. Angew Chem Int Ed 47:4986–4988
8. Kobatake S, Takami S, Muto H, Ishikawa T, Irie M (2007) Rapid and reversible shape changes of molecular crystals on photoirradiation. Nature 446:778–781

9. Fernandez-Acebes A, Lehn JM (1998) Optical switching and fluorescence modulation in photochromic metal complexes. Adv Mater 10:1519–1522
10. Volodin BL, Kippelen B, Meerholz K, Javidi B, Peyghambarian N (1996) A polymeric optical pattern-recognition system for security verification. Nature 383:58–60
11. Hampp N (2000) Bacteriorhodopsin as a photochromic retinal protein for optical memories. Chem Rev 100:1755–1776
12. Fujita K, Hatano S, Kato D, Abe J (2008) Photochromism of a radical diffusion-inhibited hexaarylbiimidazole derivative with intense coloration and fast decoloration performance. Org Lett 10:3105–3108
13. Kishimoto Y, Abe J (2009) A fast photochromic molecule that colors only under UV light. J Am Chem Soc 131:4227–4229
14. Hayashi T, Maeda K (1960) Preparation of a new phototropic substance. Bull Chem Soc Jpn 33:565–566
15. Satoh Y, Ishibashi Y, Ito S, Nagasawa Y, Miyasaka H, Chosrowjan H, Taniguchi S, Mataga N, Kato D, Kikuchi A, Abe J (2007) Ultrafast laser photolysis study on photodissociation dynamics of a hexaarylbiimidazole derivative. Chem Phys Lett 448:228–231
16. Miyasaka H, Satoh Y, Ishibashi Y, Ito S, Nagasawa Y, Taniguchi S, Chosrowjan H, Mataga N, Kato D, Kikuchi A, Abe J (2009) Ultrafast photodissociation dynamics of a hexaarylbiimidazole derivative with pyrenyl groups: dispersive reaction from femtosecond to 10 ns time regions. J Am Chem Soc 131:7256–7263
17. White DM, Sonnenberg J (1966) Oxidation of triarylimidazoles. Structures of the photochromic and piezochromic dimers of triarylimidazolyl radicals. J Am Chem Soc 88:3825–3829
18. Riem RH, MacLachlan A, Coraor GR, Urban EJ (1971) The flash photolysis of a substituted hexaarylbiimidazole and reactions of the imidazolyl radical. J Org Chem 36:2272–2275
19. Cescon LA, Coraor GR, Dessauer R, Silversmith EF, Urban EJ (1971) Some properties of triarylimidazolyl radicals and their dimer. J Org Chem 36:2262–2267
20. Qin XZ, Liu A, Trifunac AD, Krongauz VV (1991) Photodissociation of hexaarylbiimidazole. 1. Triplet-state formation. J Phys Chem 95:5822–5826
21. Kikuchi A, Iyoda T, Abe J (2002) Electronic structure of light-induced lophyl radical derived from a novel hexaarylbiimidazole with π-conjugated chromophore. Chem Commun 14:1484–1485
22. Miyamoto Y, Kikuchi A, Iwahori F, Abe J (2005) Synthesis and photochemical properties of a photochromic iron(II) complex of hexaarylbiimidazole. J Phys Chem A 109:10183–10188
23. Kikuchi A, Harada Y, Yagi M, Ubukata T, Yokoyama Y, Abe J (2010) Photoinduced diffusive mass transfer in *o*-Cl-HABI amorphous thin films. Chem Commun 46:2262–2264
24. Nakahara I, Kikuchi A, Iwahori F, Abe J (2005) Photochromism of a novel hexaarylbiimidazole derivative having azobenzene moieties. Chem Phys Lett 402:107–110
25. Kimoto A, Niitsu S, Iwahori F, Abe J (2009) Formation of hexaarylbiimidazole heterodimers via cross recombination of two lophyl radicals. New J Chemistry 33:1339–1342
26. Lenderink E, Duppen K, Wiersma DA (1992) Femtosecond fragmentation of tetraphenylhydrazine in solution. Chem Phys Lett 194:403–409
27. Meyer A, Nikowa L, Schroeder J, Schwarzer D, Thurau G (1995) Photodissociation of tetraphenylhydrazine in compressed liquid solvents. Farday Discuss 102:443–449
28. Hirata Y, Ohta M, Okada T, Mataga N (1992) Direct observation of photodissociation of tetraphenylhydrazine and its derivatives in the solution phase: picosecond study of nitrogen-nitrogen bond rupture in the fluorescence state. J Phys Chem 96:1517–1520
29. Scott TW, Liu SN (1989) Picosecond geminate recombination of phenylthiyl free-radical pairs. J Phys Chem 93:1393–1396
30. Hirata Y, Niga Y, Ohta M, Takizawa M, Okada T (1995) Photodissociation and geminate dynamics in solution phase: picosecond transient absorption studies of tetraphenylhydrazines and diphenyl disulfides. Res Chem Intermed 21:823–836
31. Bultmann T, Ernsting NP (1996) Competition between geminate recombination and solvation of polar radicals following ultrafast photodissociation of bis(*p*-aminophenyl) disulfide. J Phys Chem 100:19417–19424

32. Lochschmidt A, Eilers-König N, Heineking N, Ernsting NP (1999) Femtosecond photodissociation dynamics of bis(julolidine) disulfide in polar and apolar solvents. J Phys Chem A 103:1776–1784
33. Kawano M, Sano T, Abe J, Ohashi Y (1999) The first in situ direct observation of the light-induced radical pair from a hexaarylbiimidazolyl derivative by X-ray crystallography. J Am Chem Soc 121:8106–8107
34. Abe J, Sano T, Kawano M, Ohashi Y, Matsushita MM, Iyoda T (2001) EPR and density functional studies of light-Induced radical pairs in a single crystal of a hexaarylbiimidazolyl derivative. Angew Chem Int Ed 40:580–582
35. Kikuchi A, Iwahori F, Abe J (2004) Definitive evidence for the contribution of biradical character in a closed-shell molecule, derivative of 1,4-bis-(4,5-diphenylimidazol-2-ylidene) cyclohexa-2,5-diene. J Am Chem Soc 126:6526–6527
36. Iwahori F, Hatano S, Abe J (2007) Rational design of a new class of diffusion-inhibited HABI with fast back-reaction. J Phys Org Chem 20:857–863
37. Hatano S, Fujita K, Tamaoki N, Kaneko T, Nakashima T, Naito M, Kawai T, Abe J (2011) Reversible photogeneration of a stable chiral radical-pair from a fast photochromic molecule. J Phys Chem Lett 2:2680–2682
38. Yokoyama Y, Uchida S, Yokoyama Y, Sugawara Y, Kurita Y (1996) Diastereoselective photochromism of an (R)-binaphthol-condensed indolylfulgide. J Am Chem Soc 118:3100–3107
39. Yokoyama Y (2000) Fulgides for memories and switches. Chem Rev 100:1717–1739
40. Feringa BL, van Delden RA, Koumura N, Geertsema EM (2000) Chiral molecular switches. Chem Rev 100:1789–1816
41. Pieraccini S, Masiero S, Spada GP, Gottarelli G (2003) A new axially-chiral photochemical switch. Chem Comm 39:598–599
42. Pieraccini S, Gottarelli G, Labruto R, Masiero S, Pandoli O, Spada GP (2004) The control of the cholesteric pitch by some azo photochemical chiral switches. Chem Eur J 10:5632–5639
43. van Delden RA, Mecca T, Rosini C, Feringa BL (2004) A chiroptical molecular switch with distinct chiral and photochromic entities and its application in optical switching of a cholesteric liquid crystal. Chem Eur J 10:61–70
44. Wigglesworth TJ, Sud D, Norsten TB, Lekhi VS, Branda NR (2005) Chiral discrimination in photochromic helicenes. J Am Chem Soc 127:7272–7273
45. Hirose T, Matsuda K, Irie M (2006) Self-assembly of photochromic diarylethenes with amphiphilic side chains: reversible thermal and photochemical control. J Org Chem 71:7499–7508
46. Carreo MC, Garca I, Nez I, Merino E, Ribagorda M, Pieraccini S, Spada GP (2007) Photoinduced conformational switch of enantiopure azobenzenes controlled by a sulfoxide. J Am Chem Soc 129:7089–7100
47. Mathews M, Tamaoki N (2008) Planar chiral azobenzenophanes as chiroptic switches for photon mode reversible reflection color control in induced chiral nematic liquid crystals. J Am Chem Soc 130:11409–11416
48. Lemieux RP, Schuster GB (1993) Photochemistry of axially chiral (arylmethylene) cycloalkanes: a search for suitable photoswitchable liquid crystalline materials. J Org Chem 58:100–110
49. Mislow K, Glass MAW, O Brien RE, Rutkin P, Steinberg DH, Weiss J, Djerassi C (1962) Configuration, conformation and rotatory dispersion of optically active biaryls. J Am Chem Soc 84:1455–1478
50. Mislow K, Bunnenberg E, Records R, Wellman K, Djerassi C (1963) Inherently dissymmetric chromophores and circular dichroism. II. J Am Chem Soc 85:1342–1349
51. Scholes GD, Ghiggino KP, Oliver AM, Paddon-Row MN (1993) Through-space and through-bond effects on exciton interactions in rigidly linked dinaphthyl molecules. J Am Chem Soc 115:4345–4349
52. Gudipati MS (1994) Exciton, exchange, and through-bond interactions in multichromophoric molecules: An analysis of the electronic excited states. J Phys Chem 98:9750–9763
53. Greene FD, Berwick MA, Stowell JC (1970) Stereochemistry of free-radical recombination reactions. The cage effect in decomposition of SS-(–)-azobis-cr-phenylethane. J Am Chem Soc 92:867–874

54. Engstrom JP, Greene FD (1972) Stereochemistry of free-radical recombination reactions. The cage effect in decomposition of S-(+)-tert-butyl Z-phenylperpropionate. J Org Chem 37:968–972
55. Garst J (1975) Diffusion model analysis of cage reactions of chiral radical pairs. J Am Chem Soc 97:5062–5065
56. Lee KY, Horowitz N, Ware J, Singer LA (1977) Facile 1,3-rearrangement of ketenimines to nitriles. Stereochemical observations on a caged chiral radical pair. J Am Chem Soc 99:2622–2627
57. Step EN, Buchachenko AL, Turro NJ (1992) The cage effect in the photolysis of (S)-(+)-a-methyldeoxybenzoin: can triplet radical pairs undergo geminate recombination in nonviscous homogeneous solution? J Org Chem 57:7018–7024
58. Turro NJ, Buchachenko AL, Tarasov VF (1995) How spin stereochemistry severely complicates the formation of a carbon–carbon bond between two reactive radicals in a supercage. Acc Chem Res 28:69–80
59. Xu J, Weiss RG (2003) Enantioselectivity of prochiral radical-pair recombinations. Reaction cavity differentiation in polyethylene films. Org Lett 5:3077–3080
60. Xu J, Weiss RG (2005) Analyses of in-cage singlet radical-pair motions from irradiations of 1-naphthyl (R)-1-phenylethyl ether and 1-naphthyl (R)-2-phenylpropanoate in n-alkanes. J Org Chem 70:1243–1252
61. Resendiz MJE, Family F, Fuller K, Campos LM, Khan SI, Lebedeva NV, Forbes MDE, Garcia-Garibay MA (2009) Radical reactions with double memory of chirality (2MOC) for the enantiospecific synthesis of adjacent stereogenic quaternary centers in solution: cleavage and bonding faster than radical rotation. J Am Chem Soc 131:8425–8433
62. Tomasulo M, Sortino S, White AJP, Raymo FM (2005) Fast and stable photochromic oxazines. J Org Chem 70:8180–8189
63. Tomasulo M, Sortino S, Raymo FM (2005) A fast and stable photochromic switch based on the opening and closing of an oxazine ring. Org Lett 7:1109–1112
64. Marcus RA (1956) On the theory of oxidation-reduction reactions involving electron transfer. I. J Chem Phys 24:966–978
65. Marcus RA (1964) Chemical and electrochemical electron-transfer theory. Annu Rev Phys Chem 15:155–196
66. Marcus RA (1968) Theoretical relations among rate constants, barriers, and Brønsted slopes of chemical reactions. J Phys Chem 72:891–899
67. Marcus RA (1969) Unusual slopes of free energy plots in kinetics. J Am Chem Soc 91:7224–7225
68. Cohen AO, Marcus RA (1968) Slope of free energy plots in chemical kinetics. J Phys Chem 72:4249–4256
69. Alabugin IV, Manoharan M, Breiner B, Lewis FD (2003) Control of kinetics and thermodynamics of [1, 5]-shifts by aromaticity: a view through the prism of Marcus theory. J Am Chem Soc 125:9329–9342
70. Chen X, Brauman JI (2008) Hydrogen bonding lowers intrinsic nucleophilicity of solvated nucleophiles. J Am Chem Soc 130:15038–15046
71. Harada Y, Hatano S, Kimoto A, Abe J (2010) Remarkable acceleration for back-reaction of a fast photochromic molecule. J Phys Chem Lett 1:1112–1115
72. Mutoh K, Hatano S, Abe J (2010) An efficient strategy for enhancing the photosensitivity of photochromic [2.2]paracyclophane-bridged imidazole dimers. J Photopolym Sci Technol 23:301–306
73. Mutoh K, Abe J (2011) Comprehensive understanding of structure-photosensitivity relationships of photochromic [2.2]paracyclophane-bridged imidazole dimers. J Phys Chem A 115:4650–4656
74. Kimoto A, Tokita A, Horino T, Oshima T, Abe J (2010) Fast photochromic polymers carrying [2.2]paracyclophane-bridged imidazole dimer. Macromolecules 43:3764–3769
75. Hatano S, Sakai K, Abe J (2010) Unprecedented radical-radical reaction of a [2.2]paracyclophane derivative containing an imidazolyl radical moiety. Org Lett 12:4152–4155

76. Takizawa M, Kimoto A, Abe J (2011) Photochromic organogel based on [2.2]paracyclophane-bridged imidazole dimer with tetrapodal urea moieties. Dyes Pigm 89:254–259
77. Mutoh K, Abe J (2011) Photochromism of a water-soluble vesicular [2.2]paracyclophane-bridged imidazole dimer. Chem Comm 47:8868–8870
78. Yamashita H, Abe J (2011) Photochromic properties of [2.2]paracyclophane-bridged imidazole dimer with increased photosensitivity by introducing pyrenyl moiety. J Phys Chem A 115:13332–13337
79. Kawai S, Yamaguchi T, Kato T, Hatano S, Abe J (2012) Entropy-controlled thermal back-reaction of photochromic [2.2]paracyclophane-bridged imidazole dimer. Dyes Pigm 92:872–876

Chapter 10
Photochromic Terarylenes

Takuya Nakashima and Tsuyoshi Kawai

Abstract Properties of photochromic terarylenes are described as a novel series of photochromic switching molecules. The chemical structure of terarylenes can be recognized as an extension of diarylethenes modified at the central ethene moiety with extended π-conjugated systems. Terarylenes also show photo-induced 6π-electrocyclization reaction in a similar manner to that of diarylethenes. A variety of terarylene structures with various combinations of heteroaromatics could be achieved by metal-catalyzed conventional aryl–aryl cross-coupling methods as well as azole-forming condensation reactions from bis(heteroaryl)-α-hydroxyketones. The conformations of terarylenes are controlled both in solution and crystal with the molecular design of terarylenes by means of the combinations of heteroaromatics including their coupling directions, leading to the design of photo-switching molecules with ultimately high sensitivity. The modification of less versatile ethene moiety of diarylethenes with various π-conjugation systems has expanded the versatility as photo-switching molecular materials. In this chapter, we review the chemistry of photochromic terarylenes including synthetic approaches, photochromic properties, and various photo-switching properties.

Keywords Heteroaromatics • Photo switches • Supramolecular chemistry

10.1 Introduction

The basic structure of photochromic terarylenes is composed of three heteroaromatic rings to form a 1,3,5-hexatriene structure, which is capable of 6π-electrocyclic photochromism in a similar manner to those of fulgides [1, 2] and diarylethenes [3, 4].

T. Nakashima (✉) • T. Kawai
Graduate School of Materials Science, Nara Institute of Science and Technology, 8916-5 Takayama, Ikoma, Nara 630-0192, Japan
e-mail: ntaku@ms.naist.jp; tkawai@ms.naist.jp

M. Irie et al. (eds.), *New Frontiers in Photochromism*,
DOI 10.1007/978-4-431-54291-9_10, © Springer Japan 2013

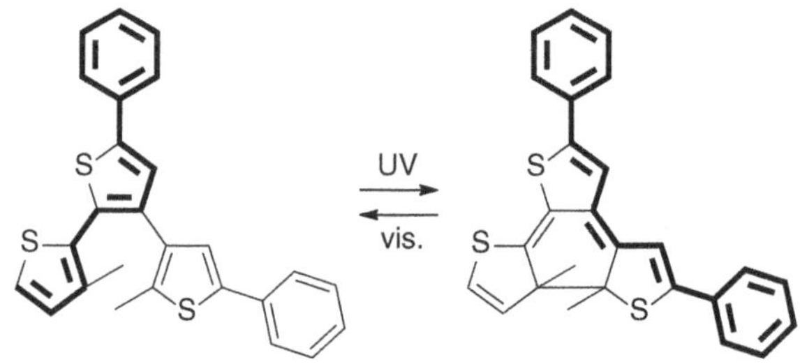

Fig. 10.1 Photochromic reaction of a terthiophene

The chemistry of photochromic terarylenes could be seen as an extension of diarylethenes modified at the central ethene moiety with aryl units. Photo-switching effects in diarylethenes have been extensively studied for controlling various chemical and physical properties such as fluorescence intensity and wavelength [5–10], refractive index [11–14], electronic conductivity [15–19], electrochemical response [20–24], magnetic interactions [25, 26], and molecular interactions [27–30]. Most of these photo-switching effects are based, at least partly, on changes in the extent of π-conjugation in the diarylethenes in the course of photochromic reactions. That is, the π-conjugation system of diarylethene extends over both sides of the molecule in the closed-ring form, while it is restricted on each side for its open-ring isomer. In this context, thiophene-bridged bis(thienyl)ethene derivatives were developed as molecular re-routers of the π-conjugation pathways [31]. As illustrated in Fig. 10.1, the π-conjugation system in the open-ring form extends from the top to the left-hand side of molecule, while it changes between the top and the right-hand side of molecule in the closed-ring isomer. Thus triangle terthiophenes have offered the switching capability of π-conjugation pathways in multiple ways.

Some diarylethene derivatives incorporating expanded π-conjugation systems such as naphthoquinone [32], phenanthroline [33], and tetraazaporphyrin [34] on the central ethene group have been proposed. The direct introduction of functional π-conjugation groups as the bridging ethene unit participating in the 6π-electrocyclization reaction would endow the significant switching effect as well as the unique photochromic reactivity, which has touched off the development of "terarylenes" as a novel series of photochromic molecules. In this chapter, we first give an overview of synthetic routes to terarylenes and then introduce their photochromic properties including the molecular design for the enhanced photochromic reactivity, the controlled thermal stability of closed-ring isomers, and various switching properties.

10.2 Synthesis

Photochromic terarylenes are prepared through two synthetic approaches. One route goes through the ring-forming reactions of azoles from diaryl-α-dions or diaryl-α-hydroxylketones to 4,5-bisarylazoles. The other involves metal-catalyzed aryl–aryl cross-coupling reactions between central and side-aryl units.

Krayushkin et al. reported the synthesis of photochromic bis(thienyl)azoles from α-hydroxyketone derivative **1** (Fig. 10.2) [35, 36]. α-Haloketone **2** and α-diketone **3**

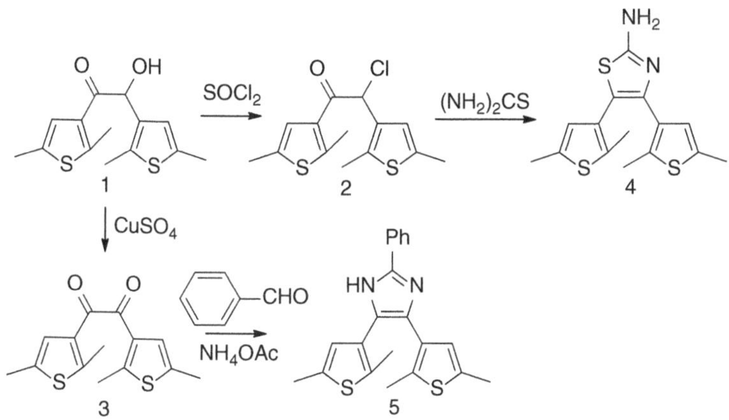

Fig. 10.2 Synthetic procedures of dithienylazoles

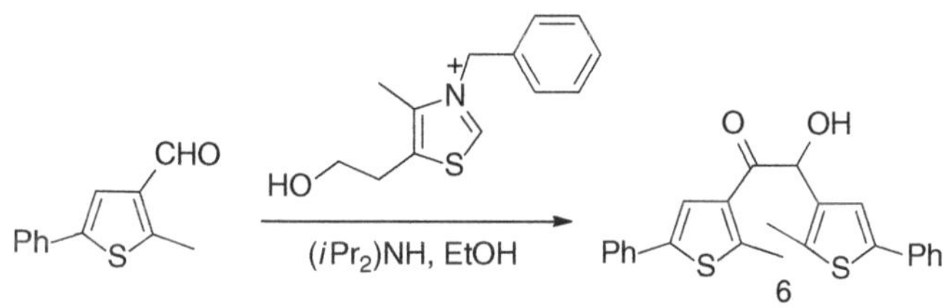

Fig. 10.3 Benzoin condensation of thienylaldehyde

were synthesized from α-hydroxyketone **1** and were converted to bis(thienyl)aminothiazole **4** and bis(thienyl)phenylimidazole **5**, respectively, by the azole-forming condensation reactions. The key substrate α-hydroxyketone **1** was synthesized starting from 2,5-dimethylthiophene through three reaction steps [35]. Alternatively, Branda et al. [37] proposed an efficient way to α-hydroxyketone **6** through a benzoin condensation reaction of thienylaldehyde (Fig. 10.3) [38].

In contrast to the route through the azole ring-formation reaction, which is limited to terarylenes with some specific azoles as a central bridging unit, a variety of terarylene molecules have been prepared from various combinations of heteroaromatics by the metal-catalyzed aryl–aryl cross-coupling reactions. Figure 10.4 shows the typical synthetic route of a terarylene with symmetrical side-aryl units. 3-Borylated thiophene as a side-aryl unit and 4,5-dibromo-2-phenylthiazole as a central unit were prepared separately, which were then combined by the Pd-catalyzed Suzuki–Miyaura cross-coupling reaction to form a photochromic bis(thienyl)thiazole **7** [39]. Tian et al. reported a one-pot synthesis of terthiophene **8** from commercially available reagents. The Grignard reagent prepared from 2-bromo-3-methylthiophene was combined with 2,3-dibromothiophene by the Ni-catalyzed cross-coupling reaction to form **8** in good yield over 90 % (Fig. 10.5) [40].

Terarylenes with asymmetric side-aryl units were synthesized by sequential cross-coupling reactions utilizing the different reactivity of α- and β-carbons of

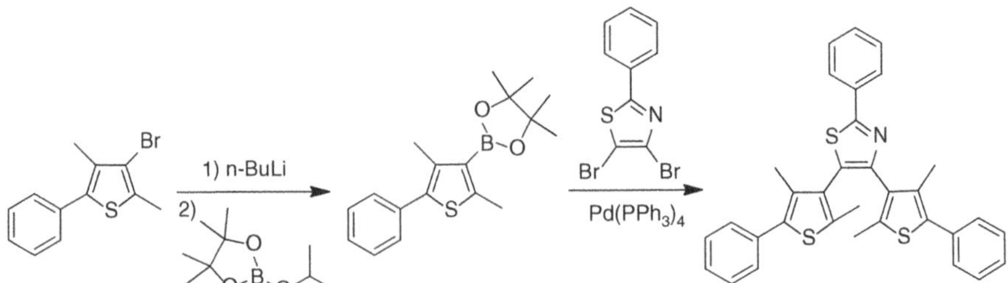

Fig. 10.4 Synthetic scheme of a terarylene with symmetric side-aryl units

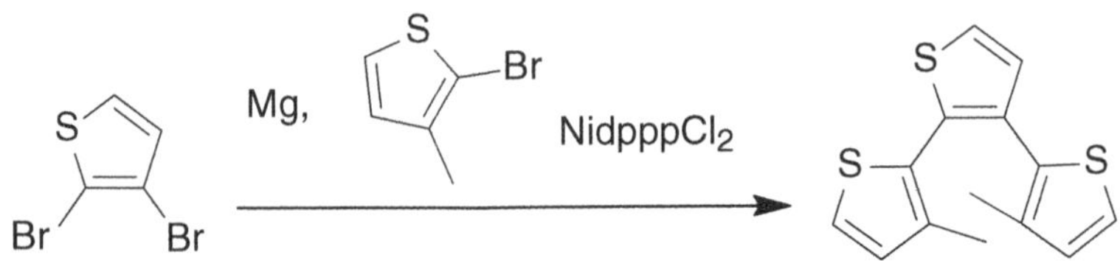

Fig. 10.5 Single-step synthesis of a terthiophene

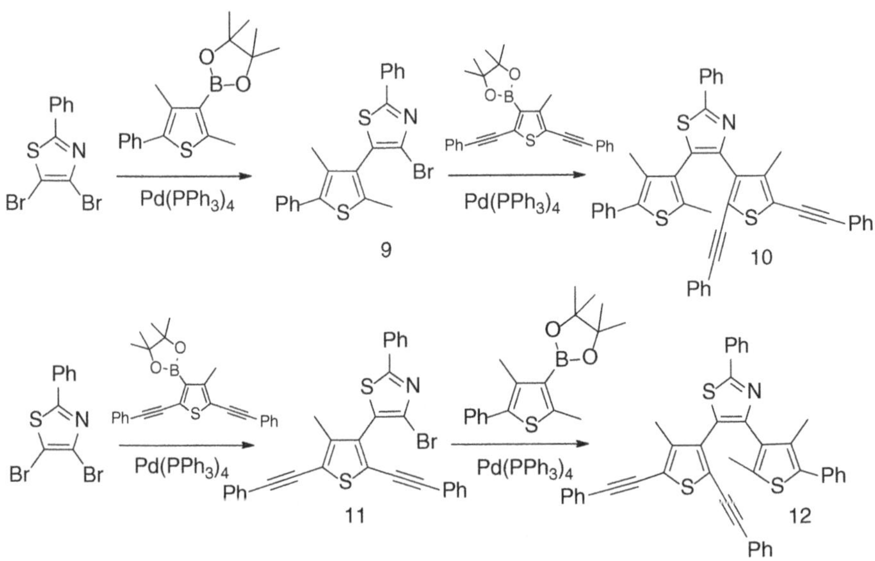

Fig. 10.6 Synthetic schemes of terarylenes with asymmetric side-aryl units

heteroaryls [41]. For instance, the nucleophilic 5-position (S-side) carbon is more reactive than 4-position (N-side) for the Suzuki–Miyaura coupling reaction of 4,5-dibromothiazole [42]. Bis(thienyl)thiazoles **10** and **12** with asymmetric side-thienyl units were synthesized according to Fig. 10.6. Since the π-conjugation pathways of closed-ring isomer of bis(thienyl)thiazoles is extended from the top to the right-hand side (N-side) of molecule, the closed-ring isomer of **10** gave an absorption maximum at longer wavelength than that of the closed-ring isomer of **12**.

10.3 Photochromic Properties

Terarylenes exhibit reversible 6π-electrocyclization reactions, photo-induced cyclization, and photo- and/or thermal cycloreversion in a similar manner to those of diarylethenes. Figure 10.7 showed a typical absorption spectral change of a terarylene **7** upon photochromic reaction. Open-ring isomer has an absorption maximum in the UV region, which shifts to longer wavelength above 600 nm by UV irradiation. The absorption spectrum of closed-ring isomer reverts to that of open-ring form by the irradiation with visible light above 400 nm. The absorption spectral change accompanies an isosbestic point at 340 nm.

Figure 10.8 summarizes the absorption maxima of each isomer of bis(aryl)thiazoles together with those of diarylethene analogs [39, 43–46]. The introduction of 2-phenylthiazole instead of hexafluorocyclopentene gives rise to the redshift in absorption peaks in both open- and closed-ring isomers because of the expansion of π-systems.

The effect of the introduction and expansion of the π-conjugation system at the central aryl unit of terarylenes on their photochemical properties has been systematically investigated by a series of bis(benzothienyl)thiazoles (Fig. 10.9, Table 10.1) [47]. The introduction and expansion of the π-conjugated system at the 2-position of the thiazole ring led to the redshift of absorption peak for both open- and closed-ring isomers. Although **18** and **17** exhibited relatively high photo-cyclization reaction quantum yields ($\Phi_{o\text{-}c}$) over 0.5, **19** with the most extended π-system showed a considerably suppressed photo-reactivity, even though it is supposed to have a similar conformation to those of the other two derivatives. Meanwhile, these molecules showed similar photo-cycloreversion quantum yields as high as 0.3. Quantum chemical calculations explained the decrease of the $\Phi_{o\text{-}c}$ and unchanged photo-cycloreversion reactivity of **19** [47].

In the following, the thermal stability of closed-ring isomers and the structure-photochromic reactivity relationship of terarylenes are introduced.

10.3.1 Thermal Stability

Although the closed-ring isomer of *cis*-stilbene quickly returns to its open form at room temperature, the closed-ring isomers of diarylethenes, which are the analogs of *cis*-stilbene with thiophene or thiazole rings instead of benzene, are highly stable in the dark. Less aromatic stabilization energies of thiophene and thiazole were reported to be responsible for the stability of closed-ring isomers of diarylethenes by reducing the degree of destabilization of closed-ring isomers to the open ring

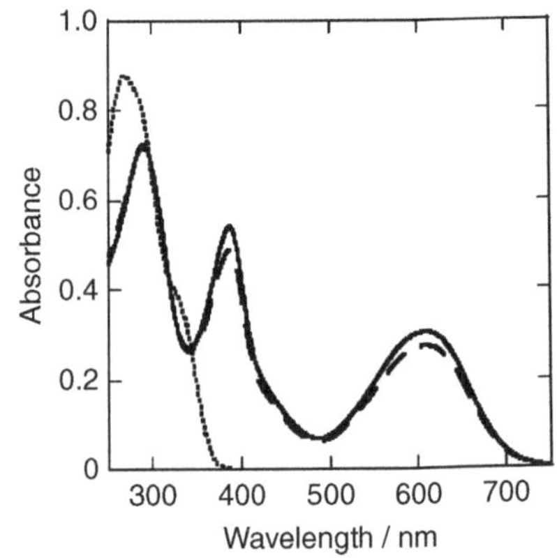

Fig. 10.7 Absorption spectra of open-ring isomer (*dotted line*), closed-ring isomer (*solid line*), and photostationary state under irradiation with 313 nm light (*dashed line*) of **7** in hexane

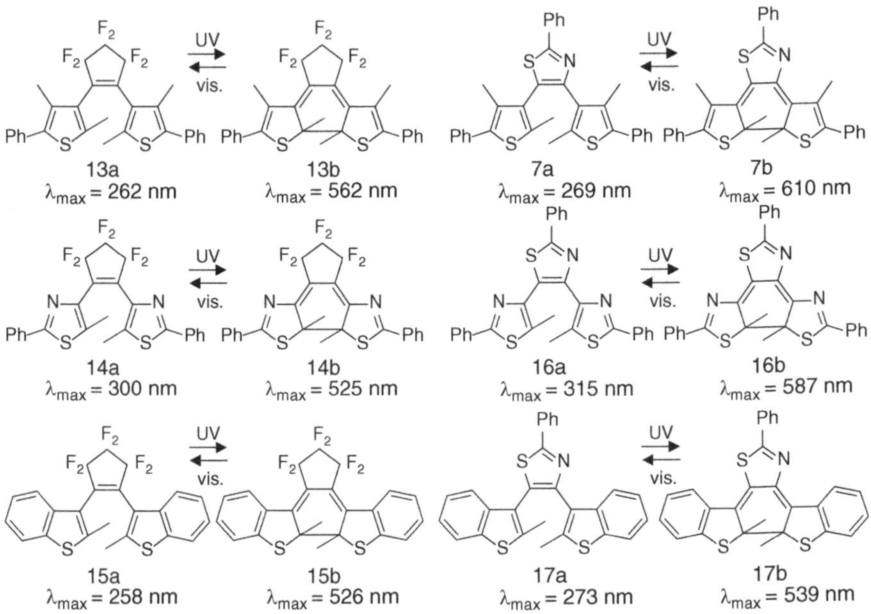

Fig. 10.8 Photochromic reactions and absorption maxima of terarylenes

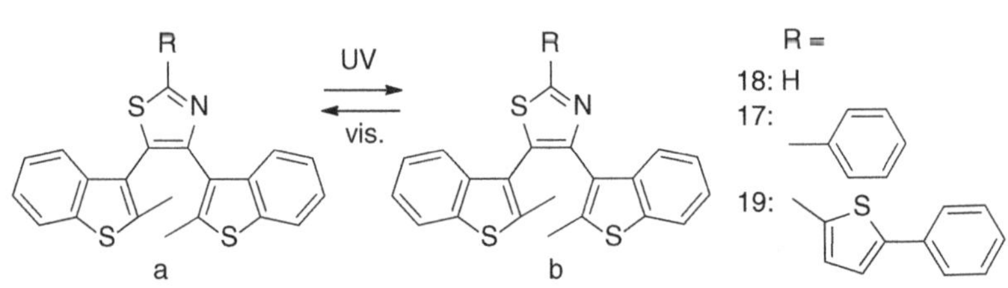

Fig. 10.9 Photochromic reaction of terarylenes with various substituents at 2-position of the central thiazole

Table 10.1 Absorption maxima and molecular extinction coefficients of the open- and closed-ring isomers of **17**, **18**, and **19**, together with their quantum yields in hexane

	λ_{max} (nm)	$\Phi_{o\text{-}c}$	$\Phi_{c\text{-}o}$
18a	295	0.56	
18b	534		0.30
17a	329	0.58	
17b	539,556		0.45
19a	370	0.078	
19b	550, 590		0.29

Table 10.2 Arrhenius parameters and thermal half-lifetime at 20 °C of cycloreversion reaction of closed-ring isomers

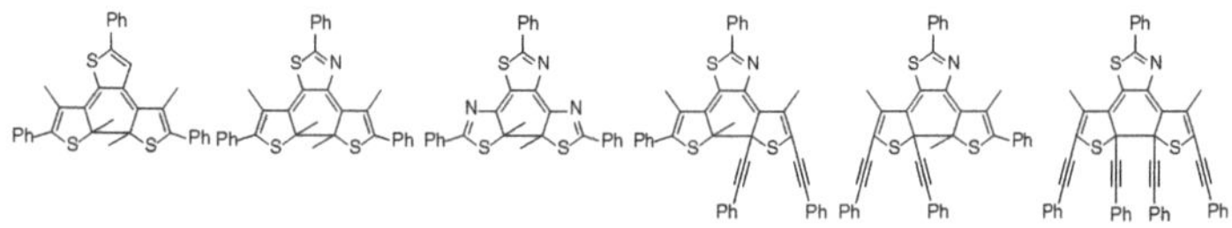

	20b	7b	16b	10b	12b	21b
E_a (kJ mol^{-1})[a]	84	92	112	83	91	83
A[a] (s^{-1})	2.0×10^{10}	1.6×10^{10}	7.1×10^{11}	1.4×10^{11}	1.8×10^{12}	9.3×10^{13}
k at 20 °C (s^{-1})	2.3×10^{-5}	6.3×10^{-7}	7.7×10^{-9}	2.2×10^{-4}	1.0×10^{-4}	1.4×10^{-1}
$t_{1/2}$ at 20 °C	11 h	14 days	3.3 years	51 min	94 min	5.4 s

[a]In toluene

form [48]. In terms of aromatic stabilization energy, the closed-ring form of terarylenes seems less stable than that of diarylethene analog because three aromatic units collapse simultaneously with the photo-cyclization. The thermal cycloreversion reaction rate constants of a series of terarylenes were studied at various temperatures, and the first-order kinetics (k) exhibited Arrhenius-type temperature dependence (Table 10.2) [39, 42].

The half-lifetime ($t_{1/2}$) of closed-ring isomer of terarylene composed of three thienyl rings **20b** was estimated to be 11 h at 20 °C [39]. While increasing the number of substitution of thienyl group with thiazole, which has a smaller aromatic stabilization energy than thiophene [49, 50], the $t_{1/2}$ values improve to 14 days for **7b** and to 3.3 years for **16b**. The contribution of lower aromaticity of thiazole units compared to thiophene was indicated by the increase of activation energy (E_a) from 84 (**20b**) to 92 and 112 kJmol^{-1} for **7b** and **16b**, respectively. Although the $t_{1/2}$ of 3.3 years seems still short in comparison with diarylethenes [3], the introduction of benzothiophene, fused heteroaromatic, or *S*,*S*-dioxide thiophene groups [51] would further improve the thermal stability of colored isomers sustainable for practical applications.

Meanwhile, the terarylene structure might have an advantage in the development of photochromic molecules with fast thermal bleaching rate. In addition to the small activation energy for the thermal cycloreversion reaction of terarylenes, the increase

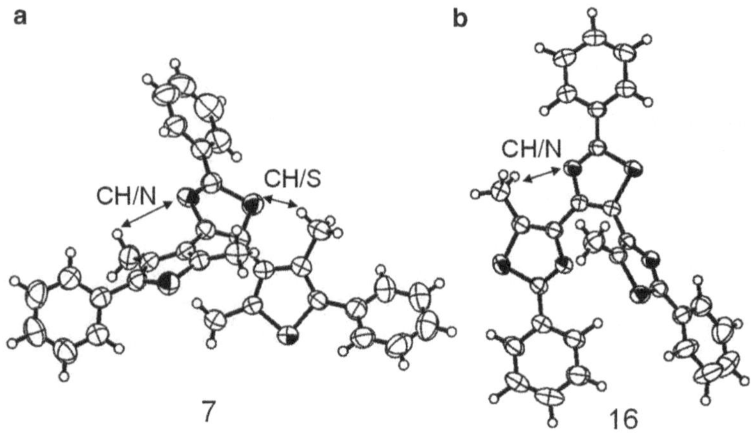

Fig. 10.10 ORTEP drawings of open-ring isomers of **7** (**a**) and **16** (**b**), showing 50 % probability displacement ellipsoids

of the frequency factor by the introduction of rigid phenylethynyl groups on the side-thienyl groups resulted in the significant acceleration of thermal bleaching rate constant from 6.3×10^{-7} for **7b** to 2.2×10^{-4}, 1.0×10^{-4}, and 0.14 s^{-1} for **10b**, **12b,** and **21b**, respectively. Thus the control of thermal cycloreversion rate constant over 7 orders of magnitude was achieved by the use of terarylene structures.

10.3.2 Structure-Photochromic Reactivity Relationship

Since the photo-electrocyclic reactions of fulgide and diarylethenes typically take place within a few ps [52] in a conrotatory manner, the photo-cyclization quantum yield is generally determined by the electronic structure in the excited state, which is directly interpreted as the molecular conformation in the ground state [3]. In contrast to diarylethenes, terarylenes offer a wide accessibility to the control of molecular folding in terms of the design of intra- and intermolecular interactions through the combination of various heteroaromatic rings.

The crystal structures of **7** and **16** exhibited the photochromic-active and photochromic-inactive molecular conformations, respectively (Fig. 10.10). The distance between the photo-reacting carbon atoms are evaluated to be 0.36 nm for **7**, which seems short enough for the electrocyclization reaction in crystal [53], whereas that of **16** was 0.49 nm, indicating nonreactive conformation. The photochromic-active conformation of **7** was seemingly supported *via* the multiple weak CH/N and CH/S hydrogen bonding interactions [54] between the methyl groups on the 4-positions of side-thienyl rings and heteroatoms (N and S) in the central thiazole. The atomic distance of CH/N and CH/S could be possible below 0.27 and 0.30 nm, respectively, which are shorter than the sums of *van der Waals* radii, 0.275 nm for H/N and 0.30 nm for H/S. On the other hand, the photochromic-inactive

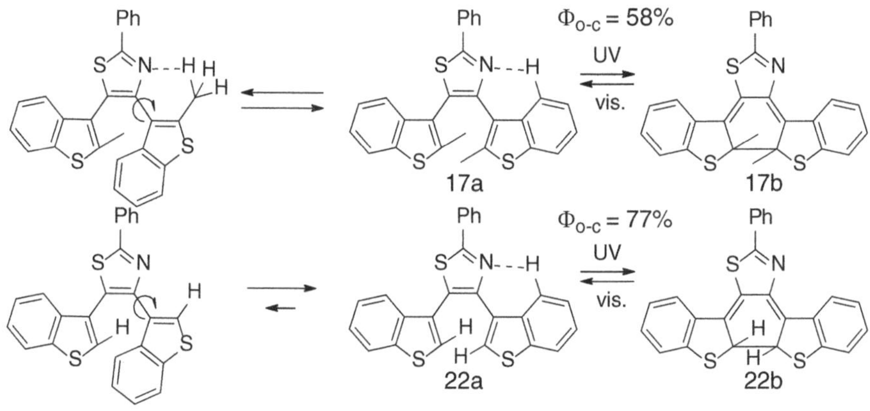

Fig. 10.11 Photochromic reactions of dibenzothienylthiazoles

conformation of **16** would be also supported by CH/N interaction between the methyl group on the reactive carbon atom of a side-aryl unit and the nitrogen of the central thiazole. The N/N repulsion between the central and side-thiazolyl rings might flip the side-thiazolyl ring, and the CH/N interaction stabilizes the flipped conformation. These intramolecular non-covalent interactions of the sort mentioned above might be responsible for the $\Phi_{o\text{-}c}$ of **7** and **16**, where **7** gave higher value of 0.6 than **16** ($\Phi_{o\text{-}c}$=0.4) in solution [39].

As another example, 2-phenyl-4,5-bis(benzothienyl)thiazole **22** (Fig. 10.11) shows $\Phi_{o\text{-}c}$ of 0.77 which is markedly larger than that of 2-phenyl-4,5-bis(2-methylbenzothienyl)thiazole **17** ($\Phi_{o\text{-}c}$=0.58). Specific intramolecular CH/N interaction is possible to stabilize the reactive conformation in these compounds. However, the nonreactive conformation of **17** is also possible to be stabilized with another CH/N interaction because of methyl protons. Therefore, the relative stability and thus population of the reactive conformation is efficiently enhanced in **22** [46].

Bis(thiazolyl)arylenes **23** [55] and **24** [56] (Fig. 10.12) were designed for further efficient photochromic reactivity in solution, whose molecular geometries were expected to be fixed in the photo-reactive conformation by multiple intramolecular interactions including CH/heteroatom and inter-heteroatom interactions. As shown in Fig. 10.12a, the CH/N and S/N [57] interactions between the central benzothiophene and both side-thiazolyl rings were indicated by the atomic contact observed in the crystal structure [55]. In combination with these interactions, CH/π interactions between the methyl groups on the reactive carbon atoms and thiazole ring at the opposite sides were suggested by the short distance between the methyl groups and the molecular plane of the thiazolyl rings (0.31 and 0.32 nm). Such multiple weak intramolecular interactions were also observed for **24** to stabilize the photoreactive conformation [56].

The conformations of **23** and **24** in solution were evaluated by temperature-dependent ^{1}H NMR spectra [55, 56]. Methyl protons on the reacting carbon atoms for both **23** and **24** gave signals at substantially high field below 2.0 ppm even

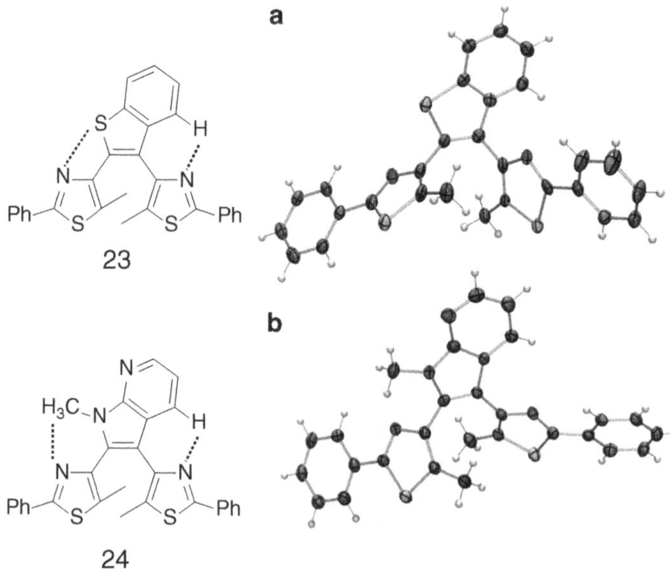

Fig. 10.12 Chemical structures and ORTEP drawings of open-ring isomers of **23** (**a**) and **24** (**b**), showing 50 % probability displacement ellipsoids

though they are attached to sp^2 carbon atoms. Such a high field chemical shift of methyl groups would originate from the ring-current effect of the neighboring thiazole ring, supporting the CH/π interactions. The downfield shift of methyl signals upon the heating suggested the weakening of CH/π interactions by increasing temperature. On the other hand, the signals of protons, which are expected to take part in CH/N interactions, showed upfield shift upon heating. The downfield shift at low temperature could be attributed to the anti-shielding effect by nitrogen atom and supported the CH/N interactions in **23** and **24**. Furthermore, the systematic DFT calculations for **23** by the rotation about the single bond between the central benzothiophene and the side-thiazolyl ring participating in S/N interaction gave the rotational energy barrier about 20 kJ/mol from photochromic-reactive conformation to an unreactive one [58]. Due to the considerable contributions of multiple intramolecular interactions, bis(thiazolyl)arylenes **23** and **24** prefer the photo-reactive conformation in solution and gave markedly high photo-cyclization quantum yields, 0.98 and 0.90, respectively.

The photochromic system integrated with molecular folding controlled by host-guest interaction has been demonstrated using terarylene **25** (Fig. 10.13) [59]. **25** has a thienopyridyl ring as a side-aryl unit. In less polar media, the N/N repulsion between the central thiazole and the pyridyl ring flips the thienopyridyl ring to form a photochromic-inactive conformation, which would be stabilized by the intramolecular CH/N interaction as shown in Fig. 10.13. Meanwhile, in a protic media such as methanol, the methanol molecule is expected to interact with both the aromatic nitrogen atoms through conventional OH/N hydrogen bonding. Binding of a

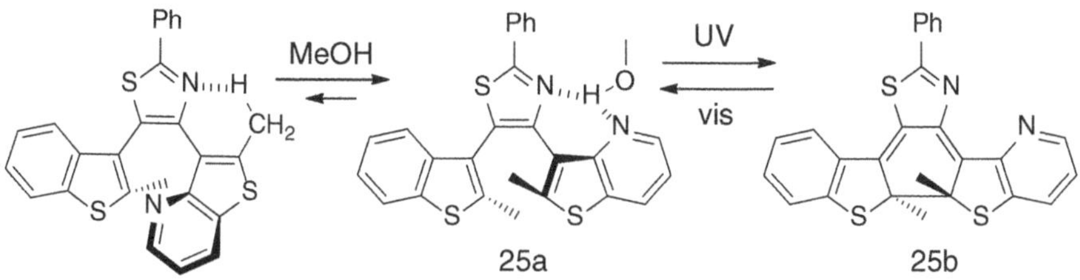

Fig. 10.13 Photochromic reaction of a terarylene with a thienopyridyl unit

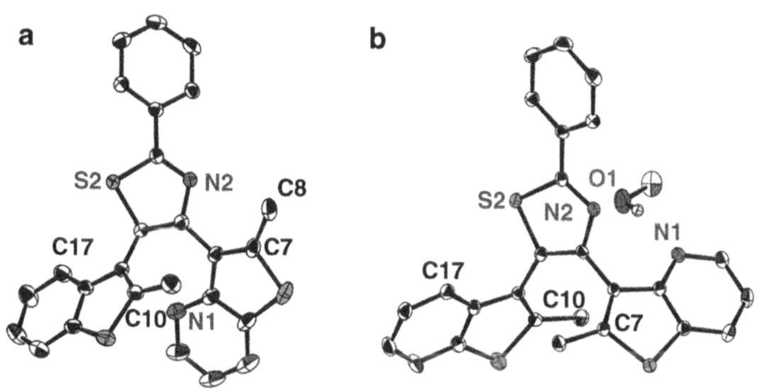

Fig. 10.14 ORTEP drawings (showing 50 % probability displacement ellipsoids) of the crystals of **25** recrystallized from (**a**) hexane and (**b**) methanol. The hydrogen atoms are omitted for clarity

methanol molecule would effectively suppress the rotation about the thiazolyl–thienopyridyl bond, therefore directing the adoption of a photochromic-active conformation. Temperature-dependent ^{1}H NMR study revealed that the photo-reactive conformation was dominant in methanol whereas the unreactive one was favored in hexane. The solvent-dependent folding of molecular conformation resulted in the photochromic cyclization quantum yields of 0.60 in methanol, which is twice of that in hexane (0.30).

Complete regulation of molecular conformation between photo-reactive and unreactive ones was achieved in the crystalline state. The geometry of **25** in the crystal from hexane is apparently unsuitable for the photo-cyclization reaction (Fig. 10.14a). The weak hydrogen bonding between (C_8)H and N_2 (about 0.24 nm) stabilizes the nonreactive form (Fig. 10.13). The atomic distance between the reacting carbons (C_{10}–C_7) of 4.24 nm was too long to exhibit the single-crystalline photochromism [59]. Interestingly, the crystal formed in the methanol solution binds a methanol molecule at the N1–N2 binding site to form an inclusion complex and adopts the photoreactive conformation (Fig. 10.14b). The distance between the reacting carbon atoms was 3.72 nm, which endows the photochromic capability in the crystalline state.

The control of photochromic reactivity of a terarylene in terms of molecular conformation was also attained by the combination with chemical reactions,

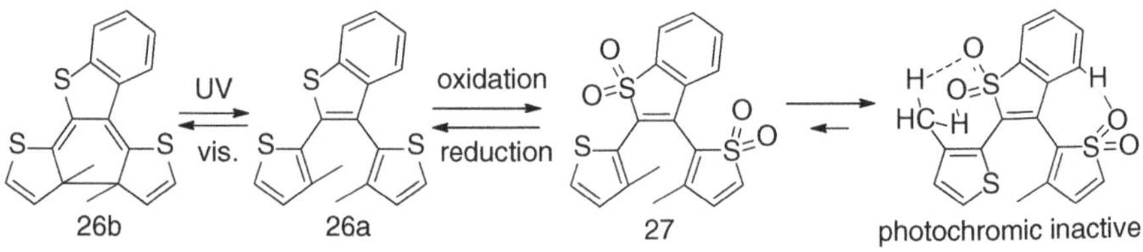

Fig. 10.15 Control of photochromic reactivity of a terthiophene with redox reactions

demonstrating a "lock and key control" system [60]. The selective oxidation of sulfur atoms in bis(thienyl)benzothiophene **26** gave a partly oxidized **27** with *S*,*S*-dioxide moieties by the oxidation using *m*-chloroperbenzoic acid (*m*CPBA). While **26** showed reversible photochromic reactions, the photochromic reactivity of **27** was strongly suppressed. X-ray crystallographic study indicated the CH/O intramolecular hydrogen bonding interactions between *S*,*S*-dioxide moieties and methyl and aryl protons as shown in Fig. 10.15, which was responsible for the suppressed photochromic reactivity of **27** to fix the unreactive conformation.

10.4 Switching Property

10.4.1 Photochromic Metal Complexes and Organometallics for Luminescence Switching

The use of extended π-system as an ethene bridging unit for diarylethenes has provided photo-responsive ligands for photochromic metal complexes and organometallics [61, 62], where the 6π-electrocyclization reaction brings about the direct modulation of π-conjugated ligand structures. Those photochromic metal complexes often provide luminescence-switching capability based on the photomodulation of metal-perturbed π–π* transition of ligand or metal to ligand charge transfer (MLCT) characters. Figure 10.16 shows typical photochromic metal complexes with photo-responsive bis(thienyl)aryl ligands [33, 34, 63–65]. Tian et al. reported preparation of tetraazaporphyrin derivative **28** having four bis(thienyl)ethene units. **28** undergoes a ring-cyclization reaction at two photochromic sites with the both UV and visible irradiation, and the near-IR emission was quenched after the cyclization reaction [34]. Yam et al. extensively studied the luminescence modulation of photochromic metal complexes as exemplified by **29**–**32** introducing coordinating units as central bridging moieties [33, 63–67].

Photochromic lanthanide complexes afford the unique mechanism for photoswitching of luminescence property, which is derived from the emission property of lanthanide ions on the basis of characteristic 4f–4f transition processes [68, 69]. The electric dipole (ED) transition on the 4f orbitals is intrinsically La Porté forbidden and is markedly sensitive to the asymmetric structure of the coordination field.

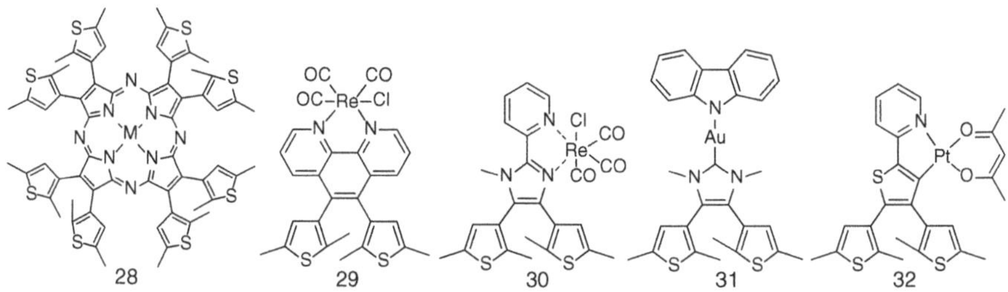

Fig. 10.16 Photochromic molecules with metal ions

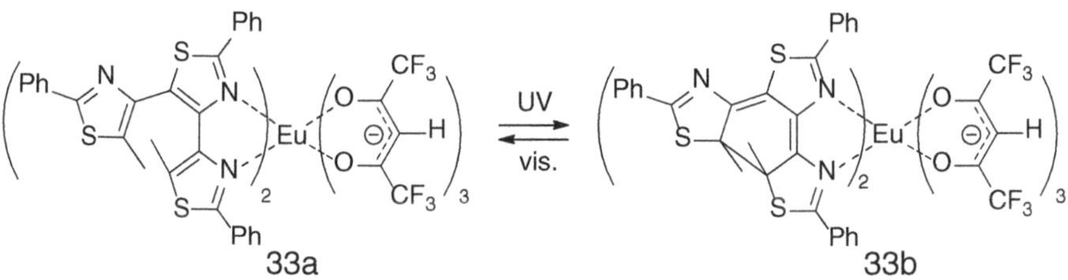

Fig. 10.17 Photochromic reaction of a terthiazole–Eu(III) complex

Thus, the probability of ED transition is expected to change upon the photochromic reaction of the ligand, which causes the modulation in the symmetry of the ligand field, or "odd parity" [61].

The terthiazole **16** was employed as a photochromic ligand to form Eu(III) complex **33** using a bidentate N–N coordination site formed between neighboring thiazolyl rings (Fig. 10.17). The reversible photochromic reaction of terthiazole ligand changed the emission profile, the relative intensity of ED transition at 614 nm to that of magnetic dipole (MD) transition at 592 nm. The relative intensity of ED transition for **33a** was larger than that of **33b**. The anisotropic polarizability of open-ring form of terthiazole **16a** was calculated to be larger than that of closed-ring isomer **16b**, which might lead to the larger ED transition probability of **33a** [69].

10.4.2 Fluorescence Turn-On Switching

In many cases, terarylenes showed fluorescence turnoff behavior, where the fluorescent open-ring form becomes nonfluorescent by the photo-cyclization reaction as also seen in most fluorescence switching diarylethenes [70]. There are a few fluorescence switching terarylenes which show the turn-on character (Fig. 10.18). While the open-form **34a** was nonfluorescent, the photo-isomer **34b** showed red fluorescence with the fluorescence quantum yield (Φ_f) of 11 % [71]. The dynamic fluorescence signal switch of compound **34** was demonstrated in live cells.

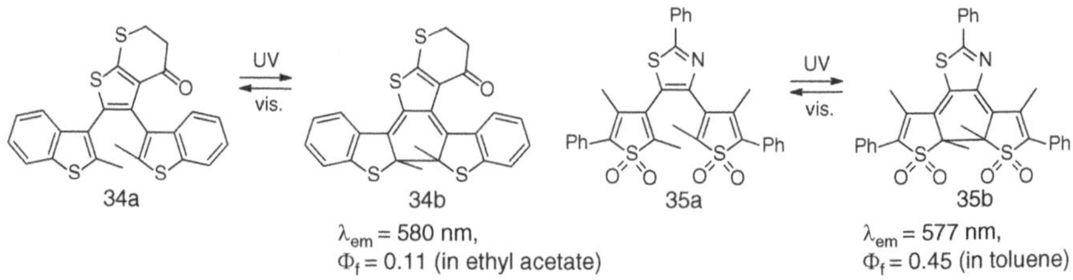

Fig. 10.18 Photochromic terarylenes with fluorescence turn-on property

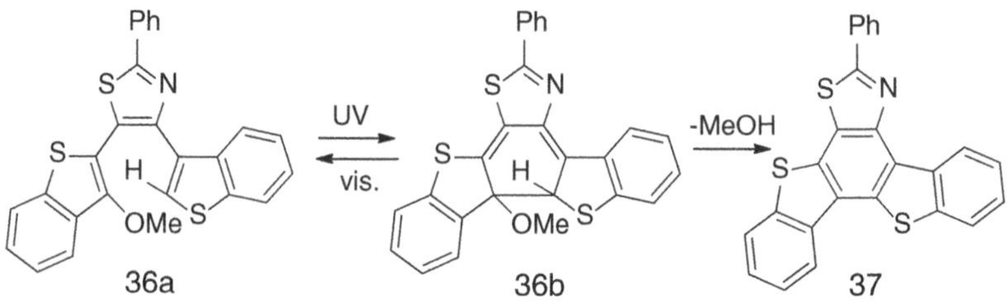

Fig. 10.19 Photochromic terarylene which shows a subsequent elimination reaction

Bis(thienyl)ethenes with *S,S*-dioxide moieties were reported to be fluorescent in the closed-ring forms [72, 73]. Terarylenes with *S,S*-dioxide moieties also showed the turn-on behavior [74]. The fluorescence quantum yield of closed-ring isomer **35b** was 45 %, which is substantially larger than that of the diarylethene analog with a perfluoro-cyclopentene bridging unit ($\Phi_f = 3$ %) [74]. The oxidation of thienyl units markedly decreases the photo-cycloreversion quantum yield as well [74]. The continuous irradiation with visible light to **35b** for hours caused no marked changes in visible absorbance and emission intensity of **35b** due to the extremely low cycloreversion quantum yield ($\Phi_{o\text{-}c} < 1.0 \times 10^{-4}$), offering a practical irreversible photo-patterning of fluorescence image.

The fluorescence turn-on system was also achieved in combination with the subsequent reaction following the photo-cyclization reaction to form a fused polycyclic compound. Terarylene **36a** has a methoxy group and a hydrogen atom at each photo-reacting carbon to release a methanol molecule from the closed-ring isomer under polar or acidic conditions (Fig. 10.19) [75]. The fused polycyclic aromatic **37** exhibited blue fluorescent with Φ_f of 10 % and never returns to **36b**, which is useful for write-once optical memory with nondestructive fluorescence readout.

10.4.3 Integration of Photochromic Reactivity with Chemical Reactivity

The physicochemical properties such as Lewis acidity and electron-donating properties of the central aromatic moieties are modulated by the photochromic reaction

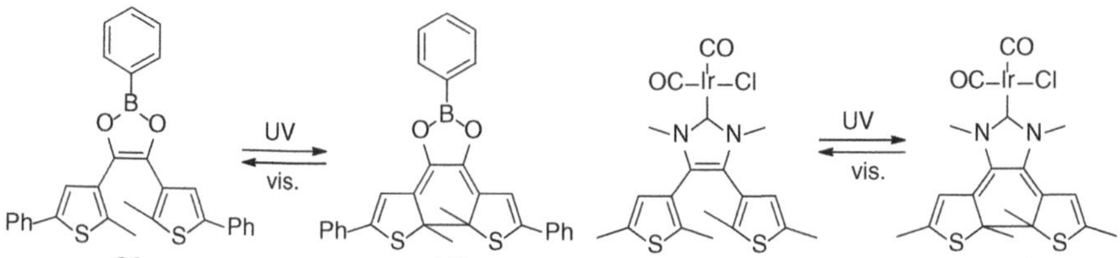

Fig. 10.20 Photochromic molecules which switch the electron-donating property

due to the reconstruction of π-systems upon the 6π-electrocyclization reactions (Fig. 10.20). Compound **38a** has a 1,3,2-dioxaborole as a central bridging unit, which has a significant aromatic character because of overlapping p-orbitals containing 4n+2π-electrons [37]. The delocalized π-electrons partially occupy the p-orbital of boron atom to reduce the Lewis acidity. UV irradiation induces the cyclization reaction to generate **38b**. The reconstruction of π-conjugation reduces the electron density of boron center and enhances the Lewis acidity. The similar modulation of electron density on the central bridging aryl unit was demonstrated using **39** [76]. Compound **39a** is composed of the same photochromic carbene ligand with the complex **31** [65]. The decrease of electron-donating property of photochromic carbene ligand was confirmed by the change in the vibration energy of carbonyl ligand coordinating to iridium.

The switching electron-donating property of *N*-heterocyclic carbene was further applied to the photo-switchable carbene catalyst [77]. Under ambient light and in the presence of base, the open-form **41a** catalyzed transesterifications as well as amidations in a manner similar to those of *N*-heterocyclic carbene precatalysts. However, upon UV irradiation to form closed-isomer **41b**, the rate of the transesterification reaction was significantly attenuated to less than a tenth (Fig. 10.21). Thus the photo-control of chemical properties such as acidity and electron-donating properties is promising for the photo-switchable catalyst whose reactivity or selectivity is controlled with light in a remote manner [78].

Some photochromic terarylenes have been involved in the gated reaction systems where the external stimuli such as chemical reaction control the photochromic reactivity or structural change by the photochromic reaction imparts chemical reactivity. Branda has extensively studied molecular systems interlocking the photochromic reactivity with chemical reaction at the central ethene moiety (Fig. 10.22). A dialkoxybenzoin structure as a photocage was introduced as the central unit bridging two benzofurans (**42**) [79]. The photo-release of acetic acid from the bridging unit generates a dimethoxybenzofuran, making up a hexatrien backbone with two side-benzofuran rings. The tris(benzofuran) derivative **43a** undergoes photo-cyclization reaction to form closed-ring isomer **43b** by irradiation with the same wavelength for the photo-release of acetic acid. Thus the photochromic reactivity was controlled with other photochemical reaction and the color change due to the formation of **43b** reports the release of acetic acid. The reversible Diels–Alder reactions of a bis(thienyl)fulvene **44** with a dienophile control the photochromic reaction between

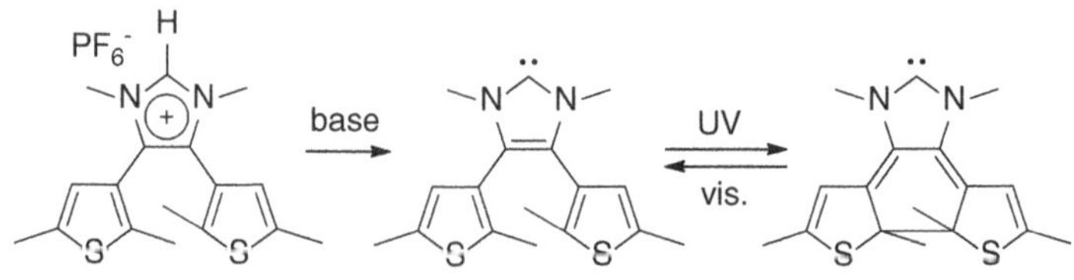

Fig. 10.21 Photoresponsive catalyst

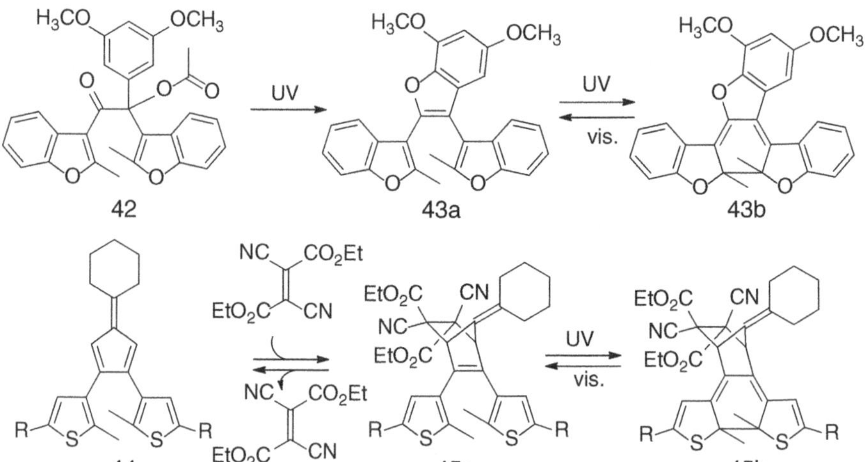

Fig. 10.22 Photochromic molecules with gated reactivities

45a and **45b** [80]. Meanwhile, the photochromic reaction controls the *retro*-Diels–Alder reaction as well. The thermally stable "armed" state **45b** is unlocked with visible light to be converted to **45a**, which thermally releases a dienophile.

Bis(thiazolyl)imidazolium **46a** (Fig. 10.23) was demonstrated to change the charge localization state of the imidazolium bridging moiety upon photochromic reactions [81]. The delocalized cationic charge in the central imidazolium ring of **46a** becomes localized at the 2-position of imidazolium ring (**46b**) by photocyclization reaction. Since the ionic interaction is dependent on the ionic radii of components and the dielectric constant of solvents, the closed-ring form **46b** showed unique solvatochromism. That is, **46b** gave two patterns of absorption peaks depending on the solvents instead of the continuous peak shift responding to the dielectric constants as is often found in conventional solvatochromism. The absorption peak of **46b** appeared at 650 nm in polar solvents such as methanol, acetonitrile, acetone, and dichloromethane, while the peak redshifted to 560 nm in less polar media with the dielectric constant below 7. In polar media, each imidazolinium cation **46b** and

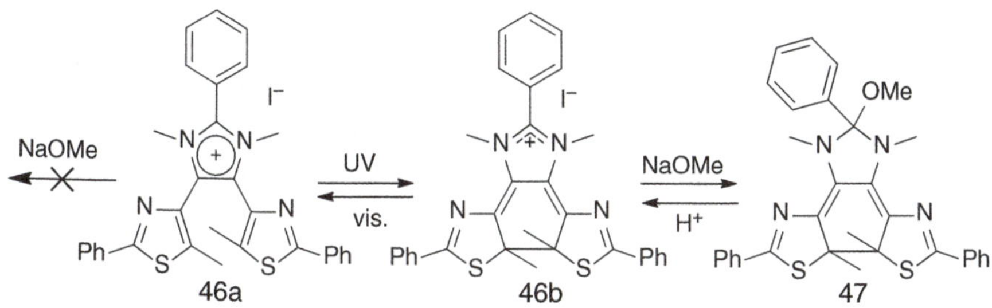

Fig. 10.23 Photo-gated reaction of a dithiazolylimidazolium salt

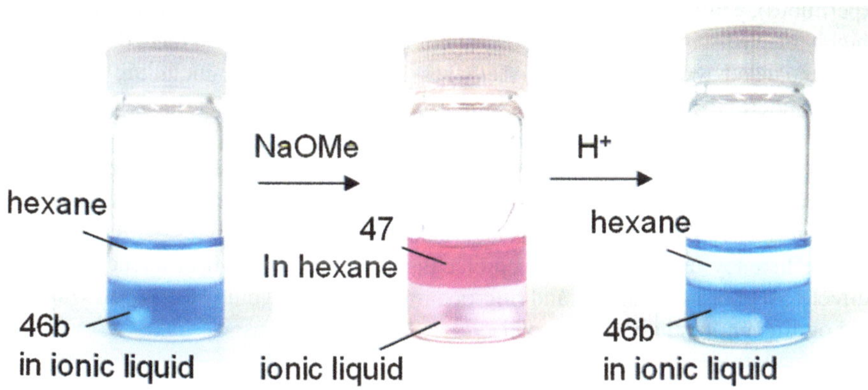

Fig. 10.24 Reversible phase transfer of **46b** by the successive addition of MeONa and TFA in the hexane/ionic liquid biphasic system

iodide anion would be well solvated with solvent molecules to result in weak electrostatic interaction. However, the insufficient solvation of ionic species in low polar media may give rise to the formation of a certain kind of contact ion pair in which the iodide anion strongly binds to the imidazolinium cation at ground state to shorten the effective π-conjugation length. Meanwhile, such solvatochromism was never found in the open-ring isomer **46a** because of the inherently weak ionic interaction between the imidazolium ring with a well delocalized cationic charge and iodine anion. In other words, as also seen in compounds **38**, **39,** and **41**, the ring-cyclization reaction reduces the electron density of the central bridging imidazolium moiety.

The photo-modulation of cationic delocalization state was further demonstrated by the "photo-gated" reactivity of **46** (Fig. 10.23). The closed-ring isomer **46b** gave an absorption band at 650 nm in methanol, which shifted to 550 nm in the presence of appropriate nucleophile such as sodium methoxide (NaOMe). The formation of methoxy adduct **47** was confirmed by ^{1}H NMR spectrum. The selective reactivity of the closed-ring isomer **46b** was demonstrated by the reaction in an imidazolium-based ionic liquid combined with reversible phase transfer between hexane and ionic liquid (Fig. 10.24). An ionic liquid solution of **46a** turned to blue with UV irradiation, and the color was corresponding to that in methanol solution. The addition of NaOMe to the ionic liquid solution changed the color to red with simultaneous phase transfer to upper hexane solution, indicating the formation of neutral compound **47**. The color reversibly changed to blue and transferred into the lower ionic liquid phase through the addition of acid (TFA) to again form ionic **46b**.

10.5 Summary and Outlook

In this chapter, we introduced photochromic terarylenes as an extension of diarylethenes modified with extended π-systems at the ethene bridging part. Substitution of (perfluoro)cyclopentene unit, which is stable against photochemical reactions, highly electro-withdrawing, and convenient for syntheses, with extended π-systems has expanded the utility of diarylethenes in many aspects, including synthetic routes, thermal stability, conformational control *via* molecular interactions, and thereby photochromic efficiency, and a variety of switching properties.

Especially, the photochromic system with high cyclization quantum yields would be beneficial for the application of light-triggered chemical reaction systems, such as photo-resist, 3D-photo molding, and related optical technologies, with extremely high efficiency. Furthermore, the molecular system with high efficiency in both directions, photo-cyclization and ring-opening reactions, would be attained with the aid of efficient oxidative or thermal cycloreversion reactions.

References

1. Heller HG, Oliver S (1981) Photochromic heterocyclic fulgides. 1. Rearrangement reactions of (E)-alpha-3-furylethylidene(isopropylidene)succinic anhydride. J Chem Soc Perkin Trans 1:197–202
2. Yokoyama Y (2000) Fulgides for memories and switches. Chem Rev 100:1717–1739
3. Irie M (2000) Diarylethenes for memories and switches. Chem Rev 100:1685–1716
4. Matsuda K, Irie M (2006) Recent development of 6 π-electrocyclic photochromic systems. Chem Lett 35:1204–1209
5. Tsivgoulis GM, Lehn JM (1995) Photonic molecular devices—reversibly photoswitchable fluorophores for nondestructive readout for optical memory. Angew Chem Int Ed Engl 34:1119–1122
6. Koshido T, Kawai T, Yoshino K (1995) Novel Photomemory effects in photochromic dye-doped conducting polymer and amorphous photochromic dye layer. Syn Met 73:257–260
7. Giordano L, Jovin TM, Irie M, Jares-Erijman EA (2002) Diheteroarylethenes as thermally stable photoswitchable acceptors in photochromic fluorescence resonance energy transfer (pcFRET). J Am Chem Soc 124:7481–7489
8. Fukaminato T, Sasaki T, Kawai T, Tamai N, Irie M (2004) Digital photoswitching of fluorescence based on the photochromism of diarylethene derivatives at a single-molecule level. J Am Chem Soc 126:14843–14849
9. Myles AJ, Branda NR (2003) Novel photochromic homopolymers based on 1,2-bis(3-thienyl) cyclopentenes. Macromolecules 36:298–303

10. Raymo FM, Tomasulo M (2005) Electron and energy transfer modulation with photochromic switches. Chem Soc Rev 34:327–336
11. Tanio N, Irie M (1994) Photooptical switching of polymer film wave-guide containing photochromic diarylethenes. Jpn J Appl Phys 1(33):1550–1553
12. Kawai T, Koshido T, Yoshino K (1995) Optical and dielectric-properties of photochromic dye in amorphous state and its application. Appl Phys Lett 67:795–797
13. Kim E, Choi YK, Lee MH (1999) Photoinduced refractive index change of a photochromic diarylethene polymer. Macromolecules 32:4855–4860
14. Kim MS, Maruyama H, Kawai T, Irie M (2003) Refractive index changes of amorphous diarylethenes containing 2,4-diphenylphenyl substituents. Chem Mater 15:4539–4543
15. Kawai T, Kunitake T, Irie M (1999) Novel photochromic conducting polymer having diarylethene derivative in the main chain. Chem Lett 905–906
16. Kawai T, Nakashima Y, Irie M (2005) A novel photoresponsive pi-conjugated polymer based on diarylethene and its photoswitching effect in electrical conductivity. Adv Mater 17:309–314
17. Katsonis N, Kudernac T, Walko M, van der Molen SJ, van Wees BJ, Feringa BL (2006) Reversible conductance switching of single diarylethenes on a gold surface. Adv Mater 18:1397–1400
18. Whalley AC, Steigerwald ML, Guo X, Nuckolls C (2007) Reversible switching in molecular electronic devices. J Am Chem Soc 129:12590–12591
19. Matsuda K, Yamaguchi H, Sakano T, Ikeda M, Tanifuji N, Irie M (2008) Conductance photoswitching of diarylethene-gold nanoparticle network induced by photochromic reaction. J Phys Chem C 112:17005–17010
20. Gilat SL, Kawai SH, Lehn JM (1995) Light-triggered molecular devices—photochemical switching of optical and electrochemical properties in molecular wire type diarylethene species. Chem Eur J 1:275–284
21. Koshido T, Kawai T, Yoshino K (1995) Optical and electrochemical properties of cis-1,2-dicyano-1,2-bis (2,4,5-trimethyl-3-thienyl)ethene. J Phys Chem 99:6110–6114
22. Gorodetsky B, Samachetty HD, Donkers RL, Workentin MS, Branda NR (2004) Reductive electrochemical cyclization of a photochromic 1,2-dithienylcyclopentene dication. Angew Chem Int Ed 43:2812–2815
23. Moriyama Y, Matsuda K, Tanifuji N, Irie S, Irie M (2005) Electrochemical cyclization/cycloreversion reactions of diarylethenes. Org Lett 7:3315–3318
24. Browne WR, de Jong JJD, Kudernac T, Walko M, Lucas LN, Uchida K, van Esch JH, Feringa BL (2005) Oxidative electrochemical switching in dithienylcyclopentenes, part 1: effect of electronic perturbation on the efficiency and direction of molecular switching. Chem Eur J 11:6414–6429
25. Matsuda K, Irie M (2001) Photochromism of diarylethenes with two nitronyl nitroxides: photoswitching of an intramolecular magnetic interaction. Chem Eur J 7:3466–3473
26. Tanifuji N, Irie M, Matsuda K (2005) New photoswitching unit for magnetic interaction: diarylethene with 2,5-bis(arylethynyl)-3-thienyl group. J Am Chem Soc 127:13344–13353
27. de Jong JJD, Lucas LN, Kellogg RM, van Esch JH, Feringa BL (2004) Reversible optical transcription of supramolecular chirality into molecular chirality. Science 304:278–281
28. Hirose T, Matsuda K, Irie M (2006) Self-assembly of photochromic diarylethenes with amphiphilic side chains: reversible thermal and photochemical control. J Org Chem 71:7499–7508
29. Nakashima T, Miyamura K, Sakai T, Kawai T (2009) Photo-, solvent-, and ion-controlled multichromism of imidazolium-substituted diarylethenes. Chem Eur J 15:1977–1984
30. Mammana A, Carroll GT, Areephong J, Feringa BL (2011) A chiroptical photoswitchable DNA complex. J Phys Chem B 115:11581–11587
31. Kawai T, Iseda T, Irie M (2004) Photochromism of triangle terthiophene derivatives as molecular re-router. Chem Commun 72–73
32. Deng XH, Liebeskind LS (2001) A contribution to the design of molecular switches: Novel acid-mediated ring-closing-photochemical ring-opening of 2,3-bis(heteroaryl)quinones (heteroaryl = thienyl, furanyl, pyrrolyl). J Am Chem Soc 123:7703–7704

33. Yam VWW, Ko CC, Zhu NY (2004) Photochromic and luminescence switching properties of a versatile diarylethene-containing 1,10-phenanthroline ligand and its rhenium(I) complex. J Am Chem Soc 126:12734–12735
34. Tian H, Chen BZ, Tu HY, Mullen K (2002) Novel bisthienylethene-based photochromic tetraazaporphyrin with photoregulating luminescence. Adv Mater 14:918–923
35. Ivanov SN, Lichitsky BV, Dudinov AA, Martynkin AY, Krayushkin MM (2001) Synthesis of substituted 1,2,4-triazines on the basis 1,2-bis-(2,5-dimethylthien-3-yl)ethanedione. Chem Heterocycl Compd 37:85–90
36. Krayushkin AM, Ivanov SN, Martynkin AY, Lichitsky BV, Dudinov AA, Uzhinov BM (2001) Photochromic dihetarylethenes-7. Synthesis of bis(thienylazoles), photochromic analogs of diarylethenes. Russ Chem Bull 50:116–121
37. Lemieux V, Spantulescu MD, Baldridge KK, Branda NR (2008) Modulating the Lewis acidity of boron using a photoswitch. Angew Chem Int Ed 47:5034–5037
38. Castells J, Lopezcalahorra F, Domingo L (1988) Postulation of bis(thiazolin-2-ylidene)s as the catalytic species in the benzoin condensation catalyzed by a thiazolium salt plus base. J Org Chem 53:4433–4436
39. Nakashima T, Atsumi K, Kawai S, Nakagawa T, Hasegawa Y, Kawai T (2007) Photochromism of thiazole-containing triangle terarylenes. Eur J Org Chem 3212–3218
40. Li XC, Tian H (2005) One-step synthesis and photochromic properties of a stable triangle terthiophene. Tetrahedron Lett 46:5409–5412
41. Shibahara F, Yamaguchi E, Murai T (2011) Direct arylation of simple azoles catalyzed by 1,10-phenanthroline containing palladium complexes: an investigation of C4 arylation of azoles and the synthesis of triarylated azoles by sequential arylation. J Org Chem 76:2680–2693
42. Kawai S, Nakashima T, Atsumi K, Sakai T, Harigai M, Imamoto Y, Kamikubo H, Kataoka M, Kawai T (2007) Novel photochromic molecules based on 4,5-dithienyl thiazole with fast thermal bleaching rate. Chem Mater 19:3479–3483
43. Irie M, Sakemura K, Okinaka M, Uchida K (1995) Photochromism of dithienylethenes with electron-donating substituents. J Org Chem 60:8305–8309
44. Uchida K, Ishikawa T, Takeshita M, Irie M (1998) Thermally irreversible photochromic systems. Reversible photocyclization of 1,2-bis(thiazolyl)perfluorocyclopentenes. Tetrahedron 54:6627–6638
45. Uchida K, Tsuchida E, Aoi Y, Nakamura S, Irie M (1999) Substitution effect on the coloration quantum yield of a photochromic bisbenzothienylethene. Chem Lett 63–64
46. Kawai S, Nakashima T, Kutsunugi Y, Nakagawa H, Nakano H, Kawai T (2009) Photochromic amorphous molecular materials based on dibenzothienylthiazole structure. J Mater Chem 19:3606–3611
47. Kutsunugi Y, Kawai S, Nakashima T, Kawai T (2009) Photochromic properties of terarylene derivatives having a pi-conjugation unit on central aromatic ring. New J Chem 33:1368–1373
48. Nakamura S, Irie M (1988) Thermally irreversible photochromic systems—a theoretical-study. J Org Chem 53:6136–6138
49. Iwata S, Ishihara Y, Qian CP, Tanaka K (1992) Thermal-stability of photochromic bis((trifluoromethyl)thiazolyl)ethene derivatives. J Org Chem 57:3726–3727
50. Katritzky AR, Karelson M, Sild S, Krygowski TM, Jug K (1998) Aromaticity as a quantitative concept. 7. Aromaticity reaffirmed as a multidimensional characteristic. J Org Chem 63:5228–5231
51. Jeong YC, Gao C, Lee IS, Yang SI, Ahn KH (2009) The considerable photostability improvement of photochromic terarylene by sulfone group. Tetrahedron Lett 50:5288–5290
52. Tamai N, Miyasaka H (2000) Ultrafast dynamics of photochromic systems. Chem Rev 100:1875–1890
53. Kobatake S, Uchida K, Tsuchida E, Irie M (2002) Single-crystalline photochromism of diarylethenes: reactivity-structure relationship. Chem Commun 2804–2805
54. Taylor R, Kennard O (1982) Crystallographic evidence for the existence of C–H...O, C–H... N, and C–H...Cl hydrogen-bonds. J Am Chem Soc 104:5063–5070

55. Fukumoto S, Nakashima T, Kawai T (2011) Photon-quantitative reaction of a dithiazolylarylene in solution. Angew Chem Int Ed 50:1565–1568
56. Fukumoto S, Nakashima T, Kawai T (2011) Intramolecular hydrogen bonding in a triangular dithiazolyl-azaindole for efficient photoreactivity in polar and nonpolar solvents. Eur J Org Chem 5047–5053
57. Karikomi M, Kitamura C, Tanaka S, Yamashita Y (1995) New narrow-bandgap polymer composed of benzobis(1,2,5-thiadiazole) and thiophenes. J Am Chem Soc 117:6791–6792
58. Song P, Gao AH, Zhou PW, Chu TS (2012) Theoretical study on photoisomerization effect with a reversible nonlinear optical switch for dithiazolylarylene. J Phys Chem A 116:5392–5397
59. Nakashima T, Fujii R, Kawai T (2011) Regulation of folding and photochromic reactivity of terarylenes through a host-guest interaction. Chem Eur J 17:10951–10957
60. Li XC, Ma YZ, Wang BC, Li GG (2008) "Lock and key control" of photochromic reactivity by controlling the oxidation/reduction state. Org Lett 10:3639–3642
61. Hasegawa Y, Nakagawa T, Kawai T (2010) Recent progress of luminescent metal complexes with photochromic units. Coord Chem Rev 254:2643–2651
62. Akita M (2011) Photochromic organometallics, A stimuli-responsive system: an approach to smart chemical systems. Organometallics 30:43–51
63. Chan JCH, Lam WH, Wong HL, Zhu NY, Wong WT, Yam VWW (2011) Diarylethene-containing cyclometalated platinum(II) complexes: tunable photochromism via metal coordination and rational ligand design. J Am Chem Soc 133:12690–12705
64. Lee PHM, Ko CC, Zhu NY, Yam VWW (2007) Metal coordination-assisted near-infrared photochromic behavior: a large perturbation on absorption wavelength properties of N, N-donor ligands containing diarylethene derivatives by coordination to the rhenium(I) metal center. J Am Chem Soc 129:6058–6059
65. Yam VWW, Lee JKW, Ko CC, Zhu NY (2009) Photochromic diarylethene-containing ionic liquids and N-heterocyclic carbenes. J Am Chem Soc 131:912–913
66. Duan GP, Wong WT, Yam VWW (2011) Synthesis and photochromic studies of eta(6)-mesitylene ruthenium(II) complexes bearing N-heterocyclic carbene ligands with the dithienylethene moiety. New J Chem 35:2267–2278
67. Wong HL, Tao CH, Zhu NY, Yam VWW (2011) Photochromic alkynes as versatile building blocks for metal alkynyl systems: design, synthesis, and photochromic studies of diarylethene-containing platinum(II) phosphine alkynyl complexes. Inorg Chem 50:471–481
68. Nakagawa T, Atsumi K, Nakashima T, Hasegawa Y, Kawai T (2007) Reversible luminescence modulation in photochromic europium(III) complex having triangle terthiazole ligands. Chem Lett 36:372–373
69. Nakagawa T, Hasegawa Y, Kawai T (2008) Photoresponsive europium(III) complex based on photochromic reaction. J Phys Chem A 112:5096–5103
70. Fukaminato T (2011) Single-molecule fluorescence photoswitching: design and synthesis of photoswitchable fluorescent molecules. J Photochem Photobiol C Photochem Rev 12:177–208
71. Pang SC, Hyun H, Lee S, Jang D, Lee MJ, Kang SH, Ahn KH (2012) Photoswitchable fluorescent diarylethene in a turn-on mode for live cell imaging. Chem Commun 48:3745–3747
72. Jeong YC, Yang SI, Ahn KH, Kim E (2005) Highly fluorescent photochromic diarylethene in the closed-ring form. Chem Commun 2503–2505
73. Uno K, Niikura H, Morimoto M, Ishibashi Y, Miyasaka H, Irie M (2011) In Situ preparation of highly fluorescent dyes upon photoirradiation. J Am Chem Soc 133:13558–13564
74. Taguchi M, Nakagawa T, Nakashima T, Kawai T (2011) Photochromic and fluorescence switching properties of oxidized triangle terarylenes in solution and in amorphous solid states. J Mater Chem 21:17425–17432
75. Nakagawa H, Kawai S, Nakashima T, Kawai T (2009) Synthesis and photochemical reactions of photochromic terarylene having a leaving methoxy group. Org Lett 11:1475–1478
76. Neilson BM, Lynch VM, Bielawski CW (2011) Photoswitchable N-heterocyclic carbenes: using light to modulate electron donating properties. Angew Chem Int Ed 50:10322–10326

77. Neilson BM, Bielawski CW (2012) Photoswitchable organocatalysis: Using light to modulate the catalytic activities of N-heterocyclic carbenes. J Am Chem Soc 134:12693–12699
78. Stoll RS, Hecht S (2010) Artificial light-gated catalyst systems. Angew Chem Int Ed 49:5054–5075
79. Wu TQ, Tang H, Bohne C, Branda NR (2012) Reporting the release of caged species by a combination of two sequential photoreactions, a molecular switch, and one color of light. Angew Chem Int Ed 51:2741–2744
80. Lemieux V, Gauthier S, Branda NR (2006) Selective and sequential photorelease using molecular switches. Angew Chem Int Ed 45:6820–6824
81. Nakashima T, Goto M, Kawai S, Kawai T (2008) Photomodulation of ionic interaction and reactivity: reversible photoconversion between imidazolium and imidazolinium. J Am Chem Soc 130:14570–14575

Chapter 11
Photochromic Organometallics: Redox-Active Iron and Ruthenium Complexes with Photochromic DTE Ligand

Takashi Koike and Munetaka Akita

Abstract This chapter focuses on our works in the area of "photochromic organometallics" with a ***M***–C (***M***: redox-active metal fragments, C: carbon atom in the photochromic unit = dithienylethene (DTE)) bond. Smart molecular systems, which are designed to bring appropriate functions in response to change of the environment, are important foundations for the development of intelligent materials. Chromic molecules can recognize external stimuli so as to trigger chemical functions required for smart chemical systems. Combination with metal components, which exhibit unique properties such as redox and photophysical properties and catalysis, should lead to more sophisticated systems. In particular, photochromic molecules with metal fragments can provide an attractive molecular system driven by light. The first part is related to photoswitchable molecular wires. The second part deals with dual stimuli-responsive system, i.e., photo- and electrochromic organometallics. The use of dithienylethene, a representative photochromic molecule, is a key for unique chemical systems in this chapter.

Keywords Dithienylethene • Dual chromism • Electrochromism • Organometallic molecular wire • Photochromism

11.1 Introduction

Stimuli–responsive systems are essential components of smart chemical systems [1]. Stimuli–responsive systems can recognize changes of the environment so as to trigger a chemical function required for the smart chemical system. Many kinds of organic and inorganic stimuli–responsive systems have been developed so far.

T. Koike • M. Akita (✉)
Chemical Resources Laboratory, Tokyo Institute of Technology,
Nagatsuta, Midori-ku, Yokohama 226-8503, Japan
e-mail: koike.t.ad@m.titech.ac.jp; makita@res.titech.ac.jp

M. Irie et al. (eds.), *New Frontiers in Photochromism*,
DOI 10.1007/978-4-431-54291-9_11, © Springer Japan 2013

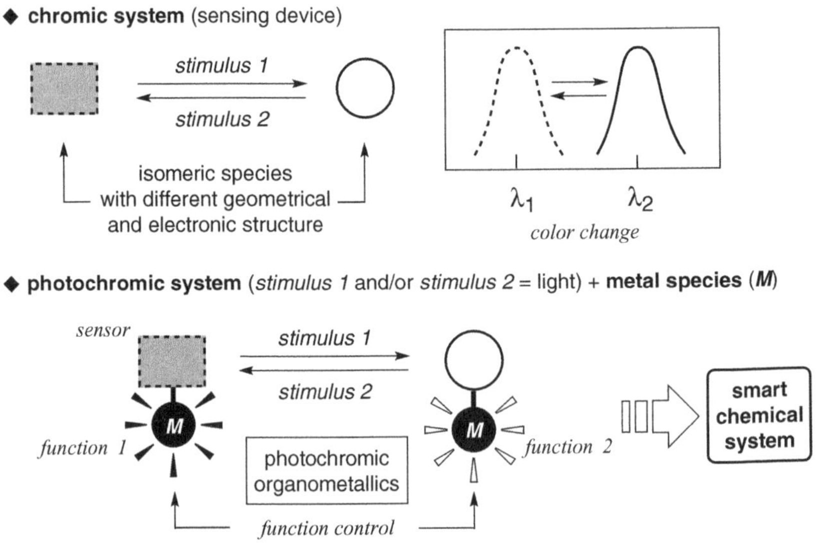

Fig. 11.1 Photochromic organometallics: combination of organic photochromic unit with metal species

Among them much attention has been focused on chromic systems [2]. Chromism is defined as a transformation of a chemical species between two forms by application of a stimulus, where the two forms have different absorption spectra, i.e., different colors, and the process is often reversible (Fig. 11.1).

The color change results from a change of the electronic structure of the chromic molecule (usually π-conjugated system), which is often associated with a change of the geometrical structure of the molecule. Combination of such chromic systems with other chemical systems should lead to the development of more sophisticated stimuli–responsive systems. One way is the combination of chromic π-conjugated organic fragments with metal species, which exhibit unique features such as redox properties, photophysical properties, and catalysis. There are many kinds of chromism, such as photo-, thermo-, iono-, halo-, electro-, solvato-, vapo-, and mechanochromism [2]. Herein photochromic system with metal complexes will be focused on, in particular, in which the photochromic unit is 1,2-dithienylethenes (DTE) [3, 4]. Over the past 10 years, photochromic metal complexes featuring other photochromes such as azobenzene, spiropyran, spirooxazine, benzopyran, and dimethyldihydropyrene derivatives have been developed from the viewpoint of modulation of photochromism or photoregulation of the redox, optical, and magnetic properties of metal units [5–10]. In addition, metal complexes themselves can also show photochromic properties usually via linkage isomerization of ligands [11]. In contrast to them, "photochromic organometallics" in this chapter refers to the DTE metal complexes, ***M***–DTE–***M*** or ***M***–DTE (***M***: redox-active metal fragments), with a ***M***–C bond. This molecular design of *photochromic organometallics* provides a promising way for creation of metal complexes with photochromic properties.

11.2 Photoswitchable Organometallic Molecular Wire

11.2.1 Introduction

Molecular electronics are regarded as one of the promising ways for miniaturization of electronic circuits leading to highly integrated electronic systems [12–25]. In order to build up molecular circuits, many molecular components such as wires, switches, transistors, and logic gates should be developed and assembled. The polyynediyldimetal complexes, ***M***–(C≡C)$_n$–***M*** (***M***: redox-active metal fragments), with a wirelike appearance are expected to display properties applicable to molecular wires [26–30], because the d-orbitals of the metal fragments at the termini may interact with each other through the π-conjugated (C≡C)$_n$ rod. For efficient communication between the two metal centers, the properties of the metal fragments must be best tuned [31–41]. Molecular electronics, however, are still in their infancy. Other molecular components need to be synthesized and their performance should be assessed [42]. Thus, we carried out the synthesis of switching systems.

11.2.2 1st Generation Dinuclear Iron and Ruthenium Complexes with Dithienylethene Ligand

Dithienylethene (DTE) has been chosen as the switching mechanism. The family of DTE is one of the versatile photochromic systems developed by Prof. Irie (Fig. 11.2) [3, 4]. Their photochromism is based on the reversible photochemical cyclization–cycloreversion processes between the open 1,3,5-hexatriene skeleton and the closed cyclohexadiene skeleton in the central part, and the forward and backward processes are triggered by UV and visible light irradiation, respectively. It is notable that the double bonds in the closed isomer **C** are fully conjugated, whereas the open isomer **O** contains cross conjugation at the thiophene–cyclopentene junctions. This dramatic structural change causes the striking color change (**O**: pale-colored, **C**: deep-colored), and DTE is superior to other photochromic systems with respect to many aspects such as quick response, fatigue resistance, and facile control of the photophysical properties (e.g., λ_{max}). If metal fragments are attached to the 5- and 5′-positions of the thiophene rings, the communication between the two metal centers through the DTE bridge may be switched.

Taking into account for the facile access to the acetylide complex-type skeleton, the DTE/Fe system with the C≡C linkers, *Fe*–C≡C–DTE–C≡C–*Fe* (*Fe* = (η^5–C_5Me_5)Fe(dppe), and dppe = $Ph_2PCH_2CH_2PPh_2$) (**1Fe*O**), was designed and readily synthesized from the corresponding 1-alkyne (Fig. 11.3) [43]. For the open isomer **1Fe*O** characterized by crystallographic methods, it is worthy of note that the DTE moiety

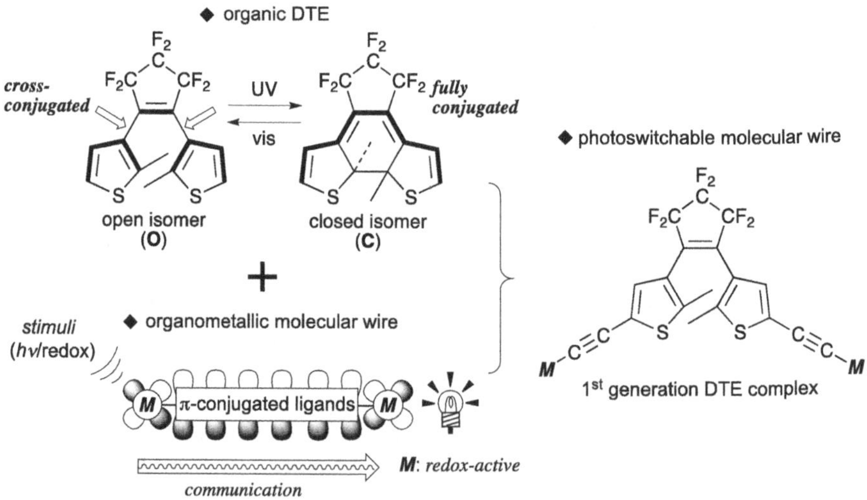

Fig. 11.2 Photoswitchable organometallic molecular wire (1st generation DTE complex)

adopts an antiparallel conformation suitable for photocyclization, i.e., the Fe groups bulkier than the central DTE part do not significantly affect the conformation of the central part in the ground state (Fig. 11.3a). The organometallic DTE derivative **1Fe*O** underwent photocyclization upon UV irradiation, in a manner similar to organic counterparts, to be converted to the closed isomer **1Fe*C**. A visible absorption at 774 nm grew progressively and reached a photostationary state with the composition of **1Fe*O/1Fe*C** = 10/90 in C_6D_6 (Fig. 11.3b). Subsequent visible light irradiation of the equilibrated mixture regenerated **1Fe*O** quantitatively. After ten photochemical cyclization–cycloreversion cycles, no noticeable deterioration was detected. Switching factor (SF) can be evaluated by comparison of wirelike performance of **1Fe*O** with that of **1Fe*C**. In general, wirelike performance of organometallic molecular wires is estimated on the basis of K_C value (comproportionation constant) obtained by electrochemical analysis. Wirelike behavior is attributed to the stability of monocationic species, that is, the mixed-valence (MV) complex. The K_C value is derived from the potential difference between the two-redox process (ΔE) according to the equation, $K_C = \exp(\Delta E \times F/RT)$ (F: Faraday constant (9.65×10^4 C mol^{-1}), R: gas constant (8.31 J K^{-1} mol^{-1}), T: temperature (K)), and represents the thermodynamic stability of the mixed-valence monocationic species against non-mixed-valence ones, indicating that the extent of delocalization of the hole over the bridging part (Fig. 11.4).

The two isomers (**1Fe*O** and **1Fe*C**) isolated by repeated recrystallization were subjected to electrochemical measurements to determine the K_C values. For the open isomer **1Fe*O**, as can be seen from the CV trace, a slightly broad single redox wave was observed indicating very weak interaction between the two metal centers with $K_C = 13$ as analyzed by simulation of the CV trace (Fig. 11.3c) and

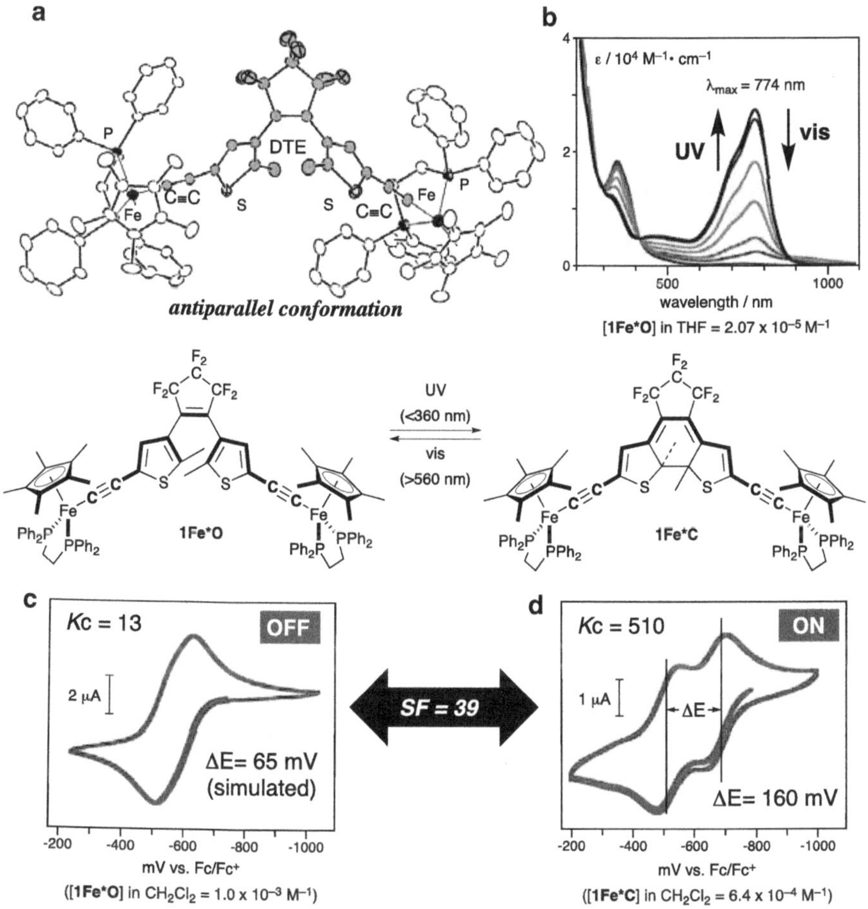

Fig. 11.3 (**a**) Molecular structure of **1Fe*O**; (**b**) UV–vis spectra for photochromism of **1Fe***; (**c**) CV trace for **1Fe*O**; (**d**) CV trace for **1Fe*C**

deconvolution analysis of a differential pulse voltammetry (DPV) trace. In sharp contrast, the closed isomer **1Fe*C** showed two separated redox waves (Fig. 11.3d). From the separation ($\Delta E = 160$ mV), the K_C value was determined to be 510, which is significantly larger than that of the open isomer. As a result, the switching factor, SF = K_C (closed)/K_C (open), turned out to be calculated at 39. It is clear that the present iron DTE complex **1Fe*** is an excellent photoswitchable molecular wire [44].

The isoelectronic complexes, *Ru*–C≡C–DTE–C≡C–*Ru* (*Ru* = (η^5–C_5Me_5)Ru(dppe)) (**1Ru*O**), were also prepared and tested for photochromic behavior and switching factor [45]. It should be noted that photochromic performance is significantly improved (**1Ru*C** was formed quantitatively under UV irradiation). Quantum yields (in toluene) were determined to be 0.0021 (**1Fe*O**; irradiated at 366 nm) and 0.38 (**1Ru*O**;

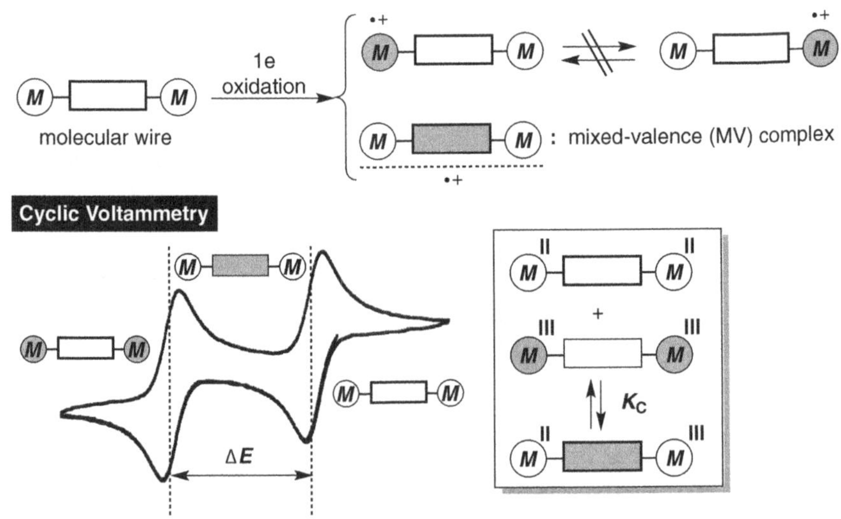

Fig. 11.4 Evaluation for organometallic molecular wires based on electrochemical analysis

irradiated at 366 nm) for the ring-closing processes and 0.00018 (**1Fe*C**; irradiated at λ_{max} of **1Fe*C** (768 nm)) and 0.00044 (**1Ru*C**; irradiated at λ_{max} of **1Ru*C** (710 nm)) for the ring-opening processes. Note that although the ring-closing quantum yield of **1Ru*O** is comparable to that of the diphenyl derivative, C_6H_5–DTE–C_6H_5 **5** (1,2-di(2-methyl-5-phenylthien-3-yl)-3,3,4,4,5,5-hexafluorocyclopentene), (0.59 for ring closure and 0.013 for ring-opening in hexane) [46], the ring-opening quantum yields are significantly even smaller than the rather small quantum yield for the ring-opening process of **5**. On the other hand, switching performance of **1Ru*** (SF=4) is inferior to that of **1Fe***. These results show that the structures of metal fragments strongly affect photochromic properties as well as redox properties.

11.3 Dual Photo- and Electrochromic Organometallics

11.3.1 Introduction

To gain further insight into the effect of metal fragments on photochromic behavior, we designed a new series of 2nd generation DTE complexes without acetylene linkers, ***M***–DTE–***M*** (***M***=(η^5–C_5R_5)ML_2, M=Fe, Ru; L=CO, phosphine), which redox-active metal fragments are directly σ-bonded to the thiophene rings at the 5,5′-positions (Fig. 11.5 and Table 11.1).

In addition, computational analysis was conducted to investigate the photochemical process in detail. Time-dependent DFT (TDDFT) calculations suggested that the photocyclization proceeds via triplet excited state, which is a different reaction pathway from the proposed process in the photocyclization of normal organic DTE.

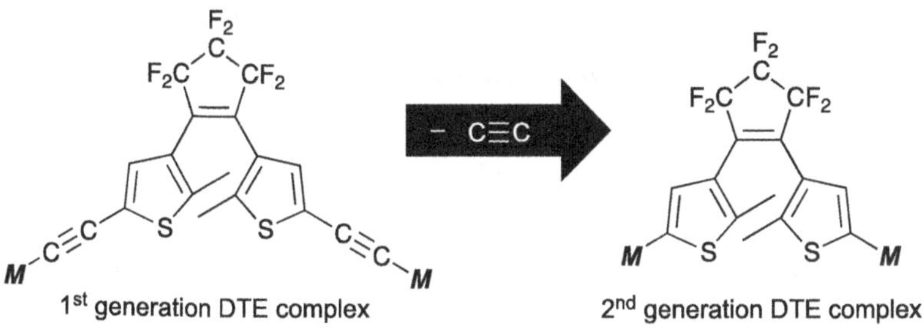

Fig. 11.5 Molecular design of 2nd generation DTE complex

Table 11.1 Photochromic behavior of DTE complexes (**1–4**)

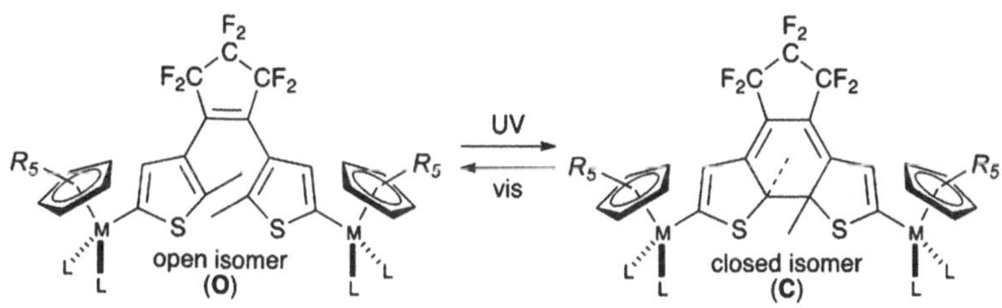

Complex	*M* ((η[5]–C$_5$R$_5$)ML$_2$)	O/C[a]	Ring closure[b] (time/min)	Ring opening[b] (time/min)	λ_{max}/nm of **C** (color)	Recycl. (times)
1Fe*	C≡C–Cp*Fe(dppe)	10/90	80	90	774 (green)	95 % (6)
1Ru*	C≡C–Cp*Ru(dppe)	~0/~100	8	60	719 (green)	88 % (10)
2Fe	CpFe(CO)$_2$	61/39	16	4	560 (brown)	Decomp.
2Ru	CpRu(CO)$_2$	36/64	24	4	549 (brown)	50 % (5)
3Fe	CpFe(CO)(PPh$_3$)	90</<10	–	–	–	Decomp.
3Ru	CpRu(CO)(PPh$_3$)	30/70	30	8	584 (blue purple)	70 % (10)
4Fe′	Cp′Fe(dppe)	100/0	–	–	–	–
4Fe*	Cp*Fe(dppe)	100/0	–	–	–	–

[a]Isomer ratios at the photostationary states in C$_6$D$_6$ determined by ^{1}H NMR
[b]Monitoring 2.0×10^{-5} M in THF by UV–vis spectra

Furthermore, these "photochromic organometallics" exhibit ring-closing reaction triggered by 2e-oxidation, i.e., the present molecular system provides dual photo- and electrochromic system. Multi-stimuli–responsive systems can be key elements for the construction of molecular devices as well as smart chemical systems, because they are expected not only to simply work as a switch but also to be applicable to logic systems. Combination of a photoresponsive DTE unit with redox-active metal fragments has potentials to produce multi-stimuli–responsive systems and multi-chromism. We also developed the mononuclear DTE complexes, ***M***–DTE (***M***=(η[5]–C$_5$H$_5$)ML$_2$, M=Fe, Ru; L=CO, phosphine), which express dual chromism in a manner different from dinuclear complexes.

11.3.2 2nd Generation Dinuclear Iron and Ruthenium Complexes with Dithienylethene Ligand

A series of 2nd generation DTE complexes, *M*–DTE–*M* (*M*=(η^5–C_5H_5)Fe(CO)$_2$ (**2Fe**), (η^5–C_5H_5)Ru(CO)$_2$ (**2Ru**), (η^5–C_5H_5)Fe(CO)(PPh$_3$) (**3Fe**), (η^5–C_5H_5)Ru(CO)(PPh$_3$) (**3Ru**), (η^5–C_5H_4(Me))Fe(dppe) (**4Fe′**), and (η^5–C_5Me_5)Fe(dppe) (**4Fe***)), was prepared by (a) metalation of the lithiated DTE using the corresponding carbonyl metal halides, (η^5–C_5R_5)MX(CO)$_2$ (M/X=Fe/I and M/X=Ru/Cl), and (b) subsequent photochemical ligand exchange reactions. Some of the derivatives are characterized by single-crystal X-ray crystallography, which reveals the open structure with the antiparallel conformation of the two thiophene rings being suitable for photochemical ring closure [47, 48]. The obtained organometallic DTE complexes exhibit photochromic behavior but the performance turns out to be dependent on the attached metal fragments, as expected from above-mentioned results (Sect. 11.2.2). The photochromic behavior on the basis of UV–vis spectra and NMR data was summarized in Table 11.1.

Complexes **2Fe**, **2Ru**, and **3Ru** undergo the reversible photochemical interconversion, whereas the other complexes **3Fe**, **4Fe′**, and **4Fe*** are virtually inert with respect to the photochromism. Isomer ratios (**O/C**) at the photostationary states in C_6D_6 are estimated to be 61/39 for **2Fe**, 36/64 for **2Ru**, 91/9 for **3Fe**, and 30/70 for **3Ru**. Dependence of the photochromic performance can be estimated by the isomer ratios at the photostationary states as follows: metal: Ru>Fe; ligand: (CO)$_2$>(CO)(PPh$_3$)≈dppe (for the iron complexes), (CO)$_2$≈(CO)(PPh$_3$) (for the ruthenium complexes); and linker: C≡C (**1**)>none (**2**–**4**). Dependence on the metal will be discussed in the next Sect. 11.3.3 on the basis of theoretical analysis. It should be noted that conversion of **2Ru** and **3Ru** to the closed isomers **C** is incomplete, whereas the acetylide derivatives **1Ru*** are converted to the closed isomers almost quantitatively. The difference could be ascribed to the absorption ranges of the closed isomers. The envelope of the UV absorption of the open isomers **O** could be extended to the absorption ranges of **2RuC**–**3RuC** (550–600 nm) but not to that of **1Ru*C** (>700 nm) being in the far longer wavelength region. Stability (fatigue resistance) of the organometallic DTE complexes is dependent on the structure of the metal auxiliaries. For the stability, the following orders are noted: metal: Ru>Fe and ligand: (CO)(PPh$_3$)>(CO)$_2$. The stability appears to be limited by photochemical decarbonylation and, therefore, is dependent on the strength of back-donation to the CO ligands. The CpM(CO)(PPh$_3$) complexes **3** are more robust than the CpM(CO)$_2$ complexes **2**, and the ruthenium complexes are more stable than the iron derivatives. In accord with this consideration, any notable deterioration was not observed for the dppe complexes without a CO ligand, **4Fe′** and **4Fe***, although they did not show photochromic behavior.

Quantum yields for the ring-closing and ring-opening processes of a representative example **2Ru** in toluene were determined to be 0.22 (irradiated at 355 nm) and 0.011 (irradiated at λ_{max} of **2RuC** (549 nm)), respectively. For **2FeC**, the ring-opening quantum yield was determined to be 0.016 (irradiated at λ_{max} of **2FeC**(560 nm)), but determination of the ring-closing quantum yield was hampered by the

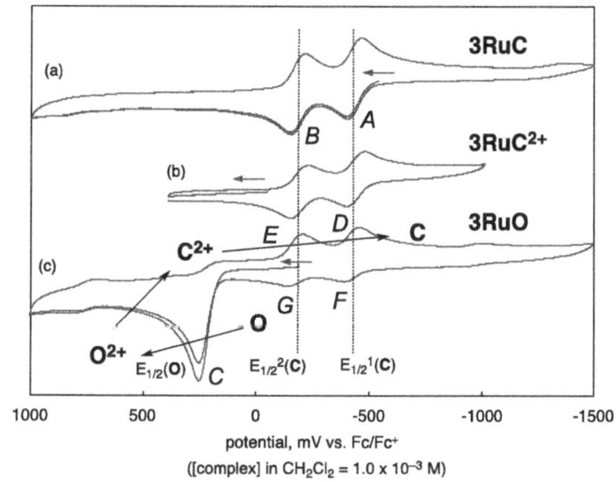

Fig. 11.6 CV traces for **3RuC**(a), **3RuC**$^{2+}$(b), and **3RuO**(c)

photochemical decomposition mentioned above proceeding at a rate comparable to that of the ring-closing process. It is notable that the ring-closing quantum yield of **2RuO** is in the same range of those of the organic diphenyl derivative, Ph-DTE-Ph **5**, (0.59 in hexane) and the ruthenium-acetylide complex **1Ru*O** (0.38) but significantly larger than that of **1Fe*O** (0.0021). On the other hand, the ring-opening quantum yields of **2FeC** and **2RuC** are virtually the same as that of **5** (0.013 in hexane) and even much larger than those of for **1Fe*C** (0.00018) and **1Ru*C** (0.00044). Thus, it turns out that photochemical reactivity of the ruthenium complexes is superior to that of the corresponding iron derivatives. The 2nd generation DTE complexes exhibit the photochromic properties in a manner similar to that of organic derivatives but the performance is dependent on the metal and ancillary ligands. With respect to central metal, it has similar results as those of 1st generation DTE complexes.

Electrochemical behavior turns out to also be dependent on the attached metal fragments. Behavior of the **2**- and **3**-series complexes is considerably different from that of the **4**-series dppe complexes. As a typical example, electrochemical behavior of **3Ru** is described in detail (Fig. 11.6). The closed isomer **3RuC** shows the reversible, two consecutive 1e-redox waves at −430 (*A*) and −174 mV (*B*) (vs. [$FeCp_2$]/[$FeCp_2$]$^+$; Fig. 11.6a). The CV trace for the open isomer **3RuO** (Fig. 11.6c) is totally different from that of **3RuC**. A 2e-oxidation wave is observed at 273 mV (*C*) but the corresponding reduction wave is not detected and, instead, two consecutive reduction waves are observed at −495 (*D*) and −217 mV (*E*). A subsequent anodic scan gives two oxidation waves (*F* and *G*) corresponding to the two reduction processes (*E* and *D*, respectively). These two redox processes observed at $E_{1/2}$=−430 and −174 mV (*D*–*G*) are superimposable on those of the closed isomer **3RuC** (*A* and *B*) mentioned above. These results suggest that 2e-oxidation induces cyclization of the open isomer. Similar electrochemical behavior is observed for the open isomers of the ruthenium complex **2RuO** and the iron complexes **2FeO** and **3FeO**. In contrast to **2O** and **3O**, the dppe complexes **4Fe′O** and **4Fe*O** show two consecutive reversible 1e-redox waves when scanned in the range of −1,500 to 500 mV.

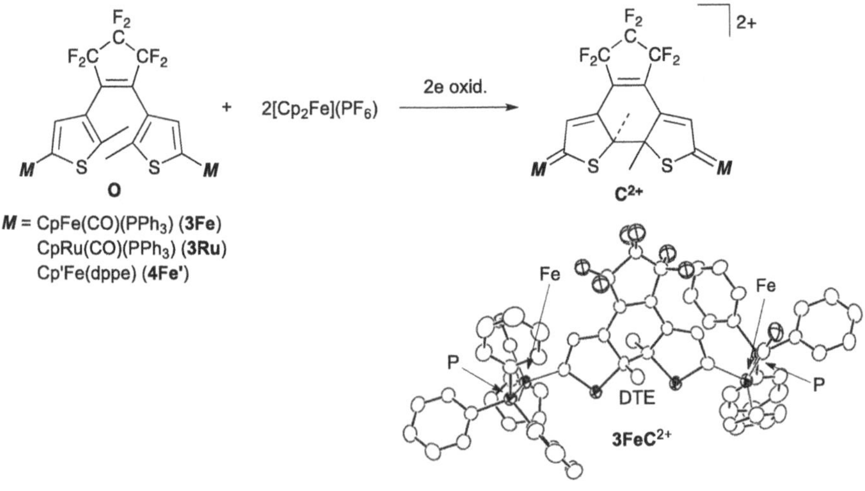

Fig. 11.7 Oxidative cyclization of **3FeO**, **3RuO**, and **4Fe′O** and ortep drawing of **3FeC**$^{2+}$

To confirm the chemical events taking place, chemical oxidation of the DTE complexes was conducted. Treatment of **3FeO**, **3RuO**, and **4Fe′O** with $[FeCp_2]PF_6$ (2 equiv.) gave the isolable, deep-green-colored diamagnetic dicationic closed species, [**3FeC**$^{2+}$]$(PF_6)_2$, [**3RuC**$^{2+}$]$(PF_6)_2$, and [**4Fe′C**$^{2+}$]$(PF_6)_2$, respectively (Fig. 11.7). Single-crystal X-ray crystallography of one of the stereoisomers of [**3FeC**$^{2+}$]$(PF_6)_2$ confirms (a) formation of a C–C single bond between the two thiophene rings at the 2- and 2′-positions (1.52(1) Å), (b) the double bond character of the Fe=C moieties (1.858(6) and 1.872(6) Å) substantially shorter than the Fe–C single bond, and (c) a change of the pattern of the bond alternation in accordance with the canonical form **C**$^{2+}$ depicted in Fig. 11.7. The CV trace for the dicationic species **3RuC**$^{2+}$ (Fig. 11.6b) identical to that of the neutral closed species (Fig. 11.6a) verifies that the two species contain the same closed carbon skeleton with different π-conjugated systems. It is significant that the Fe–phosphine complexes **3FeO** and **4Fe′O**, which do not undergo the photochemical ring closure, cyclize quantitatively upon the 2e-oxidation. As expected from the bond alternation pattern of **C**$^{2+}$, the dicationic closed isomers are so stable under daylight and do not undergo photochemical cycloreversion in contrast to the neutral closed isomers **C**. In addition, reduction of the dicationic closed species **C**$^{2+}$ by cobaltocene, $CoCp_2$, gave the corresponding neutral closed species **C**. For **2FeO** and **2RuO**, sequential in situ oxidation by CAN (cerium ammonium nitrate) and reduction with $CoCp_2$ at −78 °C afforded the neutral closed species **2C**, indicating occurrence of an analogous redox process. The cyclic voltammograms for complexes **2** similar to those of **3Ru** also suggested occurrence of analogous oxidative ring closure but the cationic carbene intermediate **2C**$^{2+}$ was too unstable to be isolated because of the lack of an electron-donating ligand (e.g., phosphine ligand) for stabilization of the electron-deficient Fischer-type carbene functional group [49, 50]. The dppe complexes **4Fe′O** and **4Fe*O** exhibit CV features different from those of **2** and **3** as mentioned above. The Cp′ complex **4Fe′O**

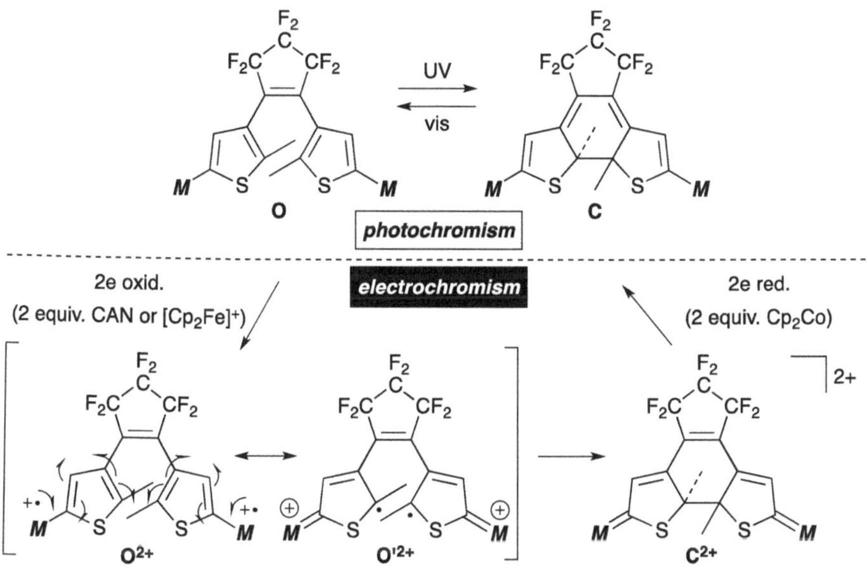

Fig. 11.8 Dual photo- and electrochromic behavior

underwent oxidative cyclization in a manner similar to **3Fe** and **3Ru** to give [**4Fe′C**$^{2+}$]$(PF_6)_2$, although reduction of the resultant dicationic species with $CoCp_2$ did not afford **4Fe′C** but a complicated mixture of products. On the other hand, 2e-oxidation of the Cp* complex **4Fe*** did not afford the cyclized product but the isolable, paramagnetic, open diradical species **4Fe*O**$^{2+}$, which shows the same CV features as those of **4Fe*O**.

The redox processes of the 2nd generation DTE complexes **2**–**4** can be interpreted in terms of the reaction sequence summarized in Fig. 11.8. The 2e-oxidation of **O** should give the dicationic diradical species **O**$^{2+}$. The radical centers in **O**$^{2+}$ should be delocalized over the thiophene rings in resonance with the thienyl radical form **O**$'^{2+}$, which undergoes radical coupling at the 2- and 2′-positions of the thiophene rings to give the closed diamagnetic species **C**$^{2+}$. Subsequent 2e-reduction gives the neutral closed isomer **C**. The lack of reduction waves for the oxidized, open dicationic species **O**$^{2+}$ (Fig. 11.6c) indicates that the ring closure (**O**$^{2+} \rightarrow$ **C**$^{2+}$) proceeds at a rate faster than the time scale of the CV measurement. A further cathodic scan gives the two reduction waves for the generated, closed dicationic species **C**$^{2+}$, which are identical to those of the neutral closed isomer **C**. The ring closure of the Cp′Fe(dppe) complex **4Fe′** should follow an analogous reaction sequence but the different CV features could be ascribed to the different time scales for the CV measurement and the chemical oxidation. The electron-donating dppe ligand should stabilize the electron-deficient diradical species **O**$^{2+}$ to elongate its lifetime. As a result, the rate of the ring-closing process becomes slower than the CV time scale but substantial with respect to the time scale of the preparative experiment. Further introduction of the electron-donating Cp* ligand (**4Fe***) finally makes the diradical

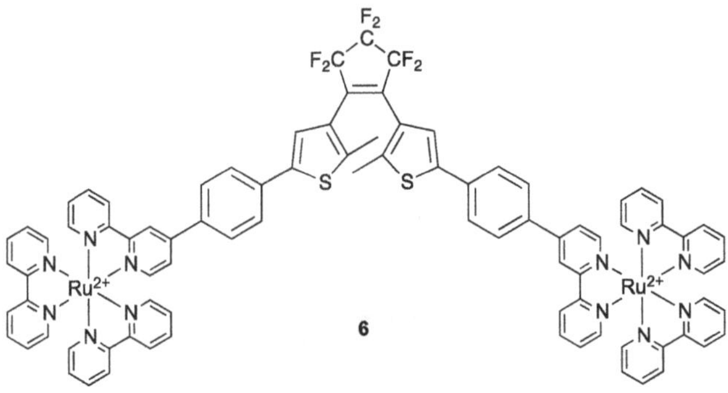

6

Fig. 11.9 A DTE complex with the $Ru(bpy)_3$ units **6**

intermediate **4Fe*O**$^{2+}$ stable so as not to undergo the cyclization. Our studies have been followed by a few related DTE–metal complexes with ring closure induced by oxidation [51, 52].

Thus, it has been revealed that the 2nd generation DTE complexes exhibit not only photochromism but also electrochromism, i.e., dual chromic behavior. Similar electrochromism is also observed in organic DTE system [53–59], however, the feature of the present "photochromic organometallics" is the involvement of isolable **C**$^{2+}$. The three colors (**C**, **O**, **C**$^{2+}$) can be expressed by a single molecule in response to different stimuli (light and redox).

11.3.3 Computational Analysis on Photochromism of Dithienylethenes with Metal Fragments

To gain further insight into the photochemical processes observed for the dinuclear DTE system, time-dependent DFT (TDDFT) analysis [60–63] was performed for singlet and triplet excited states of the $CpM(CO)(PMe_3)$ complexes **3#Fe** and **3#Ru** (simplified PMe_3-substituted analogues of **3Fe** and **3Ru**, respectively).

It has been established for organic DTE molecules that the ring closure occurs via the lowest singlet excited state (corresponding to S_9 for **3#Fe** in Fig. 11.10a) [64, 65]. By contrast, De Cola and her coworkers recently studied photophysical properties of a transition metal complex **6**, where the photochemically active $Ru(bpy)_3$ fragments were attached to DTE (Fig. 11.9), and proposed that the ring closure of **6** proceeds not only via the ligand(DTE)-centered singlet state but also via the ligand-centered triplet state resulting from energy transfer processes by way of the metal-based excited states [66]. Taking into account for these two extreme cases, we have carried out TDDFT analysis. To be explained is the following photocyclization tendency observed for the central metal: Ru > Fe.

Energy levels for Franck–Condon states for the two complexes are summarized in Fig. 11.10. Singlet and triplet states are denoted by bold lines and plain lines,

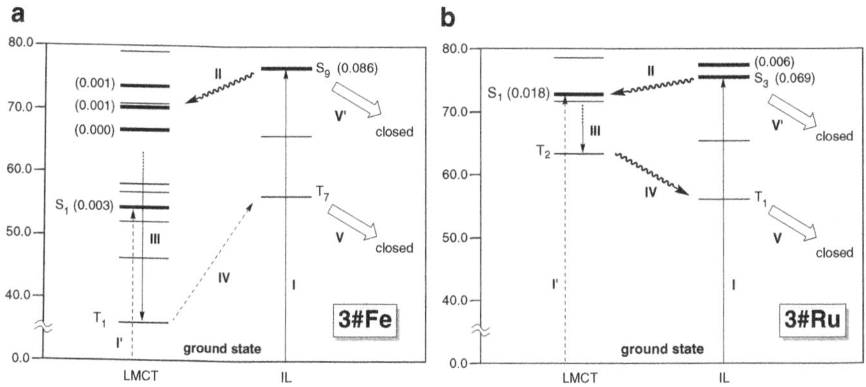

Fig. 11.10 Energy diagrams for Frank–Condon excited states of **3#Fe** (**a**) and **3#Ru** (**b**). *Bold lines*: singlet states; *plain lines*: triplet states

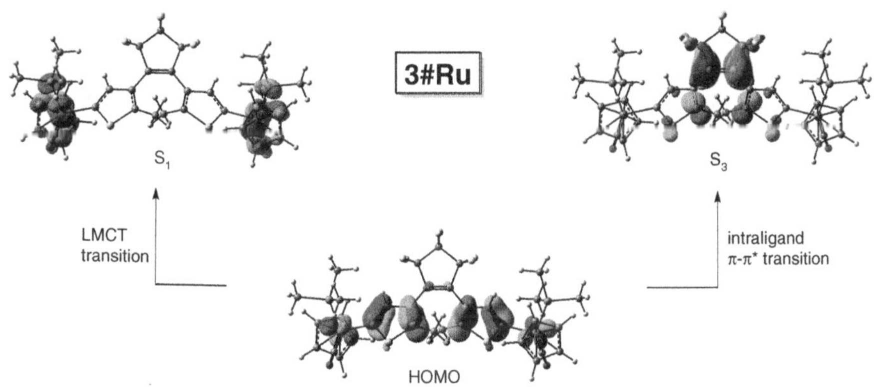

Fig. 11.11 HOMO and the lowest metal-(S_1) and ligand-based singlet excited states (S_3) for **3#Ru**

respectively, and labels for the excited states of higher energies are omitted for clarity. The excited states can be divided into two categories, LMCT and IL (intraligand). The excited states of the LMCT series are metal d-orbital-based, whereas those of the IL series are DTE-based, as can be seen, for example, from the S_1 and S_3 states of the CpRu(CO)(PMe_3) complex **3#Ru** (Fig. 11.11). In addition, the metal-based LMCT series orbitals contain antibonding combinations between the metal d-orbital and the ligand orbitals (Cp, CO, and PR_3). These features are common to these complexes discussed herein.

Let us consider the chemical processes of **3#Ru** (Fig. 11.10b). Irradiation causes excitation of a HOMO electron to singlet orbitals of higher energies. The S_1 state resulting from LMCT transition lies substantially lower in energy than the S_3 state resulting from IL transition, suggesting that the initial excitation may occur toward the S_1 state preferentially. The oscillator strength shown in parentheses, which indicates relative transition probability, clearly demonstrates that the GS(S_0)→S_3 transition (**I**) is the most probable and major initial photochemical event; the oscillator strength for the GS→S_3 transition (0.069) is substantially larger than the negligible oscillator strength for other transitions (<0.018). The energy level of S_3 is in the UV region in accordance with the lack of a visible band for **3Ru**. Subsequent energy transfer to a LMCT excited state (**II**) followed by intersystem crossing and relaxation according to the Kasha's rule (**III**) leads to the lowest LMCT triplet state T_2. Further back energy transfer toward the DTE part leads to the lowest excited state T_1 with the significant DTE-based character (**IV**). Let us point out that the MO features for the DTE moiety in T_1 (MO is the same as that for S_3 depicted in Fig. 11.11), i.e., (a) antibonding combination of the p-orbitals of the C=C moiety in the cyclopentene ring, (b) π-bonding interactions at the thiophene–cyclopentene junctions, and (c) in-phase combination of the p-orbitals of the 2- and 2′-carbon atoms in the thiophene rings to be connected in the closed isomer, are reminiscent of the molecular orbital of a closed cyclohexadiene skeleton. These features tell us that the final ring-closing process from T_1 (**V**) is a likely process.

Then we examined the effects of the metal centers. For the iron analogue **3#Fe**, an energy diagram similar to that for **3#Ru** is obtained (Fig. 11.10a) but a significant difference is noted for the relative energy gap between the excited states associated with step **IV**. The DTE-based T_7 state lies significantly higher in energy than the lowest metal-based triplet state T_1 so that the final endothermic process **IV** is not viable and, as a result, the iron complex **3#Fe** eventually deactivates to the ground state without cyclization. The dependence of the efficiency of the photocyclization on the central metal (Ru>Fe) has been successfully interpreted in terms of the mechanism involving the crucial metal-to-DTE energy transfer step **IV**. The inertness of **3#Fe** reveals that the ring closure of the organometallic DTE compounds may proceed via triplet excited states, because, otherwise, photocyclization should occur directly via the initial singlet state (S_9) in a manner similar to organic derivatives. The inertness suggests that energy transfer from S_9 to LMCT triplet states is much faster than direct ring closure from S_9.

More recently, our collaborators, the group of Dr. Nakamura, reported on explanation for ancillary ligand effect on the photocyclization of **2Fe** and **3#Fe** [67]. However, the origin of the dependence on the metal fragment is still open to further discussion and investigation.

11.3.4 Mononuclear Iron and Ruthenium Complexes with Dithienylethene Ligand

A series of mononuclear DTE complexes, ***M***–DTE (***M***=(η^5–C_5H_5)Fe(CO)$_2$ (**7Fe**), (η^5–C_5H_5)Ru(CO)$_2$ (**7Ru**), (η^5–C_5H_5)Fe(CO)(PPh_3) (**8Fe**), (η^5–C_5H_5)Ru(CO)(PPh_3) (**8Ru**), and (η^5–C_5H_4(Me))Fe(dppe) (**9Fe**)), was prepared in a manner similar to the synthesis of dinuclear complexes.

Table 11.2 Photochromic behavior of the mononuclear DTE complexes (**7**–**9**)

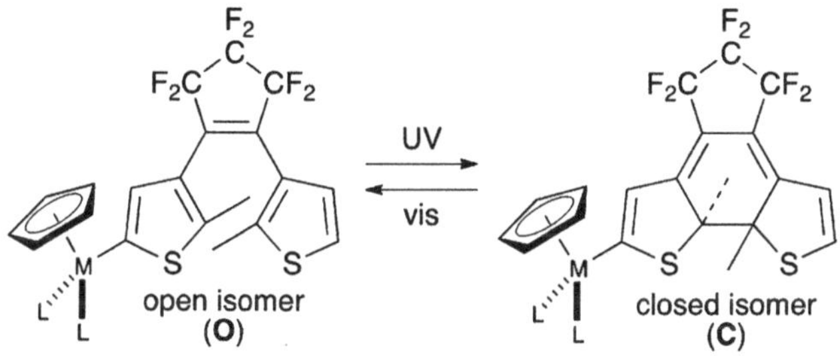

Complex	*M* ((η^5-C_5H_5) ML_2)	**O**/**C**[a]	Ring closure[b] (time/min)	Ring opening[b] (time/min)	λ_{max}/nm of **C** (color)	Recycl. (times)
7Fe	$CpFe(CO)_2$	–	–	–	–	Decomp.
7Ru	$CpRu(CO)_2$	62/38	2.5	2	547 (reddish purple)	70 % (2)
8Fe	CpFe(CO) (PPh_3)	–	–	–	–	Decomp.
8Ru	CpRu(CO) (PPh_3)	58/42	3	3	570 (purple)	>90 % (7)
9Fe	CpFe(dppe)	100/0	–	–	–	–

[a]Isomer ratios at the photostationary states in C_6D_6 determined by ^1H NMR
[b]Monitoring 2.0×10^{-5} M in THF by UV–vis spectra

Photochromic behavior of ***M***–DTE is summarized in Table 11.2. It is revealed that the structure of the metal fragments dramatically affected the photochromic properties as observed in dinuclear systems. In the photochromic performance, the mononuclear complexes exhibited lower than the corresponding dinuclear complexes did (Table 11.1).

The mononuclear DTE complexes are also redox-active. CV traces of **9Fe** and **8Ru** apparently contained single reversible redox waves at 450 and 335 mV, respectively. These results suggested that the one-electron redox processes might induce reversible color change of the mononuclear DTE complexes because transition metal fragments can serve as chromophores originated in d–d, MLCT, and LMCT transitions. In fact, radical species $\mathbf{O}^{+\bullet}$ derived from 1e-oxidation of redox-active metal fragments have a characteristic absorption band assigned to LMCT around visible light region (**9Fe**: 676 nm, **8Ru**: 500 nm). As mentioned above, 2e-oxidation of the DTE derivatives with two redox-active organometallic fragments causes the ring closure giving the dicationic closed isomer $\mathbf{C}^{2+}$. That 2e-redox process is not reversible as shown in Figs. 11.7 and 11.8. It is notable, in the mononuclear systems, the color of the open isomer **O** can be changed into two directions in reversible manners (dual chromism), i.e., photochemical ring closure/opening process of DTE moiety between **O** and **C** and 1e-redox process of the organometallic fragment between **O** and $\mathbf{O}^{+\bullet}$ as shown in Fig. 11.12 [68].

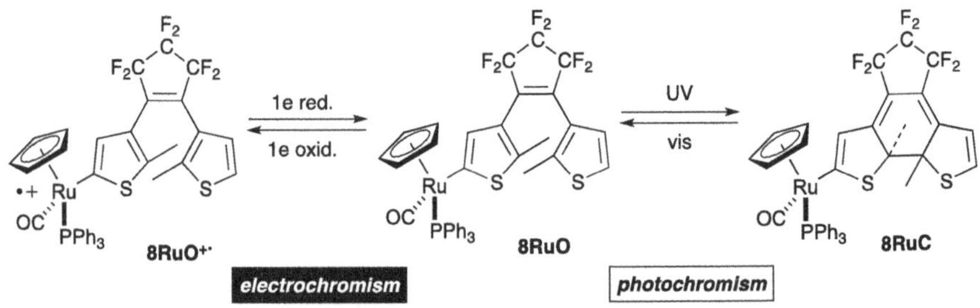

Fig. 11.12 Dual chromism of the mononuclear DTE complexes

These results suggest that combination of a redox-active metal fragment with a DTE makes it dual chromic system (photo- and electrochromism), leading to plural colors and states. Recently, Humphrey and coworkers showed the *multi-stimuli–responsive systems* composed with DTE, redox-active metal fragments, and pH-responsive C≡C units [69]. They have six different states, which can be logically switched by application of the three different stimuli (light, redox, and pH) and detected by a single technique (NLO).

11.4 Conclusion

Photochromic organometallics with a ***M***–C (***M***: redox-active metal fragments, C: carbon atom in the photochromic unit = dithienylethene (DTE)) bond show intriguing stimuli–responsive functions as described above. The functions can lead to molecular devices and sophisticated logic systems.

11.4.1 Molecular Devices

DTE is an excellent switch for π-conjugated systems. Organometallic molecular wires are not also exceptions. Redox properties can be switched photochemically. Switching factors and photochemical process are strongly affected by the structure of metal fragments (Fe or Ru, auxiliary ligands). The basic information on molecular design of metal complexes with photochromic properties has been collected.

11.4.2 Multimodal Stimuli–Responsive Systems

Attachment of a redox-active metal fragment to a photochromic system can make it dual chromic system (photo- and electrochromism). Organic photochromic ligand and redox-active organometallic fragments can serve as photochrome and electrochrome, respectively. Molecular design is important to construct such stimuli–responsive systems. In particular, the number of metal fragments changes their chromic mechanism. Further combination with other chromic systems and metal fragments should contribute to the development of sophisticated logic systems.

Photochromic organometallics are promising molecular systems for molecular devices and logic systems, although a series of systematic studies is needed for realization. Integration of these stimuli–responsive systems would lead to smart chemical systems.

Acknowledgments This research was financially supported by the Ministry of Education, Culture, Sports, Science and Technology of the Japanese Government (the Grant-in-Aid for Scientific Research on Priority Areas, "New Frontiers in Photochromism (No. 471)" (T.K.), which is gratefully acknowledged. We are also grateful to the Japan Society for Promotion of Science for the Grant-in-Aid for Scientific Research (No. 22350026) (M.A.). We thank Dr Shinichiro Nakamura (RIKEN) for his valuable comments and assistance for computational analysis. A generous gift of perfluorocyclopentene, the starting compound for DTE, from ZEON Corporation is gratefully acknowledged.

References

1. Schwartz M (ed) (2002) Encyclopedia of smart materials. Wiley, New York
2. Bamfield P, Hutchings MG (eds) (2010) Chromic phenomena, technological applications of color chemistry, 2nd ed. Royal Society of Chemistry, Cambridge
3. Irie M (2000) Diarylethenes for memories and switches. Chem Rev 965:1685–1716
4. Irie M, Uchida K (1998) Synthesis and properties of photochromic diarylethenes with heterocyclic aryl groups. Bull Chem Soc Jpn 71:985–996
5. Wang M-S, Xu G, Zhang Z-J, Guo G-C (2010) Inorganic-organic hybrid photochromic material. Chem Commun 46:361–376
6. Tian H, Yang S (2004) Recent progress on diarylethene based photochromic switches. Chem Soc Rev 33:85–97
7. Duerchais V, Le Bozec H (2010) Metal complexes featuring photochromic ligands. Top Organomet Chem 28:171–225 and references therein
8. Ko C-C, Yam VW-W (2010) Transition metal complexes with photochromic ligands-photosensitization and photoswitchable properties. J Mater Chem 20:2063–2070
9. Hasegawa Y, Nakagawa T, Kawai T (2010) Recent progress of luminescent metal complexes with photochromic units. Coord Chem Rev 254:2643–2651
10. Akita M (2011) Photochromic organometallics, a stimuli-responsive system: an approach to smart chemical systems. Organometallics 30:43–51 and references therein
11. Nakai H, Isobe K (2010) Photochromism of organometallic compounds with structural rearrangement. Coord Chem Rev 254:2652–2662 and references therein
12. Jortner J, Ratner M (1997) Molecular electronics. Blackwell, Oxford
13. Aviram A, Ratner M (1998) Molecular electronics: science and technology. Ann N Y Acad Sci
14. Ratner M (2000) Molecular electronics: pushing electrons around. Nature 404:137–138
15. Tour JM (2000) Molecular electronics. Synthesis and testing of components. Acc Chem Res 33:791–804
16. Hipps KW (2001) It's all about contacts. Science 294:536
17. Cahen D, Hodes G (2002) Molecules and electronic materials. Adv Mater 14:789–798
18. Caroll RL, Gorman CB (2002) The genesis of molecular electronics. Angew Chem Int Ed 41:4378–4400

19. Robertson N, McGowan GA (2003) A comparison of potential molecular wires as components for molecular electronics. Chem Soc Rev 32:96–103
20. Reed MA, Lee T (eds) (2003) Molecular nanoelectronics. American Scientific Publishers, Stevenson Ranch
21. Balzani V, Credi A, Venturi M (2003) Molecular devices and machines. Wiley, Weinheim
22. Flood AH, Stoddart JF, Steuerman DW, Heath JR (2004) Where molecular electronics? Science 306:2055–2056
23. Nørgaard K, Bjørnholm T (2005) Supramolecular chemistry on water–towards self-assembling molecular electronic circuitry. Chem Commun 1812–1823
24. Low PJ (2005) Metal complexes in molecular electronics: progress and possibilities. Dalton Trans 2821–2824
25. Benniston AC (2004) Pushing around electrons: towards 2-D and 3-D molecular switches. Chem Soc Rev 33:573–578
26. Schwab PFH, Levin MD, Michl J (1999) Molecular rods. 1. Simple axial rods. Chem Rev 99:1863–1934
27. Schwab PFH, Smith JR, Michl J (2005) Synthesis and properties of molecular rods. 2. Zig–zag rods. Chem Rev 105:1197–1280
28. Zheng QL, Gladysz JA (2005) A synthetic breakthrough into an unanticipated stability regime: readily isolable complexes in which C_{16}–C_{28} polyynediyl chains span two platinum atoms. J Am Chem Soc 127:10508–10509
29. Touchard D, Dixneuf PH (1998) A new class of carbon-rich organometallics. The C_3, C_4 and C_5 metallacumulenes $Ru{=}(C{=})_nCR_2$. Coord Chem Rev 178–180:409–429
30. Diederich F, Stang PJ, Tykwinski RR (2004) Acetylene chemistry. Wiley, Weinheim
31. Paul F, Lapinte C (1998) Organometallic molecular wires and other nanoscale-sized devices: an approach using the organoiron (dppe)Cp*Fe building block. Coord Chem Rev 178–180:431–509
32. Ghazala SI, Paul F, Toupet L, Roisnel T, Hapiot P, Lapinte C (2006) Di-organoiron mixed valent complexes featuring "(η^2–dppe)(η^5–Me_5)Fe" endgroups: smooth class-III to class-II transition induced by successive insertion of 1,4-phenylene units in a butadiyne-diyl bridge. J Am Chem Soc 128:2463–2476
33. De Montigny F, Aegrouarch G, Costuas K, Halet J-F, Roisnel T, Toupet L, Lapinte C (2005) Electron transfer and electron exchange between $[Cp^*(dppe)Fe]n^+$ (n=0, 1) building blocks mediated by the 9,10-bis(ethynyl)anthracene bridge. Organometallics 24:4558–4572
34. Le Narvoir N, Lapinte C (1995) 1,4-Diethynylbenzene bridged Fe(Cp*)(dppe) units: mixed-valence 35-electron and bisiron(III) 34-electron complexes. Organometallics 14:634–639
35. Le Stang S, Paul F, Lapinte C (2000) Molecular wires: synthesis and properties of the new mixed-valence complex $[Cp^*(dppe)Fe{-}C{\equiv}C{-}X{-}C{\equiv}C{-}Fe(dppe)Cp^*][PF_6]$ (X=2,5-C_4H_2S) and comparison of its properties with those of the related all-carbon-bridged complex (X=–C_4–). Organometallics 19:1035–1043
36. Antonova AB, Bruce MI, Ellis BG, Gaudio M, Humphrey PA, Jevric M, Melino G, Nicholson GK, Perkins GJ, Skelton BW, Stapleton B, White AH, Zaitseva NN (2004) A novel methodology for the synthesis of complexes containing long carbon chains linking metal centres: molecular structures of $\{Ru(dppe)Cp^*\}_2(\mu\text{-}C_{14})$ and $\{Co_3(\mu\text{-dppm})(CO)_7\}_2(\mu_3{:}\mu_3\text{-}C_{16})$. Chem Commun 960–961
37. Bruce MI, Costuas K, Ellis BG, Halet J-F, Low PJ, Moubaraki B, Murray KS, Ouddai N, Perkins GJ, Skelton BW, White AH (2007) Redox-active complexes containing group 8 metal centers linked by C_2 bridges. Organometallics 26:3735–3745
38. Dembinski R, Bartik T, Bartik B, Jaeger M, Gladysz JA (2000) Toward metal-capped one-dimensional carbon allotropes: wirelike C_6–C_{20} polyynediyl chains that span two redox-active (η^5–Me_5)Re(NO)(PPh_3) endgroups. J Am Chem Soc 122:810–822
39. Stahl J, Mohr W, De Quadras L, Peters TB, Bohling JC, Martín-Alvarez JM, Owen GR, Hampel F, Gladysz JA (2007) sp Carbon chains surrounded by sp^3 carbon double helices: coordination-driven self-assembly of wirelike Pt(C≡C)nPt moieties that are spanned by two $P(CH_2)mP$ linkages. J Am Chem Soc 129:8282–8295

40. De Quadras L, Bauer EB, Mohr W, Bohling JC, Peters TB, Martín-Alvarez JM, Hampel F, Gladysz JA (2007) sp Carbon chains surrounded by sp^3 carbon double helices: directed syntheses of wirelike Pt(C≡C)nPt moieties that are spanned by two $P(CH_2)mP$ linkages via alkene metathesis. J Am Chem Soc 129:8296–8309
41. De Quadras L, Bauer EB, Stahl J, Zhuravlev F, Hampel F, Gladysz JA (2007) sp Carbon chains surrounded by sp^3 carbon double helices: wire-like Pt(C≡C)nPt moieties that are spanned by two α, ω-diphosphines that bear heteroatoms or alkyl substituents. New J Chem 31:1594–1604
42. Akita M, Koike T (2008) Organometallic chemistry of polycarbon species: from clusters to molecular devices. Dalton Trans 3523–3530
43. Tanaka Y, Inagaki A, Akita M (2007) A Photoswitchable molecular wire with the dithienylethene (DTE) linker, (dppe)(η^5–C_5Me_5)Fe–C≡C–DTE–C≡C–Fe(η^5–C_5Me_5)(dppe). Chem Commun 1169–1171
44. Fraysse S, Coudret C, Launay J-P (2000) Synthesis and properties of dinuclear complexes with a photochromic bridge: an intervalence electron transfer switching "On" and "Off". Eur J Inorg Chem 1581–1590
45. Tanaka Y, Ishisaka T, Inagaki A, Koike T, Lapinte C, Akita M (2010) Photochromic organometallics with a dithienylethene (DTE) bridge, [Y–C≡C–DTE–C≡C–Y] (Y={MCp*(dppe)}): photoswitchable molecular wire (M=Fe) versus dual photo- and electrochromism (M=Ru). Chem Eur J 16:4762–4776
46. Irie M, Lifka T, Kobatake S, Kato N (2000) Photochromism of 1,2-bis(2-methyl-5-phenyl-3-thienyl)perfluorocyclopentene in a single-crystalline phase. J Am Chem Soc 122:4871–4876
47. Motoyama K, Koike T, Akita M (2008) Remarkable switching behavior of bimodally stimuli-responsive photochromic dithienylethenes with redox-active organometallic attachments. Chem Commun 5812–5814
48. Motoyama K, Li H, Koike T, Hatakeyama M, Yokojima S, Nakamura S, Akita M (2011) Photo- and electro-chromic organometallics with dithienylethene (DTE) linker, L_2CpM-DTE-$MCpL_2$: dually stimuli-responsive molecular switch. Dalton Trans 40:10643–10657
49. Petz W (1992) Iron–carbene complexes. Springer, Berlin
50. Brookhart M, Studabaker WB (1987) Cyclopropanes from reactions of transition metal carbene complexes with olefins. Chem Rev 87:411–432
51. Liu YF, Lagrost C, Constuas K, Touchar N, Le Bozac H, Rigaut S (2008) A multifunctional organometallic switch with carbon-rich ruthenium and diarylethene units. Chem Commun 6117–6118
52. Lin Y, Yuan JJ, Hu M, Yin J, Jin S, Liu SH (2009) Syntheses and properties of binuclear ruthenium vinyl complexes with dithienylethene units as multifunction switches. Organometallics 28:6402–6409
53. Kawai SH, Gilat SL, Ponsinet R, Lehn J-M (1995) A dual-mode molecular switching device: bisphenolic diarylethenes with integrated photochromic and electrochromic properties. Chem Eur J 1:285–293
54. Koshido T, Kawai T, Yoshino K (1995) Optical and electrochemical properties of cis-1,2-dicyano-1,2-bis(2,4,5-trimethyl-3-thienyl)ethane. J Phys Chem 99:6110–6114
55. Peters A, Branda NR (2003) Electrochromism in photochromic dithienylcyclopentenes. J Am Chem Soc 125:3404–3405
56. Peters A, Branda NR (2003) Electrochemically induced ring-closing of photochromic 1,2-dithienylcyclopentenes. Chem Commun 954–955
57. Zhou X-H, Zhang F-S, Yuan P, Sun F, Pu S-Z, Zhao F-Q et al (2004) Photoelectrochromic dithienylperfluorocyclopentene derivatives. Chem Lett 33:1006–1007
58. Moriyama Y, Matsuda K, Tanifuji N, Irie S, Irie M (2005) Electrochemical cyclization/cycloreversion reactions of diarylethenes. Org Lett 7:3315–3318
59. Guirado G, Coudret C, Hliwa M, Launay J-P (2005) Understanding electrochromic processes initiated by dithienylcyclopentene cation-radicals. J Phys Chem B 109:17445–17459

60. Grabom T, Petersilka M, Gross EKU (2000) Molecular excitation energies from time-dependent density functional theory. THEOCHEM 501:353–367
61. Furche F (2001) On the density matrix based approach to time-dependent density functional response theory. J Chem Phys 114:5982–5992
62. Becke AD (1993) Density functional thermochemistry. III. The role of exact exchange. J Chem Phys 98:5648–5652
63. Lee C, Yang W, Parr RG (1998) Development of the colle-salvetti correlation-energy formula into a functional of the electron density. Phys Rev B 37:785–789
64. Hania PR, Telesca R, Lucas LN, Pugzlys A, Van Esch J, Feringa BL, Snijders JG, Duppen K (2002) An optical and theoretical investigation of the ultrafast dynamics of a bisthienylethene-based photochromic switch. J Phys Chem A 106:8498–8507
65. Guillaumont D, Kobayashi T, Kanda K, Miyasaka H, Uchida K, Kobatake S, Shibata K, Nakamura S, Irie M (2002) An ab Initio MO study of the photochromic reaction of dithienylethenes. J Phys Chem A 106:7222–7227
66. Jukes RTF, Adamo V, Hartl F, Belser P, De Cola L (2004) Photochromic dithienylethene derivatives containing Ru(II) or Os(II) metal units. Sensitized photocyclization from a triplet state. Inorg Chem 43:2779–2792
67. Hatakeyama M, Yokojima S, Shinoda K, Koike T, Akita M, Nakamura S (2012) Theoretical study on the photocyclization mechanism of diarylethenes with transition-metal substituents. Bull Chem Soc Jpn 85:679–686
68. Li H, Koike T, Akita M (2012) Color change of redox-active organometallic dithienylethene complexes by photochemical and redox processes. Dyes Pigm 92:854–860
69. Green KA, Cifuentes MP, Corkery TC, Samoc M, Humphrey MG (2009) Switching the cubic nonlinear optical properties of an electro-, halo-, and photochromic ruthenium alkynyl complex across six states. Angew Chem Int Ed 48:7867–7870

Chapter 12
Ultrafast Dynamics and Mechanisms of One-Photon and Multiphoton Photochromic Reactions

Yukihide Ishibashi, Tetsuro Katayama, and Hiroshi Miyasaka

Abstract Dynamics and mechanisms of photochromic reactions of organic molecular systems, such as diarylethene, fulgide, and hexaarylbiimidazole derivatives, were investigated by time-resolved measurements in femtosecond to picosecond regions. Ultrafast cyclization and cycloreversion reactions in diarylethene derivatives were directly detected by femtosecond laser photolysis. By integrating the temperature dependence of faction of conformers, reaction yields, and dynamic behaviors, the mechanisms of these reactions in the excited states were discussed on the basis of the adiabatic potential surfaces. For the photodissociation dynamics of hexaarylbiimidazole derivatives, the role of the local excited state in the rapid reaction on the dissociative potential surface was discussed by introducing the ultrafast dynamics in these systems with different aromatic substituents. In addition to the elucidation of dynamic behaviors of cyclization, cycloreversion, and photodissociation in these systems by one-photon excitation, nonlinear photochromic reactions were also studied. Stepwise two-photon absorption leading to a specific higher excited state, which was not accessible via one-photon absorption from the ground state, induced drastic enhancement of the cycloreversion reaction (>2,500 times compared to the one-photon process). In addition, one-color reversible control of both directions of photochromic reactions was attained for a diarylethene derivative through simultaneous three-photon cyclization and two-photon cycloreversion using femtosecond near-IR laser pulse at 1.28 μm.

Keywords Diarylethene • Fulgide • Multiphoton-gated photochromic reaction • One-color reversible control • Ultrafast transient absorption spectroscopy

---

Y. Ishibashi
Graduate School of Science and Engineering, Ehime University, Matsuyama, Japan

T. Katayama • H. Miyasaka (✉)
Division of Frontier Materials Science, Graduate School of Engineering Science, Osaka University, Toyonaka, Japan
e-mail: miyasaka@chem.es.osaka-u.ac.jp

M. Irie et al. (eds.), *New Frontiers in Photochromism*,
DOI 10.1007/978-4-431-54291-9_12, © Springer Japan 2013

12.1 Introduction

Various kinds of photochromic molecules have been developed so far, such as azobenzene, spiropyran, spirooxazine, hexaarylbiimidazole, fulgide, diarylethene, and so forth [1–6]. Some of these systems are covered in other chapters in this book in detail. The first four derivatives undergo the colorization reactions upon UV light irradiation, while the decolorization reactions can occur without light irradiation in these systems (T-type). On the other hand, fulgide and diarylethene derivatives have large activation barrier between the colored and uncolored isomers in the ground state, so that both colorization and decolorization reactions need the photoexcitation at an appropriate wavelength (P-type).

Photochemical reactions take place in competition with various processes such as radiative and nonradiative transitions in a finite lifetime of the excited state. In addition, rapid reactions in unrelaxed excited states with excess vibrational energy and in high electronically excited states sometimes contribute to the overall reaction processes. Accordingly, direct detection of these processes with high temporal resolution is quite important to acquire the information on the dynamics and mechanism of the reaction profiles in the excited state and leads to the detailed elucidation of photofunctional responses and the rational principles for the advanced molecular systems [7]. Moreover, pulsed laser excitation can open a new response of the photochromic systems via nonlinear processes. In the present chapter, we discuss the reaction dynamics and mechanisms of several organic photochromic molecules as revealed by ultrafast laser spectroscopy. In addition, we introduce the photochromic reactions induced by multiphoton processes under the pulsed laser excitation, such as multiphoton-gated photochromic reactions and one-color reversible control of both colorization and decolorization through higher-order simultaneous multiphoton absorption processes.

12.2 Photochromic Reaction Dynamics of Diarylethene Derivatives

Diarylethene derivatives, which were originally invented by Irie et al. [3, 4], have various excellent properties, such as thermal stability of both isomers and fatigue resistance. In addition, some of the derivatives can undergo photochromic reactions even in the crystalline phase. These properties have been attracting much attention not only from the viewpoint of the application to various optical and photonic devices and photo-actuators but also from the basic viewpoint of the elucidation of photoinduced chemical reactions [3, 4, 8–11].

The photochromism of diarylethene derivatives is based on the photoinduced cyclization (ring-closing) and cycloreversion (ring opening) reactions as shown in Fig. 12.1a. The open-ring isomer (left-hand side) has in general absorption bands only in UV region, while the closed-ring isomer has absorption bands also in visible

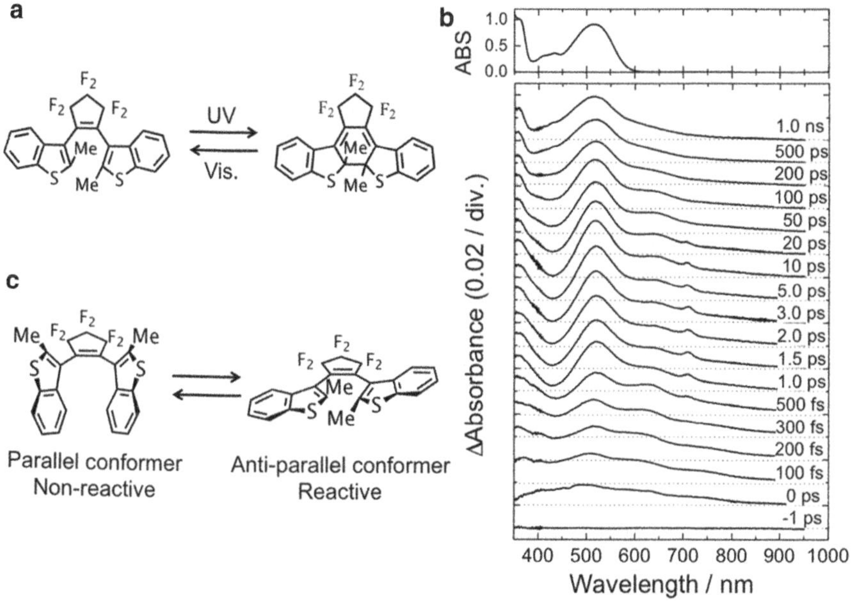

Fig. 12.1 (**a**) Molecular structures and photochromic reactions of a diarylethene derivative. (**b**) Transient absorption spectra of **BT(o)** in *n*-hexane solution, excited with a fs 320-nm laser pulse. The spectrum in the top column is a steady-state absorption spectrum of **BT(c)**. (**c**) Conformers of **BT(o)** in the ground state

region. Accordingly, the UV light irradiation can induce the colorization, while the selective excitation in the visible light for the closed-ring isomer leads to the decolorization via the cycloreversion reaction.

12.2.1 Direct Detection of Cyclization Dynamics

Figure 12.1b exhibits time-resolved transient absorption spectra of the open-ring isomer of **BT**, **BT(o)**, in *n*-hexane solution, excited with a fs 320-nm laser pulse. Broad absorption band around 500 nm appears immediately after the excitation, followed by the gradual evolution into a new band at 520 nm in a few ps time range. Spectral band shape and the absorption maximum are in agreement with those of the closed-ring isomer, **BT(c)**, as shown in the top column of Fig. 12.1b, indicating that the cyclization takes place very rapidly. From the analysis of time profiles of the transient absorbance, the time constant of the cyclization reaction was obtained to be 450 fs. After this rapid cyclization reaction with the 450-fs time constant, one can still find spectral evolution in a few tens of ps to a few ns time region. The time constant for this slow evolution was 150 ps, of which value was in agreement with the decay time constant of the fluorescence of **BT(o)** in *n*-hexane solution.

For various diarylethene derivatives, it has been reported [3, 4, 8, 9] that at least two conformers exist in the open-ring isomer in the ground state, as shown in Fig. 12.1c. One of the two conformers has a molecular structure with C_{2V} symmetry (antiparallel conformer, AP), while the other one is in the C_S symmetry (parallel conformer, P). These two conformers are generally distinguishable in the steady-state NMR spectroscopy, indicating that the interconversion between the two conformers takes place in the time region longer than ms in the ground state. As shown in Fig. 12.1c, cyclization reaction can occur in the AP conformer, while it seems difficult for the P conformer to undergo the cyclization reaction accompanied with the large geometrical rearrangement in the short lifetime of the excited state.

In order to elucidate the role of the 150-ps component in the cyclization reaction, the solvent viscosity effect of the fluorescence lifetime and the cyclization reaction yield was investigated [12]. Because the geometrical rearrangement of solute molecules is usually affected by the viscosity of the solvent, the lifetime and the reaction yield may depend on the solvent viscosity in the case that the 150-ps component participates in the cyclization reaction. The experimental result, however, did not show any solvent viscosity effect for the lifetime or reaction yield. In addition, no apparent rise of the closed-ring isomer was observed in the time evolution of the transient absorption spectra. From these results, it was concluded that the 150-ps component did not take a significant role in the cyclization reaction [12] and this species was assigned to the excited singlet state of the P conformer of **BT(o)**. After the decay of the 150-ps component, the transient absorption spectra of Fig. 12.1b in ns time region still show some broad absorption band in the wavelength region >600 nm where **BT(c)** has no absorption. This broad absorption was attributed to the triplet state (T_1) of **BT(o)** in the P conformer, which is produced through the decay of the excited singlet state of **BT(o)** with the 150-ps time constant. From the laser spectroscopy covering μs time region and the effect of the oxygen on the cyclization yield, it was concluded that the T_1 state of the P conformer does not contribute to the cyclization reaction. That is, the cyclization reaction takes place only through the rapid reaction channel with the 450-fs time constant for the present derivative.

Rapid cyclization reactions within a few ps were confirmed also in other diarylethene derivatives in solutions as well as in the crystalline phase [13–23]. From the theoretical investigations on the potential energy surfaces of diarylethene derivatives both in the ground and excited states [24–28], it was predicted for the AP conformer of the open-ring isomer that the reaction path of the cyclization had no activation barrier between the excited Franck–Condon (FC) state and the reaction point. This theoretical prediction is in accordance with the rapid cyclization reaction as observed in the experimental investigations.

Although most of the cyclization processes of diarylethene derivatives occur within a few ps following the UV excitation, it was reported for some diarylethene derivatives that slow cyclization processes with time constants of a few hundreds of ps could be observed in addition to the rapid reaction channel within a few ps [17]. In this system, it was suggested that the excited Franck–Condon state of the AP conformer undergoes the relaxation of the molecular geometry in competition with the rapid cyclization, leading to the structure with skewed thiophene rings owing to the charge transfer in the excited state [12, 17]. The slow cyclization process, which is strongly dependent on the solvent viscosity [17], was attributed to the reaction from the excited state with the relaxed molecular geometry. It is also worth noting that the triplet state of the diarylethene derivative undergoes the cyclization reaction in some cases [29–31]. From recent experimental investigation [31], it was revealed that the sensitization (T_1 to S_0 energy transfer) yielding the T_1 state of the AP conformer of **BT(o)** led to the cyclization but no cyclization occurred through the triplet state of the P conformer of **BT(o)**. Usual one-photon excitation prepares both P and AP conformers in the excited singlet state but the excited singlet state of the AP conformer undergoes the cyclization very rapidly. Hence, the T_1 state of the AP conformer is not produced [12] in the normal excitation condition. The result on the triplet sensitization confirms the experimental results on the usual excitation of the S_0 state and adds new channel of the cyclization via the sensitization. At any rate, the interconversion of the AP and P conformers is difficult to occur even in the triplet lifetime of a few tens of μs. This slow interconversion is consistent with the result of the NMR spectroscopy for the ground state. Cyclization in the triplet manifold was reported also for the intramolecular dyad systems with a triplet sensitizer and a diarylethene derivative [29, 30]. The formation of the triplet state of the AP conformer seems an important factor for the triplet cyclization to occur also in these dyad systems.

12.2.2 Direct Detection of Cycloreversion Dynamics

Cycloreversion processes of diarylethene derivatives have been also investigated by ultrafast laser photolysis methods [26, 27, 32–42]. The excited state lifetimes of the closed-ring isomer are in the range of a few ps to a few tens of ps, as shown in Table 12.1. These lifetimes are two to three orders shorter than the fluorescent lifetimes of typical organic dyes but clear correlation between the excited state lifetime and the cycloreversion reaction yield was not obtained. This result suggests the presence of several nonradiative decay channels competing with the cycloreversion reaction. In general, it has been observed that the cycloreversion reaction yield of the closed-ring isomer is temperature-dependent and it increases with an increase in the temperature.

12.2.3 Experimental Studies on the Potential Energy Surfaces of Cycloreversion and Cyclization Reactions

From a number of experimental results by steady-state and time-resolved measurements on the cycloreversion reactions, it has been predicted that the activation barrier exists in the adiabatic potential surface from the minimum position of the

Table 12.1 Excited-state lifetime and cycloreversion reaction yield of several diarylethene derivatives in the closed-ring isomer

Molecule	Excited-state lifetime (ps)	Cycloreversion reaction yield	Environment (solvent)	References
1	0.7	10^{-5}	*n*-Hexane	[32]
2	2.1	0.001	1,2-Dichloroethane	[33]
3	2.0	0.0013	Chloroform	[34]
4	10	0.013	*n*-Hexane	[35]
	15	≦0.01	PMMA	[36]
	25	0.017	Polycrystal	[23, 37]
5	13	0.04	1,2-Dichloroethane	[26]
6	1.3	0.077	*n*-Hexane	[38]
7	8	0.077	*n*-Hexane	[27]
8	25	0.35	*n*-Hexane	[39, 40, 42]

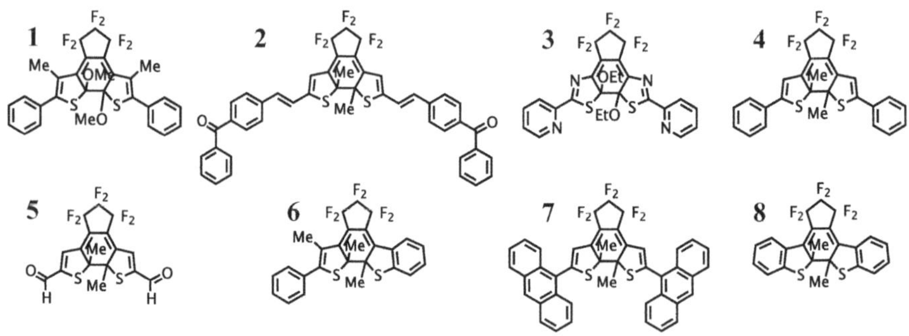

excited state of the closed-ring isomer to the conical intersection (CI) where the fate of the excited molecules is determined. In addition, it has been proposed from the theoretical investigations [24–28] that the cyclization and cycloreversion in the excited state occur in the adiabatic potential surface with the same CI. In order to experimentally obtain the detailed information on the potential surfaces responsible for the cyclization and cycloreversion reactions, temperature dependence of the excited state dynamics and reaction yields of **BT** in Fig. 12.1 was investigated in *n*-hexane solution [42].

For the cyclization reaction of **BT(o)**, it was confirmed that the reaction yield was independent of the temperature in the range of the investigation (253–343 K). In addition, no temperature dependence of the fraction of the AP conformer in **BT(o)**, 0.65, was observed in this temperature range from the steady-state NMR spectroscopy. These two results led to the conclusion that the cyclization reaction is actually temperature-independent. It should be noted that the apparent reaction yield of the cyclization is 0.3 and the fraction of the AP conformer is 0.65. Hence, the cyclization reaction yield could be estimated to be 0.45 for the excited state of the AP conformer provided that the extinction coefficients of both AP and P conformers are almost the same at the excitation wavelength. In addition, the time constant of the cyclization reaction was quite short, 450 fs, as was introduced in Fig. 12.1. From these results, it was concluded that the excited state of the AP

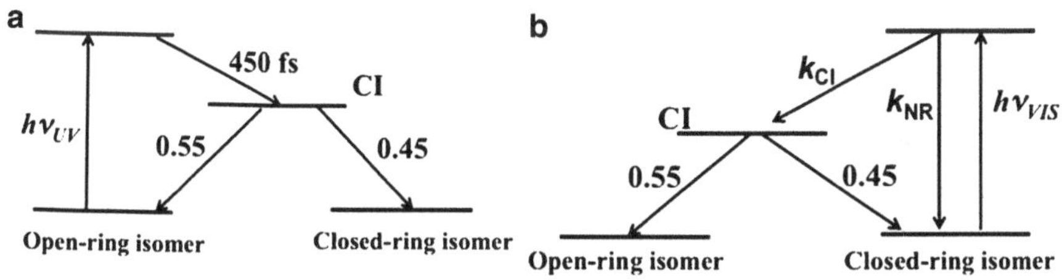

Fig. 12.2 (**a**) Reaction scheme for the cyclization the AP conformer of **BT(o)** and (**b**) the scheme for the cycloreversion based on the adiabatic potential surface

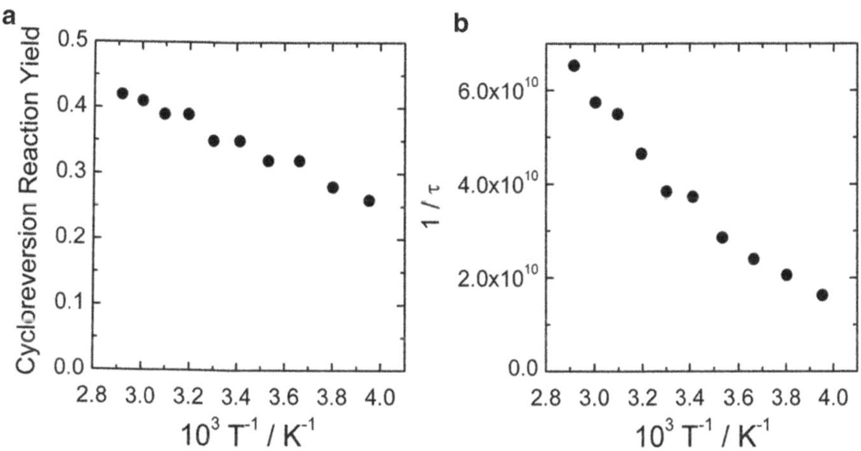

Fig. 12.3 (**a**) Temperature dependence of cycloreversion reaction quantum yield of **BT(c)** in *n*-hexane solution. (**b**) Temperature dependence of the excited state lifetime of **BT(c)** in *n*-hexane solution

conformer of **BT(o)** undergoes the rapid relaxation into the CI at which the branching of the reactions, such as the formation of the closed-ring isomer or the deactivation to the ground state of **BT(o)**, is determined. In the present case, the fraction to the closed-ring isomer is 0.45 and 0.55 is to the deactivation as summarized in Fig. 12.2a.

On the other hand, the cycloreversion reaction yield and the excited state lifetime of **BT(c)** are strongly dependent on the temperature. As shown in Fig. 12.3, the cyclization reaction yield increases and the lifetime of the excited state becomes short with an increase in the temperature. These temperature dependencies clearly show that the very rapid nonradiative process and the cycloreversion reaction take place and these two processes have activation energies different from each other. In order to verify whether the adiabatic potential model could rationally explain the experimental results, the cycloreversion process were analyzed on the basis of the scheme shown in Fig. 12.2b where the same CI is involved both for the cycloreversion and cyclization reactions. In this hypothesis, the branching ratio at the CI point is temperature-independent because the cyclization reaction was temperature-independent in the temperature region examined here. Hence, the short lifetime of

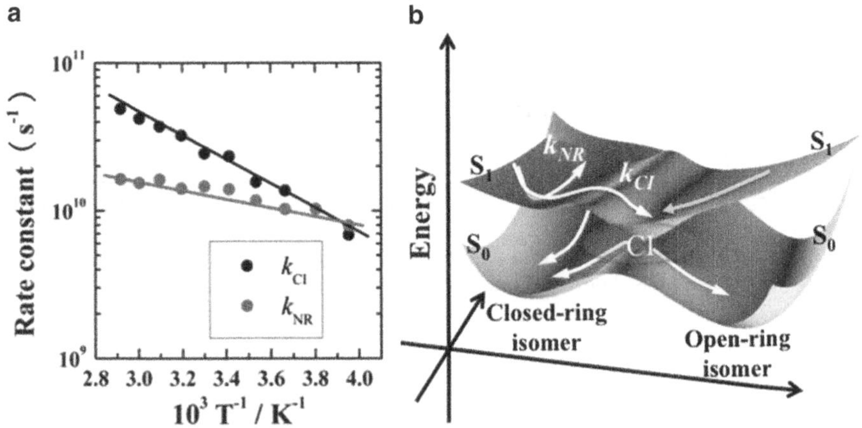

Fig. 12.4 (**a**) Temperature dependence of k_{CI} and k_{NR} in the decay process of the excited state of **BT(c)** in *n*-hexane solution. (**b**) Potential energy surfaces employed for the analysis of the temperature dependencies of cyclization and cycloreversion reactions of BT in *n*-hexane

the excited state of the closed-ring isomer and the cycloreversion reaction yield <0.55 are mainly due to the nonradiative decay to the ground state, k_{NR}, from the excited state as shown in Fig. 12.2b.

Figure 12.4a shows the temperature dependencies of k_{NR} and k_{CI}, which were obtained by the analysis based on Fig. 12.2b, indicating that the activation energy of k_{CI} is higher than that of k_{NR} and thus the contribution of k_{CI} increases with increasing temperature. Because the cycloreversion reaction yield increases in the present temperature range but it keeps less than 0.55 as shown in Fig. 12.3a, the adiabatic potential surface as shown in Fig. 12.4b could rationally explain the experimental results. To obtain more rigorous conclusion on the reaction profiles, and potential surfaces, however, more direct experimental results are needed. In anyhow, the present experimental results indicate that the rapid nonradiative deactivation from the excited state takes an important role in the temperature dependence of the cycloreversion reactions. Such rapid cyclization and cycloreversion reactions were observed also for fulgide derivatives [43–50], which can be regarded as 6-π electron systems as diarylethene derivatives and similar factors may regulate photochromic reactions also in fulgide derivatives.

12.3 Photodissociation Dynamics of Hexaarylbiimidazole Derivatives

Hexaarylbiimidazoles (HABIs) are one of the photochromic molecules, of which original derivative was developed by Hayashi and Maeda in 1970 [5]. Photochemical reaction of a typical HABI derivative (**h1**) is shown in Fig. 12.5a. The photochromic behavior of HABIs is attributed to the homolytic cleavage of the C–N bond

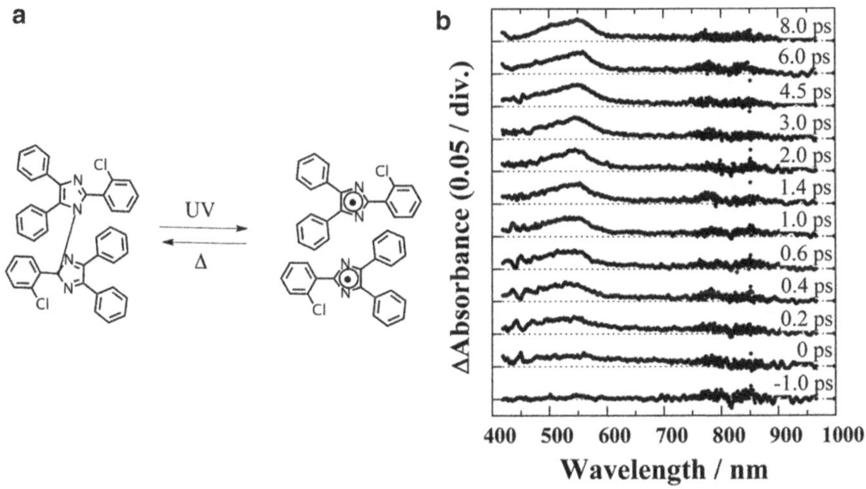

Fig. 12.5 (**a**) Photochromic reaction of a representative HABI derivative. (**b**)Transient absorption spectra of **h1** on benzene solution, excited with a fs 400-nm laser pulse

between the imidazole rings, leading to the formation of a pair of triphenylimidazolyle radicals (TPIRs). TPIRs usually have absorption bands in the visible region, while the parent HABI has no or little absorption bands in the visible region. Two radicals thermally recombine to reform its parent dimer without any other reactions in inert environments. This recombination occurs in bimolecular manner in seconds to minutes time regions (dependent on the concentration of radicals). This extremely slow recombination rate was attributed to the large activation entropy for the C–N bond reconstruction. The high yield of radicals in solid matrices and their low sensitivity to the presence of oxygen stimulated industrial interest in the use of HABIs as a photoinitiator for a wide variety of materials. In addition, new HABI materials where two imidazole groups are attached to one cyclophane or naphthalene group were recently developed [51, 52]. In these compounds, the independent translational diffusion of two radicals (radical dissociation) is prohibited and, as a result, very rapid colorization and depolarization are attained, of which property has been attracting much attention from the viewpoint of application to the imaging materials.

Figure 12.5b shows time-resolved transient absorption spectra of the HABI derivative, (**h1**), in benzene solution excited with a fs 400-nm laser pulse [53]. With an increase in the delay time after the excitation, one can find a growth of the absorption band at 530 nm due to the radical (**h2**). From the analysis of the time evolution of the transient absorption spectra, the dissociation time constant of the parent HABI, **h1**, was obtained to be 80 fs. This short time constant and the large reaction yield (almost unity) strongly suggest that the radical is produced along with the dissociative potential surface after the photoexcitation.

It is worth noting that the geminate recombination process in the nascent radical pair after the bond fission has been observed in the homolytic photodissociation of

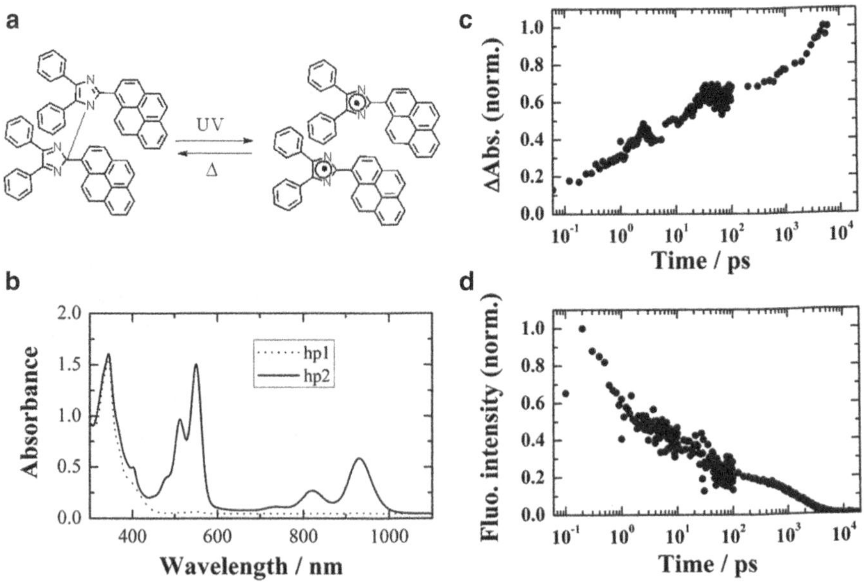

Fig. 12.6 (**a**) Photochromic reaction of a HABI derivative (**hp1**). (**b**) Absorption spectra of **hp1** (parent molecule, *smooth line*) and **hp2** (produced radical, *dotted line*). (**c**) Time profile of transient absorbance of **hp1** in benzene solution, excited with a fs 400-nm laser pulse and monitored at 560 nm (the absorption maximum of **hp2**). (**d**) Time profile of fluorescence at 480 nm, observed after the excitation of **hp1** in benzene solution with a fs 400-nm laser pulse

tetraphenylhydrazine and diphenyl disulfide derivatives [54–59]. This geminate recombination process in a few ps time region occurs in competition with the dissociation of the radical pair to free radicals in solutions and reduces the yield of free radicals [56–59]. On the other hand, such a rapid decay due to the geminate recombination was not detected for the photodissociation of the HABI derivative, of which result was attributed to the high geometrical restriction for the reconstruction of the C–N chemical bond as mentioned in previous sections. In actual, the formation yield of the radicals is almost unity for many HABI derivatives, which is consistent with the absence of the rapid geminate recombination reducing the reaction yield.

Because the photodissociation yield of HABI derivatrives is usually high as shown above, larger absorption coefficient in long wavelength region is an important factor to increase the total sensitivity of the molecule. Introduction of larger aromatic groups, instead of phenyl rings, is one of efficient methods to improve the photochromic properties of HABI derivatives. However, the local attractive potential surface of the large aromatic moiety in lower energy area may affect the crossing of the surface to the repulsive one, leading to the decrease of the reaction yield. Hence, the detailed information on the dissociation involving surface hopping between the two intersecting diabatic potential energy surfaces is indispensable for the rational designing of the improved derivatives and the control of the reaction speed. Figure 12.6a shows a HABI derivative, **hp1**, synthesized along with the above ideas. Introduction of two pyrenyl groups increases the extinction coefficient of the parent molecule in the wavelength region >350 nm, together with an increase in the absorption intensity of radicals in the visible region as shown in Fig. 12.6b.

Photodissociation dynamics of **hp1** in benzene solution is shown in Fig. 12.6c, where the time profile of the transient absorbance at 550 nm, the absorption maximum of the radical, was plotted. As this figure shows, the photodissociation takes place in the wide time region of <100 fs to several ns [60].

To elucidate the dynamic behaviors in detail, time profiles of fluorescence signals of **hp1** were also studied. In **hp1** system, weak fluorescence band with an maximum around 470 nm can be detected in the steady-state measurement. The band shape and its maximum of the fluorescence spectrum in solutions are almost the same with those observed in the crystalline phase where two pyrenyl groups have face-to-face orientation [60]. In addition, these fluorescence band shapes and their maximum are close to those of the excimer of pyrene and no emission was observed for **h1**. From these results, the fluorescence of **hp1** was attributed to the pyrenyl moieties with mutual interaction [60]. In Fig. 12.6d, the time profile of the fluorescence at 480 nm is exhibited. As observed in the rise profile of the radical, the decay of the fluorescence shows complicated temporal profiles in sub-ps to ns time region, for which at least triple-exponential function with time constants of 0.26 ps, 3.3 ps, and 1.7 ns was necessary to analyze. In addition, the contribution from the two short time constants of the fluorescence was pronounced in shorter wavelength region, indicating that the fluorescence maximum redshifts with increasing time in the region of initial ca. 10 ps after the excitation and the 1.7-ns decay of the fluorescence follows while keeping the fluorescence maximum around 470 nm. These time dependencies of the fluorescence indicate that the some geometrical relaxation leading to the excimer of pyrenyl groups takes place in the initial 10-ps time region.

In order to more precisely elucidate these behaviors, time profiles of the transient absorbance and the fluorescence were analyzed in the following manner. The fluorescent intensity $I_F(t)$ at a delay time, t, is proportional to the radiative rate constant, k_F, as represented by Eq. 12.1.

$$I_F(t) = k_F[R(t)]. \tag{12.1}$$

Here, $[R(t)]$ is the concentration of the fluorescent state. On the other hand, the formation of the radical is written as

$$\frac{d[P(t)]}{dt} = 2k_{diss}\left[R(t)\right], \tag{12.2}$$

where $[P(t)]$ and k_{diss} are, respectively, the concentration of the radical and the rate constant (or the probability of the reaction in a unit time) of photodissociation. Because the quantum yield of the photodissociation of **hp1** is almost unity [60], k_{diss} could be much larger than k_F. On the basis of above equations and the experimental results, the rate constant of photodissociation (the probability of the reaction in a unit time), k_{diss}, could be estimated by Eq. 12.3

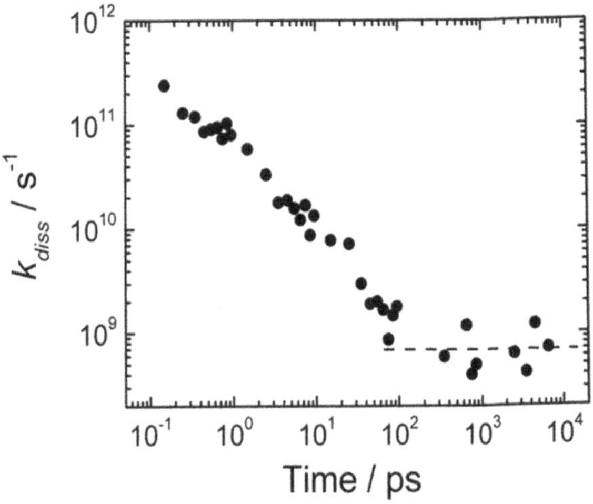

Fig. 12.7 Time dependence of the photodissociation probability of **hp1** in benzene solution. *Dashed line* indicates the rate constant of 6.1×10^8 s^{-1}

$$k_{diss}(t) \propto \frac{\Delta A_{550}(t+\Delta t) - \Delta A_{550}(t)}{2 \bullet \Delta t \bullet I_F(t)}. \tag{12.3}$$

Here, $\Delta A_{550}(t)$ is the transient absorbance at 550 nm at time, t. In the actual calculation, the averaged value, $(I_F(t)+I_F(t+\Delta t))/2$, was used. The time dependence of k_{diss} thus obtained is shown in Fig. 12.7. In this figure, the value of k_{diss} in ns time region was set 6.1×10^8 s^{-1} (1.7 ns^{-1}) because the quantum yield of photodissociation was almost unity and the averaged rate constant of the rise and the decay of the fluorescence were ca. 1.7 ns (6.1×10^8 s^{-1}) in the ns time region. Figure 12.7 clearly shows that the reaction probability, k_{diss}, is strongly dependent on the delay time after the excitation. In the time region of 0.1–100 ps after the excitation, it drastically decreases with an increase in the delay time.

The above result of the time dependence of the rate constant (reaction probability in a unit time) implies following two possible mechanisms of the photodissociation. The first mechanism is based on the coexistence of multistates where the conformation of **hp1** just after the excitation has wide distribution and the reaction rate constant is dependent on the conformation. That is, the molecule with the conformation favorable for the reaction undergoes rapid photodissociation, while that with unfavorable geometry may have small reaction rate constant. In other words, the inhomogeneity of the conformation leads to the dispersive kinetics. In this case, the dynamic behaviors can be regarded as the integration of molecules with different rate constants. The second mechanism is based on the dynamic relaxation model, where some relaxation process of the excited **hp1** competes with the photodissociation reaction and the rate constant decreases with an increase in the time following the excitation. That is, the apparent activation energy of the photodissociation increases during this relaxation and the rate constant of the photodissociation is intrinsically time-dependent.

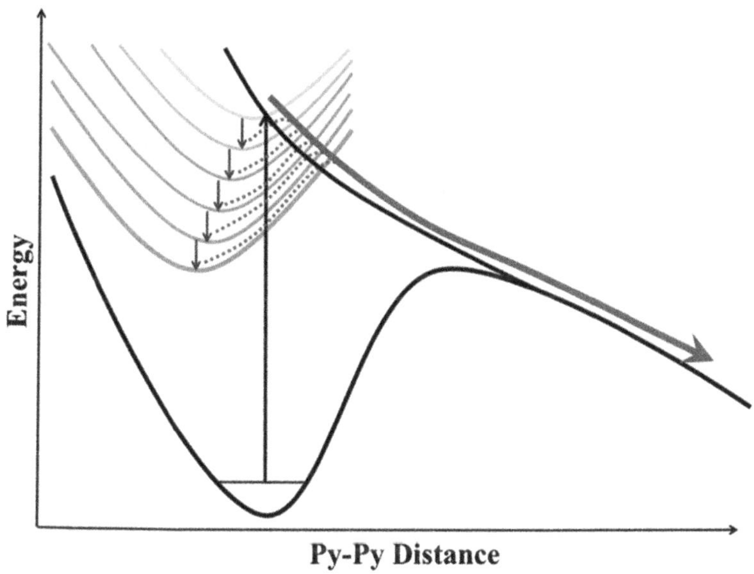

Fig. 12.8 Schematic representation of photodissociation dynamics of **hp1** and the role of the pyrene excimer formation

In the beginning of the discussion on the two mechanisms, it is worth mentioning that the photodissociation rate constant in the early stage after the excitation in Fig. 12.7 is rather small ($<5 \times 10^{11}$ s^{-1}) compared to the fluorescence decay. The rate constant of the fluorescence decay at 480 nm immediately after the excitation was ca. 4.5×10^{12} s^{-1} (220 fs) and this value is almost ten times larger than the rate constant of the photodissociation in the early stage after the excitation. This large difference indicates that the main process of the fluorescence decay in the early stage after the excitation is not the photodissociation. In addition, the internal conversion to the ground state is not the main deactivation of the fluorescence decay because the photodissociation yield was almost unity by the steady-state irradiation. In addition, large Stokes-shift of the absorption and the fluorescence in the steady-state spectra, ca. 3,000 cm^{-1}, also supports the idea that some geometrical relaxation process takes place in the excited state. Although the contribution from the first mechanism of the coexistence of various conformations cannot be excluded entirely, these results strongly suggest that the dynamic relaxation process takes an essential role in the complex dynamic behaviors in the fluorescence decay and the formation of **hp2**.

By summarizing above results and discussion, the schematic potential energy curves could be represented as shown in Fig. 12.8 where the abscissa is given as the distance between two pyrenyl moieties. After the photoexcitation, the energy minimum of the locally excited state of pyrenyl moiety gradually lowers its energy with increasing time after the excitation. During this relaxation process, the minimum position of the locally excited state may shift toward left position (decrease in the distance between pyrenyl moieties) because of the increase in the interaction of two pyrenyl moieties. The lowering of the energy of the locally excited state as well as the slight shift of the potential curve toward the left-hand side in the abscissa may increase the apparent activation energy of the photodissociation to result in the decrease in the reaction probability with an increase in the delay time. For *o*-Cl-HABI with phenyl groups (**h1** in Fig. 12.5), the energy level of the locally excited state is higher than the potential curve of the dissociation. In addition, the interaction between phenyl groups is rather small, as the equilibrium constant of intermolecular benzene excimer is less than unity. Owing to these two effects, the photodissociation of **h1** is rather simple, while that of **hp1** shows the complex behaviors.

12.4 Multiphoton Photochromic Reactions

Although the application of ultrafast laser pulses is indispensable to the direct detection of the photochromic reaction processes as shown above, high peak power and large number of photons in the ultrafast laser pulse easily induce nonlinear phenomena such as multiphoton absorption processes. These nonlinear phenomena sometimes inhibit the precise measurements in the ultrafast time regions. The application of these nonlinear phenomena, however, can open new responses of photochromic molecules, which cannot be easily attained by the one-photon absorption process. In this section, we introduce the multiphoton-gated cycloreversion reaction induced by picosecond laser pulse and the one-color reversible control of photochromic reactions using higher-order nonresonant multiphoton absorption process as examples of nonlinear photochromic reactions.

12.4.1 Multiphoton-Gated Cycloreversion Photochromic Reaction

Diarylethene and fulgide derivatives undergo the cycloreversion reaction upon irradiation of the visible light. The pulsed excitation using ps lasers, however, induces the drastic enhancement of the apparent reaction yield in these systems. From the detailed investigation on the excitation intensity dependence of the cycloreversion reaction, it was revealed that this enhancement was due to the successive two-photon absorption process [32, 34, 35, 38, 44, 45, 61, 62]. That is, the molecules excited by the first photon absorption are pumped up into higher excited states by the second photon absorption, as shown in Fig. 12.9a. Large cycloreversion reaction yield in the higher excited state is a preferable condition for the enhancement of the cycloreversion reaction. However, even in the case that the reaction yield is small in the higher excited state, the $S_1 \rightarrow S_n$ excitation and $S_n \rightarrow S_1$ deactivation, in competition with the cycloreversion, can be repeated during the pulse duration of the laser,

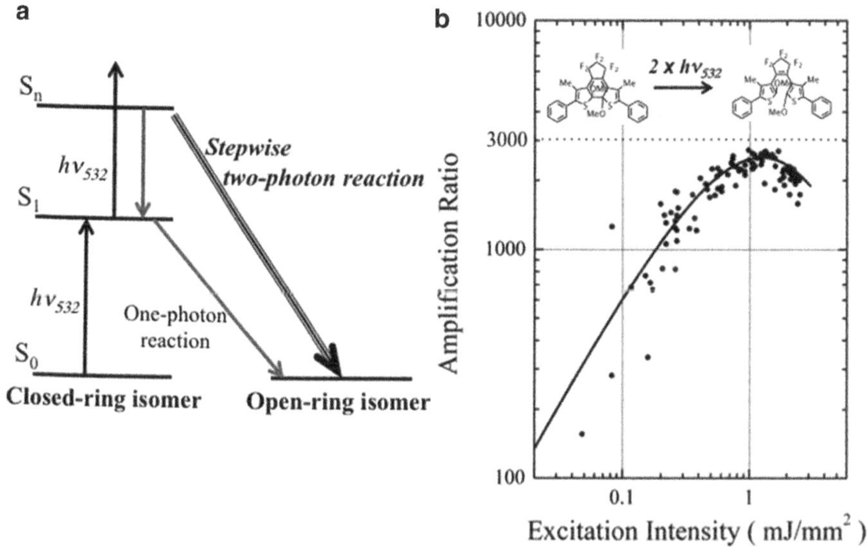

Fig. 12.9 (**a**) Schematic representation of stepwise two-photon gated cycloreversion reaction. (**b**) Excitation intensity dependence of the amplification ratio of the cycloreversion reaction under the picosecond 532-nm laser pulse with 15-ps laser pulse

because the lifetime of the higher excited state is usually very short (<1 ps). As a result of these iterative processes, large amount of the molecules can be led to the open-ring isomer.

Figure 12.9b shows the excitation intensity dependence of the number of the molecules led to the open-isomer under the ps 532-nm laser pulse. The ordinate is given as the amplification ratio defined by $N_{RM}/(\Phi_{CO} \times N_p)$. Here, N_{RM}, Φ_{CO}, and N_p are respectively the number of the reacted molecules, one-photon cycloreversion reaction yield, and the number of photons exposed into the system. In the case that all the exposed photons are absorbed by one-photon absorption process and that the cycloreversion reaction occurs with Φ_{CO}, the amplification ratio is unity. As shown in Fig. 12.9b, this value increases with an increase in the excitation intensity with a slope of ca. 2. This result indicates that the two-photon absorption process is responsible for the enhancement of the cycloreversion reaction. After it reaches the maximum value, it decreases with further increase in the excitation intensity. This decrease is due to the depletion of the closed-ring isomer. That is, almost all of the closed-ring isomer in the exposed volume was converted into the open-ring isomer, and hence, further increase in the intensity of the ps 532-nm laser pulse did not lead to the increase of the open-isomer. As this figure shows, the amplification ratio more than 2,500 could be attained for the diarylethene derivative with small Φ_{CO} [32], and almost all of the molecules in the exposed volume could be converted into the open-ring isomer by one laser pulse.

For the property of the higher excited state leading to the effective cycloreversion reaction, it is worth noting that the UV one-photon absorption to the energy level similar to that attained by the visible two-photon absorption does not lead to the effective cycloreversion [23, 32, 34–36, 44, 45, 61, 62] In general, the electronic state attained by the two-photon absorption is different from that by allowed one-photon transition. The difference in the reactivity depending on the mode of the excitation indicates that the character of the excited state takes an important role and that the internal conversion between the one-photon allowed excited state and that allowed by the two-photon transition is not fast compared to the effective cycloreversion from the latter excited state.

Moreover, it is also worth adding here the reason why ps laser pulse is employed for the multiphoton-gated cycloreversion reaction. In the successive two-photon absorption process under the one-color laser excitation, the second photon absorption via the S_1 state occurs in competition with the first photon absorption of the S_0 state. Hence, the large number of photons in a laser pulse is necessary to prepare the large number of S_1 molecules. (Otherwise, most of the incident photons are absorbed by S_0 molecules.) In addition to the number of photons, the pulse duration of the laser is an important factor to keep the large number of the S_1 state molecules. Because the lifetime of the S_1 state of the closed-ring isomers is in the order of a few to a few tens of ps as shown in Table 12.1, ns pulsed laser excitation usually cannot prepare a number of S_1 state molecules to induce the successive two-photon enhancement of the cycloreversion [62]. On the other hand, femtosecond pulsed laser has a time duration short enough to prepare the large number of the S_1 state in the laser pulse. However, extremely short laser pulse with a number of photons easily induces nonlinear processes other than the successive two-photon absorption. So as to avoid these nonlinear phenomena, the intensity of the femtosecond laser has to be kept small, and as a result, it is rather difficult to lead the effective enhancement of the stepwise two-photon process. The above multiphoton-gated reaction could be applied to the realization of the photoerasable optical memory system with nondestructive readout capability and the site-selective colorization/decolorization with high spatial resolution.

12.4.2 One-Color Reversible Control of Photochromic Reactions in a Diarylethene Derivative Using Higher-Order Multiphoton Absorption

The above reaction switching is attained by the stepwise two-photon absorption process via the actual intermediate state with a finite lifetime. Hence, it is necessary to employ the laser pulse which is resonant to the absorption of the closed-ring isomer for the reaction switching and the other laser light with different color is needed to induce the colorization reaction. As stated in Sect. 12.2, the open-isomer of diarylethene derivatives has absorption bands only in the UV region, while the closed-ring isomer has absorption bands both in the visible and UV regions. Accordingly, using a nonresonant laser pulse of which three-photon energy can excite the open-ring isomer and two-photon energy corresponds to the energy level of the closed-ring isomer, we can expect one-color reversible control of the cyclization and cycloreversion reactions. Although the number of the experimental results is very small for the cross section of the simultaneous three-photon absorption, it was reported that it is as small as 10^{-32} to 10^{-34} of those for the two-photon absorption process [63]. Hence, the fs laser system with high output energies was needed to induce the simultaneous three-photon absorption process [64].

The focusing of the laser beam with an objective lens with high NA is an effective method to increase the peak power of the light in a tiny spot. Such an objective lens, however, leads to the broadening of the pulse duration of the fs laser owing to the dispersion of the wide spectrum in the fs laser pulse, resulting in the decrease in the peak power owing to the increase in the pulse duration. To overcome these factors and to keep the short pulse duration to induce the higher-order simultaneous multiphoton absorption process, the laser microscope system with a cavity-dumped Cr:forsterite laser [65, 66] was applied for the one-color control of the photochromic reactions. Because the output of the Cr:forsterite laser locates at 1.28 μm, the short pulse duration of 35 fs is obtained under the objective with high NA (0.95) owing to small dispersion of the lens materials in the near-infrared region. As a result of the short pulse duration, we can attain the high peak power that can easily induce three- and four-photon fluorescence emissions with the laser energy of ≤1 nJ/pulse [65, 66].

Figure 12.10b shows the transmission images of an amorphous solid film of the diarylethene derivative, whose molecular structure is shown in Fig. 12.10a, irradiated with 1.28-μm laser pulse under the microscope [67]. The open-ring isomer of this diarylethene derivative has absorption bands in the wavelength region <450 nm, while the closed-ring isomer has a visible band around 600 nm. This figure shows the irradiation with the 1.28-μm laser pulse for 50 ms (0.7 nJ/pulse and 100 kHz repletion rate) induces the colorization due to the cyclization. From the excitation intensity dependence, this colorization was found to be due to the simultaneous three-photon absorption process. On the other hand, the irradiation of the same laser pulse with a weak intensity(48 pJ/pulse)for 180 s diminishes the color via the simultaneous two-photon cycloreversion reaction. This colorization and decolorization cycle could be repeated at least five to ten times. Figure 12.10c shows the transmission image after the irradiation of the 1.28-μm laser pulse to the amorphous film of the diarylethene derivative. In this case, the film was exposed to the steady-state UV light before the laser excitation. The number of the closed-ring isomer, N_C, and that of the open-ring isomer, N_O, is almost the same in this film. This figure shows that, with an increase in the exposure time of the laser pulse at 1.28 μm, decolorization is induced in the peripheral region of the focusing spot of the laser but the center area keeps the color due to the closed-ring isomer. This difference of the colorization and decolorization depending on the position in the laser spot is attributable to the difference of the light intensity in the beam profile of the laser. That is, the high intensity of the 1.28-μm laser pulse in the center area induces the colorization via the three-photon cyclization, while the decolorization by the two-photon cycloreversion is dominant in the peripheral area with weak intensity of the light. This result

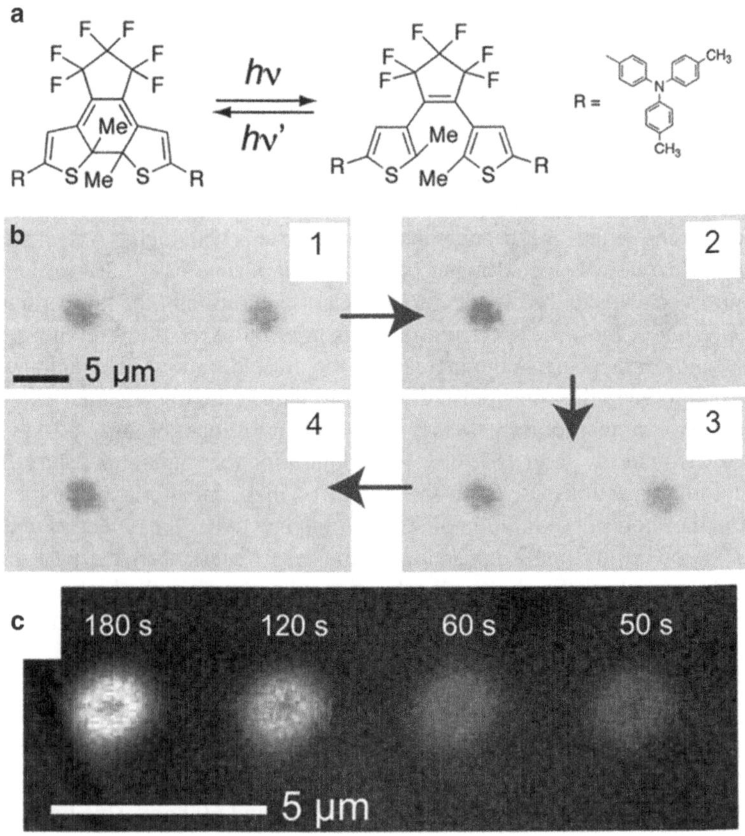

Fig. 12.10 (**a**) Molecular structure and reactions of a diarylethene derivative used. (**b**) Transmittance images of the amorphous film of the compound above, excited with Cr:Forfsterite laser centered at 1.28 μm. Irradiation of the NIR laser with the intensity of 0.7 nJ/pulse for 50 ms (100 kHz) induces the colorization and the exposure with weaker intensity of 48 pJ/pulse (100 kHz) for 180 s leads to decolorization (**c**) Transmittance images of the amorphous film (at N_C/N_O = ca. 1), irradiated by Cr:F laser with the intensity of 66 pJ/pulse at 100 kHz. Exposure time is shown in the figure

demonstrates that the one-color control of the photochromic reactions can be potentially applicable to the maskless patterning by optimizing various parameters such as pulse duration, intensity, exposure time, and so on.

12.4.3 Cooperative Photochromic Reactions in Molecular Assemblies

Pulsed laser excitation can induce the dense population of the excited state in the molecular assembly such as solid systems, especially in the case that the allowed one-photon absorption band is excited. Temporally and spatially simultaneous production of a number of the excited state can induce the photochromic reactions even in the environmental condition that the weak excitation does not lead to the isomerization reaction owing to the small free volume to change the molecular geometry [68–71]. In these cases, the reaction in the molecular level is not related to the multiphoton process but the response of the molecular assembly is nonlinear with the light intensity. These responses can be utilized to realize the photoresponse system with a threshold for the exposed light intensity.

Acknowledgements This work was supported by Grand-in-Aids for Research in Priority Areas "New Frontiers in Photochromism (No. 471)" from the Ministry of Education, Culture, Sports, Science, and Technology (MEXT), Japan. The authors appreciate the collaboration of Professors Masahiro Irie, Seiya Kobatake, Yasushi Yokoyama, Kingo Uchida, Shin-ichiro Nakamura, and Jiro Abe.

References

1. Brown GH (ed) (1971) Photochromism. Wiley-Interscience, New York
2. Dürr H, Bouas-Laurent H (eds) (1990) Photochromism: molecules and systems. Elsevier, Amsterdam
3. Irie M (2000) Diarylethenes for memories and switches. Chem Rev 100:1685–1716
4. Kobatake S, Irie M (2003) Photochromism. Annu Rep Prog Chem C 99:277–313
5. Hayashi T, Maeda K (1970) The mechanism of photochromism, thermochromism and piezochromism of dimers of triarylimidazolyl. Bull Chem Soc Jpn 43:429–438
6. Yokoyama Y (2000) Fulgides for memories and switches. Chem Rev 100:1717–1739
7. Tamai N, Miyasaka H (2000) Ultrafast dynamics of photochromic systems. Chem Rev 100:1875–1890
8. Kobatake S, Irie M (2004) Single-crystalline photochromism of diarylethenes. Bull Chem Soc Jpn 77:195–210
9. Irie M (2008) Photochromism and molecular mechanical devices. Bull Chem Soc Jpn 81:917–926
10. Irie M, Kobatake S, Horichi M (2001) Reversible surface morphology changes of a photochromic diarylethene single crystal by photoirradaition. Science 291:1769
11. Kobatake S, Takami S, Muto H, Ishikawa T, Irie M (2007) Rapid and reversible shape changes of molecular crystals on photoirradaition. Nature 446:778–781
12. Ishibashi Y, Fujiwara M, Umesato T, Saito H, Kobatake S, Irie M, Miyasaka H (2011) Cyclization reaction dynamics of a photochromic diarylethene derivative as revealed by femtosecond to microsecond time-resolved spectroscopy. J Phys Chem C 115:4265–4272
13. Miyasaka H, Araki S, Tabata A, Nobuto T, Mataga N, Irie M (1994) Picosecond laser photolysis studies on photochromic reactions of 1,2-bis(2,4,5-trimethyl-3-thienyl)maleic anhydride in solutions. Chem Phys Lett 230:249–254
14. Miyasaka H, Nobuto T, Itaya A, Tamai N, Irie M (1997) Picosecond laser photolysis studies on a photochromic dithienylethene in solution and in crystalline phases. Chem Phys Lett 269:281–285
15. Kaieda T, Kobatake S, Miyasaka H, Murakami M, Iwai N, Nagata Y, Itaya A, Irie M (2002) Efficient photocyclization of dithienylethene dimer, trimer, and tetramer: Quantum yield and reaction dynamics. J Am Chem Soc 124:2015–2024
16. Tamai N, Saika T, Shimidzu T, Irie M (1996) Femtosecond dynamics of a thiophene oligomer with a photoswitch by transient absorption spectroscopy. J Phys Chem 100:4689–4692
17. Miyasaka H, Nobuto T, Murakami M, Itaya A, Tamai N, Irie M (2002) Solvent viscosity effects on photochromic reactions of a diarylethene derivative as revealed by picosecond laser spectroscopy. J Phys Chem A 106:8096–8102

18. Ern J, Bens AT, Martin HD, Kuldova K, Trommsdorff HP, Kryschi C (2002) Ring-opening and -closure reaction dynamics of a photochromic dithienylethene derivative. J Phys Chem A 106:1654–1660
19. Hania PR, Telesca R, Lucas LN, Pugzlys A, van Esch J, Feringa BL, Snijders JG, Duppen K (2002) An optical and theoretical investigation of the ultrafast dynamics of a bisthienylethene-based photochromic switch. J Phys Chem A 106:8498–8507
20. Hania PR, Pugzlys A, Lucas LN, de Jong JJD, Feringa BL, van Esch JH, Jonkman HT, Duppen K (2005) Ring closure dynamics of BTE-based photochromic switches: Perfluoro- versus perhydrocyclopentene derivatives. J Phys Chem A 109:9437–9442
21. Shim S, Eom I, Joo T, Kim E, Kim KS (2007) Ring closure reaction dynamics of diarylethene derivatives in solution. J Phys Chem A 111:8910–8917
22. Elsner C, Cordes T, Dietrich P, Zastrow M, Herzog TT, Rück-Braun K, Zinth W (2009) Photochromic bis(thiophen-3-yl)maleimides studied with time-resolved spectroscopy. J Phys Chem A 113:1033–1039
23. Tani K, Ishibashi Y, Miyasaka H, Kobatake S, Irie M (2008) Dynamics of cyclization, cycloreversion, and multiphoton-gated reaction of a photochromic diarylethene derivative in crystalline phase. J Phys Chem C 112:11150–11157
24. Guillaumont D, Kobayashi K, Kanda K, Miyasaka H, Uchida K, Kobatake S, Shibata K, Nakamura S, Irie M (2002) An ab initio MO study of the photochromic reaction of dithienylethenes. J Phys Chem A 106:7222–7227
25. Nakamura S, Kobayashi T, Takata A, Uchida K, Asano Y, Murakami A, Goldberg A, Guillaumont D, Yokojima S, Kobatake S, Irie M (2007) Quantum yields and potential energy surfaces: a theoretical study. J Phys Org Chem 20:821–829
26. Ern J, Bens AT, Martin HD, Mukamel S, Schmid D, Tretiak S, Tsiper E, Kryschi C (1999) Reaction dynamics of photochromic dithienylethene derivatives. Chem Phys 246:115–125
27. Ern J, Bens AT, Martin HD, Mukamel S, Schimid D, Tretiak S, Tsiper T, Kryschi C (2001) Reaction dynamics of a photochromic fluorescing dithienylethene. J Phys Chem A 105: 1741–1749
28. Boggio-Pasqua M, Ravaglia M, Bearpark MJ, Gavavelli M, Robb MA (2003) Can diarylethene photochromism be explained by a reaction path alone? A CASSCF study with model MMVB dynamics. J Phys Chem A 107:11139–11152
29. Indelli MT, Carli S, Ghirotti M, Chiorboli C, Ravaglia M, Garavelli M, Scandola F (2008) Triplet pathways in diarylethene photochromism: photophysical and computational study of dyads containing ruthenium(II) polypyridine and 1,2-bis(2-methylbenzothiophene-3-yl) maleimide units. J Am Chem Soc 130:7286–7299
30. Fukaminato T, Doi T, Tanaka M, Irie M (2009) Photocyclization reaction of diarylethene–perylenebisimide dyads upon irradiation with visible (>500 nm) light. J Phys Chem C 113:11623–11627
31. Murata R, Yago T, Wakasa M (2011) Cyclization reaction of diarylethene through the triplet excited state. Bull Chem Soc Jpn 84:1336–1338
32. Ishibashi Y, Okuno K, Ota C, Umesato T, Katayama T, Murakami M, Kobatake S, Irie M, Miyasaka H (2010) Multiphoton-gated cycloreversion reactions of photochromic diarylethene derivatives with low reaction yields upon one-photon visible excitation. Photochem Photobiol Sci 9:172–180
33. Ern J, Bens AT, Bock A, Martin HD, Kryschi C (1998) Femtosecond transient absorption studies on photochromism of dithienylethene derivates. J Lumin 76&77:90–94
34. Piard J, Ishibashi Y, Saito H, Métivier R, Nakatani K, Gavrel G, Yu P, Miyasaka H (2012) Multiphoton-gated cycloreversion reaction of a photochromic 1,2-bis(hiazolyl)- perfluorocyclopentene diarylethene derivative. J Photochem Photobiol A: Chem 234:57–65
35. Miyasaka H, Murakami M, Itaya A, Guillaumont D, Nakamura S, Irie M (2001) Multiphoton gated photochromic reaction in a diarylethene derivative. J Am Chem Soc 123:753–754
36. Ryo S, Ishibashi Y, Murakami M, Miyasaka H, Kobatake S, Irie M (2007) Multiphoton-gated photochromic reaction of diarylethene derivatives in PMMA solid film. J Phys Org Chem 20:953–959

37. Ishibashi Y, Tani K, Miyasaka H, Kobatake S, Irie M (2007) Picosecond laser photolysis study of cycloreversion reaction of a diarylethene derivative in polycrystals: multiphoton-gated reaction. Chem Phys Lett 437:243–247
38. Ishibashi Y, Mukaida M, Falkenström M, Miyasaka H, Kobatake S, Irie M (2009) One- and multi-photon cycloreversion reaction dynamics of diarylethene derivative with asymmetrical structure, as revealed by ultrafast laser spectroscopy. Phys Chem Chem Phys 11:2640–2648
39. Miyasaka H, Murakami M, Okada T, Nagata Y, Itaya A, Kobatake S, Irie M (2003) Picosecond and femtosecond laser photolysis studies of a photochromic diarylethene derivative: multiphoton gated reaction. Chem Phys Lett 371:40–48
40. Shim S, Joo T, Bae SC, Kim KS, Kim E (2003) Ring opening dynamics of a photochromic diarylethene derivative in solution. J Phys Chem A 107:8106–8110
41. Kuldová K, Tsyganenkoa K, Corvala A, Trommsdorff HP, Bens AT, Kryschi C (2000) Photoswitchable dithienylethenes: threshold of the photoreactivity. Synth Metal 115:163–166
42. Ishibashi Y, Umesato T, Kobatake S, Irie M, Miyasaka H (2012) Femtosecond laser photolysis studies on temperature dependence of cyclization and cycloreversion reactions of a photochromic diarylethene derivative. J Phys Chem C 116:4862–4859
43. Kurita S, Kashiwagi A, Kurita Y, Miyasaka H, Mataga N (1990) Picosecond laser photolysis studies on the photochromism of a furylfulgide. Chem Phys Lett 171:553–557
44. Ishibashi Y, Murakami M, Miyasaka H, Kobatake S, Irie M, Yokoyama Y (2007) Laser multiphoton-gated photochromic reaction of a fulgide derivative. J Phys Chem C 111:2730–2737
45. Ishibashi Y, Katayama T, Ota C, Kobatake S, Irie M, Yokoyama Y, Miyasaka H (2009) Ultrafast laser spectroscopic study on photochromic cycloreversion dynamics in fulgide derivatives: one-photon and multiphoton-gated reactions. New J Chem 33:1409–1419
46. Handschuh M, Seibold M, Port H, Wolf HC (1997) Dynamics of the cyclization reaction in photochromic furyl fulgides. J Phys Chem A 101:502–506
47. Koller FO, Schreier WJ, Schrader TE, Sieg A, Malkmus S, Schulz C, Dietrich S, Ruck-Braun K, Zinth W, Braun M (2006) Ultrafast structural dynamics of photochromic indolylfulgimides studied by vibrational spectroscopy and DFT calculations. J Phys Chem A 110:12769–12776
48. Malkmus S, Koller FO, Heinz B, Schreier WJ, Schrader TE, Zinth W, Schulz C, Dietrich S, Rück-Braun K, Braun M (2006) Ultrafast ring opening reaction of a photochromic indolylfulgimide. Chem Phys Lett 417:266–271
49. Heinz B, Malkmus S, Laimgruber S, Dietrich S, Schulz C, Rück-Braun K, Braun M, Zinth W, Gilch P (2007) Comparing a photoinduced pericyclic ring opening and closure: differences in the excited state pathways. J Am Chem Soc 129:8577–8584
50. Cordes T, Malkmus S, DiGirolamo JA, Lees WJ, Nenov A, de Vivie-Riedle R, Braun M, Zinth W (2008) Accelerated and efficient photochemistry from higher excited electronic states in fulgide molecules. J Phys Chem A 112:13364–13371
51. Fujita K, Hatano S, Kato D, Abe J (2008) Photochromism of a radical diffusion-inhibited hexaarylbiimidazole derivative with intense coloration and fast decoloration performance. Org Lett 10:3105–3108
52. Kishimoto Y, Abe J (2009) A fast photochromic molecule that colors only under UV light. J Am Chem Soc 131:4227–4229
53. Satoh Y, Ishibashi Y, Ito S, Nagasawa Y, Miyasaka H, Chosrowjan H, Taniguchi S, Mataga N, Kato D, Kikuchi A, Abe J (2007) Ultrafast laser photolysis study on photodissociation dynamics of a hexaarylbiimidazole derivative. Chem Phys Lett 448:228–231
54. Lenderink E, Duppen K, Wiersma DA (1992) Femtosecond fragmentation of tetraphenylhydrazine in solution. Chem Phys Lett 194:403–409
55. Hirata Y, Ohta M, Okada T, Mataga N (1992) Direct observation of photodissociation of tetraphenylhydrazine and its derivatives in the solution phase: picosecond study of nitrogen-nitrogen bond rupture in the fluorescence state. J Phys Chem 96:1517–1520
56. Hirata Y, Niga Y, Ohta M, Takizawa M, Okada T (1995) Photodissociation and geminate dynamics in solution phase: picosecond transient absorption studies of tetraphenylhydrazines and diphenyl disulfides. Res Chem Intermed 21:823–836

57. Hirata Y, Niga Y, Makita SI, Okada T (1997) Geminate recombination of the p-aminophenylthiyl radical pair produced by the photodissociation of p-aminophenyl disulfide in nonpolar solvents. J Phys Chem A 101:561–565
58. Bultmann T, Ernsting NP (1996) Competition between geminate recombination and solvation of polar radicals following ultrafast photodissociation of bis(p-aminophenyl) disulfide. J Phys Chem 100:19417–19424
59. Lochschmidt A, Eilers-König N, Heuneking N, Ernsting NP (1999) Femtosecond photodissociation dynamics of bis(julolidine) disulfide in polar and apolar solvents. J Phys Chem A 103:1776–1784
60. Miyasaka H, Satoh Y, Ishibashi Y, Ito S, Nagasawa Y, Taniguchi S, Chosrowjan H, Mataga N, Kato D, Kikuchi A, Abe J (2009) Ultrafast photodissociation dynamics of a hexaarylbiimidazole derivative with pyrenyl groups: dispersive reaction from femtosecond to 10 ns time regions. J Am Chem Soc 131:7256–7263
61. Murakami M, Miyasaka H, Okada T, Kobatake S, Irie M (2004) Dynamics and mechanisms of the multiphoton gated photochromic reaction of diarylethene derivatives. J Am Chem Soc 126:14764–14772
62. Uchida K, Takata A, Ryo S, Saito M, Murakami M, Ishibashi Y, Miyasaka H, Irie M (2005) Picosecond laser photolysis studies on a photochromic oxidation polymer film consisting of diarylethene molecules. J Mater Chem 15:2128–2133
63. Birks JB (ed) (1970) Photophysics of aromatic molecules. Wiley-Interscience, Amsterdam
64. Yatsuhashi T, Nakahagi Y, Okamoto H, Nakashima N (2010) Linear response of multiphoton reaction: three-photon cycloreversion of anthracene biplanemer in solution by intense femtosecond laser pulses. J Phys Chem A 114:10475–10480
65. Matsuda H, Fujimoto Y, Ito S, Nagasawa Y, Miyasaka H, Asahi T, Masuhara H (2006) Development of near-infrared 35 fs laser microscope and its application to the detection of three- and four-photon fluorescence of organic microcrystals. J Phys Chem B 110:1091–1094
66. Matsuda H, Ito S, Nagasawa Y, Asahi T, Masuhara H, Kobatake S, Irie M, Miyasaka H (2006) Higher-order multiphoton imaging by femtosecond near-infrared laser microscope system. J Photochem Photobio A 183:261–266
67. Mori K, Ishibashi Y, Matsuda H, Ito S, Nagasawa Y, Nakagawa H, Uchida K, Yokojima S, Nakamura S, Irie M, Miyasaka H (2011) One-color reversible control of photochromic reactions in a diarylethene derivative: three-photon cyclization and two-photon cycloreversion by, a near-infrared femtosecond laser pulse at 1.28 mm. J Am Chem Soc 133:2621–2625
68. Suzuki M, Asahi T, Masuhara H (2002) Photochromic reactions of crystalline spiropyrans and spirooxazines induced by intense femtosecond laser excitation. Phys Chem Chem Phys 4:185–192
69. Asahi T, Suzuki M, Masuhara H (2002) Cooperative photochemical reaction in molecular crystal induced by intense femtosecond laser excitation: photochromism of spironaphthooxazine. J Phys Chem A 106:2335–2340
70. Suzuki M, Asahi T, Takahashi K, Masuhara H (2003) Ultrafast dynamics of photoinduced ring-opening and the subsequent ring-closure reactions of spirooxazines in crystalline state. Chem Phys Lett 368:384–392
71. Uchida K, Yamaguchi S, Yamada H, Akazawa M, Katayama T, Ishibashi Y, Miyasaka H (2009) Photoisomerization of an azobenzene gel by pulsed laser irradiation. Chem Commun 29:4420–4422

Chapter 13
Photochromism of Organic Compounds Undergoing Isomerization Around the C═C Double Bond and/or Intramolecular Hydrogen Atom Transfer: Small and Large Structural Change

Kayoko Kataoka, Takuya Kobayashi, and Tatsuo Arai

Abstract Photochromism by way of isomerization around the C═C double bond and/or intramolecular hydrogen atom transfer is discussed from the viewpoint of how large the structural change taking place is for the photochemical reaction processes of organic compounds. The large structural change could be achieved by introduction of highly rigid large substituent such as polyphenylene at the photoresponsive core or by introduction of hydrophilic substituent at the peripheries of large dendritic molecules. The occurrence of large structural change can be monitored by fluorescence spectroscopy as well as transient spectroscopy and GPC analysis. The intramolecular hydrogen bonding affects the electronic state to change the color between the isomer having intramolecular hydrogen bonding and the isomer not having intramolecular hydrogen bonding. These hydrogen-bonded compounds underwent intramolecular hydrogen atom transfer, which is a small structural change, to give the tautomer with different color from the normal form. The large structural change of dendritic molecules and the color change of hydrogen-bonded compounds are summarized in this chapter.

Keywords Charge transfer • Dendrimers • Fluorescence • Hydrogen atom transfer • Isomerization • Structural change

13.1 Introduction

Photochromism is defined as reversible color change by photochemical processes. There have been many examples of photochromic compounds [1–3]. In this chapter, we would like to focus on the phenomena of structural change upon the

K. Kataoka • T. Kobayashi • T. Arai (✉)
Graduate School of Pure and Applied Science, University of Tsukuba,
Tsukuba, Ibaraki 305-8571, Japan
e-mail: arai@chem.tsukuba.ac.jp

M. Irie et al. (eds.), *New Frontiers in Photochromism*,
DOI 10.1007/978-4-431-54291-9_13, © Springer Japan 2013

photochemical processes of photochromic compounds and their related compounds, especially how large of a structural change we can expect upon the photoreversible photochromic processes. In this respect, the photoisomerization around the C=C double bond and photoinduced hydrogen atom transfer processes of dendritic molecules as well as their parent molecules were studied. In addition, cooperative effect of photoreactive C=C double bond and intramolecular hydrogen bonding on the photochemistry of these compounds has been studied.

13.2 Dendritic Molecules

The approach to amplify structural changes of the chromophore with a small molecular weight to a large structural change of the whole structure of large molecules is discussed. In order to create such system and investigate their basic photochemical properties, the use of dendritic molecules could be a suitable way for research. Dendrimers are well-defined, highly branched molecules which have uniform molecular weight and size dependent on the number of generations. There are three sites for the functional groups to be placed in dendritic molecules, i.e., the core, branching units, and surface groups of dendrimers [4]. We wish to introduce here several types of dendrimers having photoresponsive core and their specific photochemical properties, especially the photoisomerization around C=C double bond, from the viewpoint of how large of a structural change one can expect on photoirradiation of the molecules.

13.2.1 Stilbene Dendrimers

We have chosen the stilbene structure for the study to induce large structural change upon photoirradiation from several reasons. On direct irradiation, stilbene undergoes mutual isomerization between the *cis*- and *trans*-isomers, deactivating from the perpendicular conformation ($^1p^*$) of the excited singlet state (Fig. 13.1) [5]. The *trans*-isomer ($^1t^*$) gives fluorescence emission, with a quantum yield of 0.04 and a fluorescence lifetime (τ_f) of 70 ps. The *cis*-isomer ($^1c^*$) gives practically no fluorescence and its singlet excited state lifetime is less than 1 ps.

In order to construct the target dendrimer molecules one should modify the core structure to 3,3′,5,5′-tetramethoxystilbene (TMST). Thus, the photochemical properties of TMST have also been studied [6]. While *cis*-TMST does not give fluorescence as is the case of *cis*-stilbene, *trans*-TMST gives fluorescence *ca.* 10 times more efficiently than *trans*-stilbene. In addition, the excited singlet state of *trans-TMST* has a charge-transfer (CT) character, and the fluorescence lifetime of *trans-TMST* experienced a large solvent effect increasing from 2.3 ns in cyclohexane to 16.6 ns in acetonitrile [6b] (Fig. 13.2).

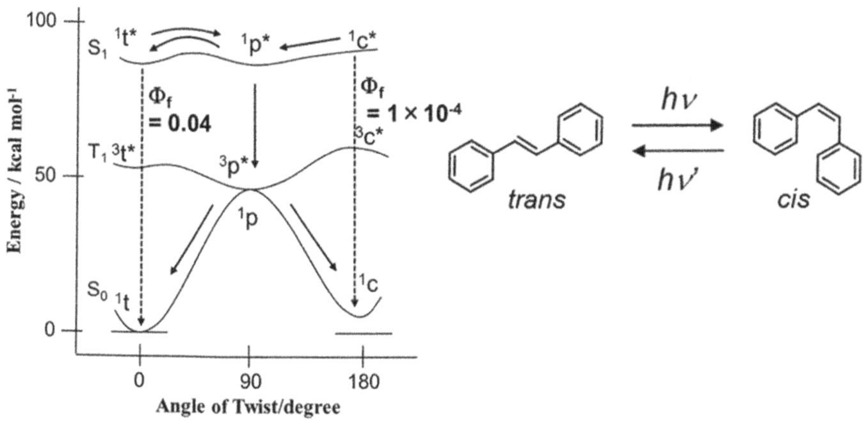

Fig. 13.1 Potential energy surfaces of *trans–cis* isomerization of stilbene

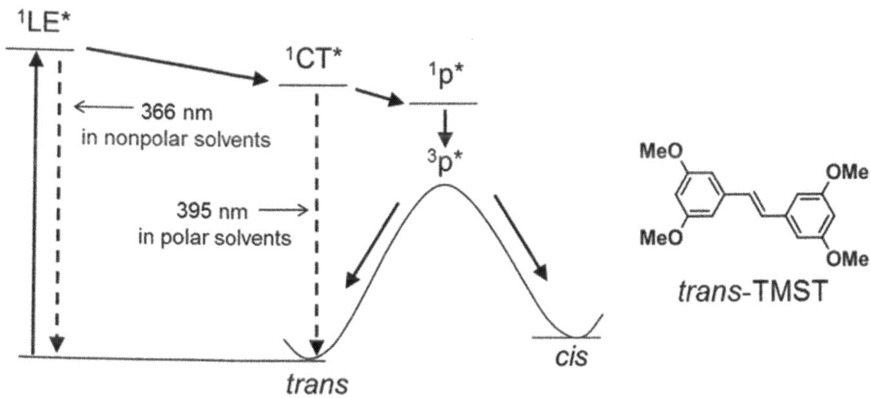

Fig. 13.2 Scheme for the formation and decay of the fluorescent ^{1}CT* state and ^{1}LE* state and isomerization of the *trans*-TMST

On the basis of previous researches for stilbene and TMST, we have investigated photochemical properties of several types of stilbene-cored dendrimers. In these studies, the fluorescence and photoisomerization experiments could be a monitor of the environmental and isomerization properties of the core stilbene or TMST.

13.2.1.1 Benzyl Ether-Type Stilbene Dendrimers (1)

The excited singlet lifetimes (*ca.* 10 ns) and fluorescence quantum yields (Φ_f=0.4) of the stilbene dendrimers (*trans*-**1** G1~G4) (Fig. 13.3) in THF were almost the same as those of *trans*-TMST in THF [7]. This result indicates that solvent THF molecules can penetrate into the inner part of the dendrimers and can interact with stilbene core regardless of the generation of dendrimers.

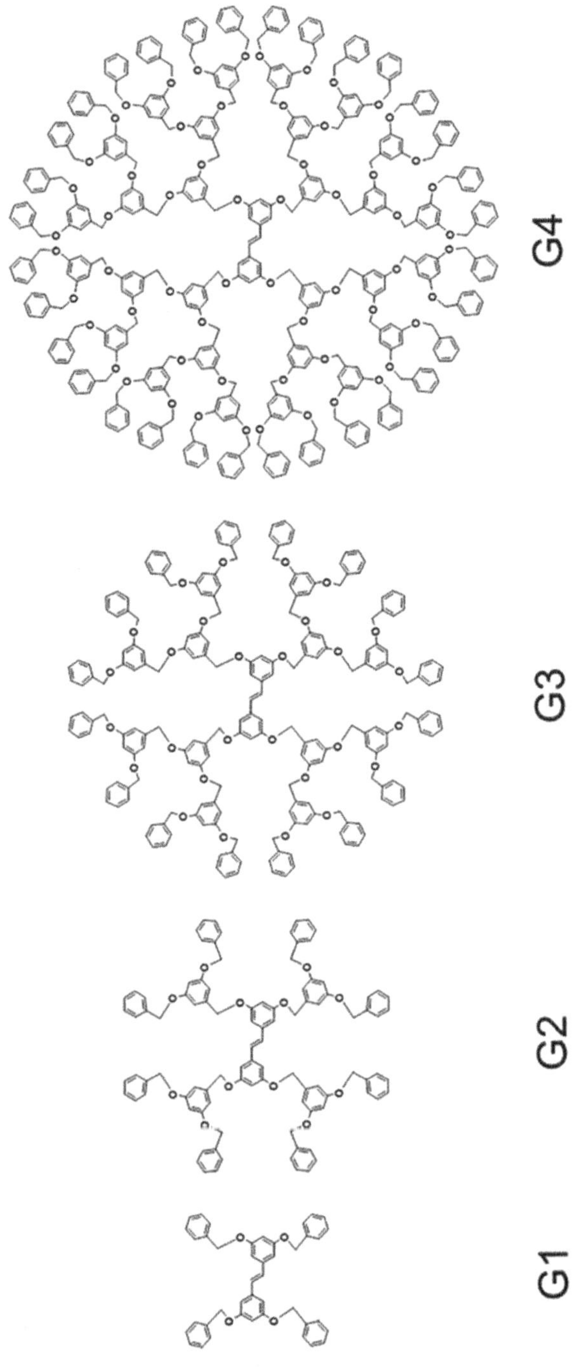

Fig. 13.3 Benzyl ether-type stilbene dendrimer **1** (G1–G4)

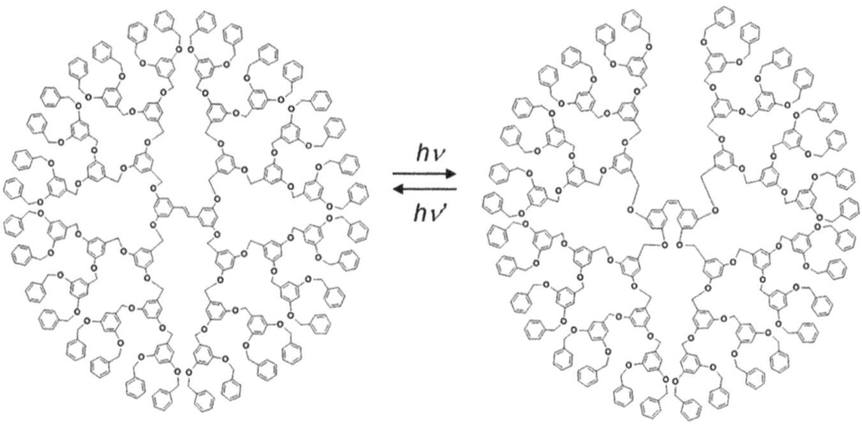

Fig. 13.4 Photoisomerization of benzyl ether-type stilbene dendrimer **1**(G4)

As to the photochemical properties, even the stilbene dendrimer (*trans*-**1** G4) with molecular weight over 6,500 underwent *trans–cis* isomerization in the excited singlet state within the lifetime of 10 ns (Fig. 13.4). Since we could not observe the effect of large dendron substituent on the photochemistry of stilbene core, one can assume that the isomerization may proceed by a volume-conserving mechanism such as the Hula-Twist mechanism rather than conventional 180° rotation around the C=C double bond [8]. Otherwise, the flexibility of the surrounding benzyl ether group may allow the isomerization of the core stilbene without the large structural change of the surrounding dendron group.

13.2.1.2 Ester- or Carboxylate-Terminated Dendrimers

As we observed almost no effect of benzyl ether-type dendrons on the photochemical and photophysical properties of stilbene-cored dendrimers, we have changed the dendron group by introducing electron-accepting substituents and water-soluble carboxylate anions at the peripheries of the benzyl ether-type dendrons to study the effect of substitution at the peripheries.

Ester-Terminated Dendrimers

Stilbene dendrimers with ester-terminated dendron groups as electron-accepting group exhibited quenching of excited singlet state of stilbene core which is caused by charge-transfer interactions with peripheral ester groups (Fig. 13.5) [9, 10]. This is supported by the decrease of the excited singlet lifetime with increasing number of peripheral ester functional groups (Table 13.1). If the quenching of the S_1 state of

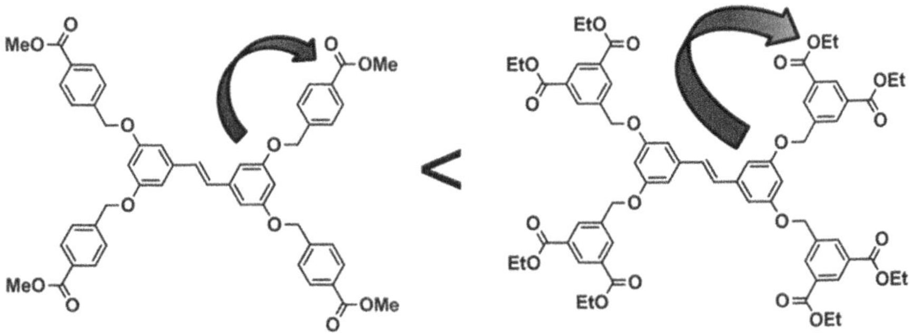

trans-**2** G1 *trans*-**3** G1

Fig. 13.5 Intramolecular charge-transfer interaction between the excited state core and the peripheral electron-accepting group of ester-terminated dendrimers

Table 13.1 Excited singlet lifetimes and *trans*-to-*cis* photoisomerization quantum yields of TMST and ester-terminated stilbene dendrimers

	trans-TMST	*trans*-**1** G1	*trans*-**2** G1	*trans*-**3** G1
τ_s (ns)	10.0	9.7	4.6	2.6
$\Phi_{t\rightarrow c}$	0.38	0.32	0.27	0.30

the core stilbene leads to no isomerization, the quantum yield of isomerization $\Phi_{t\rightarrow c}$ should be assumed to be 0.15 and 0.086 for *trans*-**2** G1 and *trans*-**3** G1, respectively. However, this is not the case. The experimental values are far more efficient: 0.27 and 0.30 for *trans*-**2** G1 and *trans*-**3** G1, respectively. These considerations indicate that the quenching by electron-accepting substituent at the remote peripheral position affects the electronic state of the core stilbene to quench and accelerate the intersystem crossing of the core to increase the isomerization efficiency by forming exciplex or serving as an electron-accepting spherical environment [10].

Water-Soluble (Carboxylate-Terminated) Dendrimers (**4**)

Unlike hydrophobic stilbene dendrimers, the environment around the stilbene core is different among the generations of **4** (WG1–WG3) (Fig. 13.6). With increasing the generation of the dendrimers, Stokes shifts tended to become smaller (Table 13.2). The Stokes shift of **4**WG3 was almost the same as that of hydrophobic dendrimer **2**G3 in THF [11]. In case of **4**WG1, solvent water should exist around the stilbene core and stabilize the excited state of stilbene, and the fluorescence spectrum of **4**WG1 appeared at the longest wavelength region among the compounds. With increasing the generation of dendrimer, the interaction between the stilbene core and water decreased due to the bulky hydrophobic dendron group and the increase of the number of hydrophilic carboxylate anions at the peripheries. The similarity of

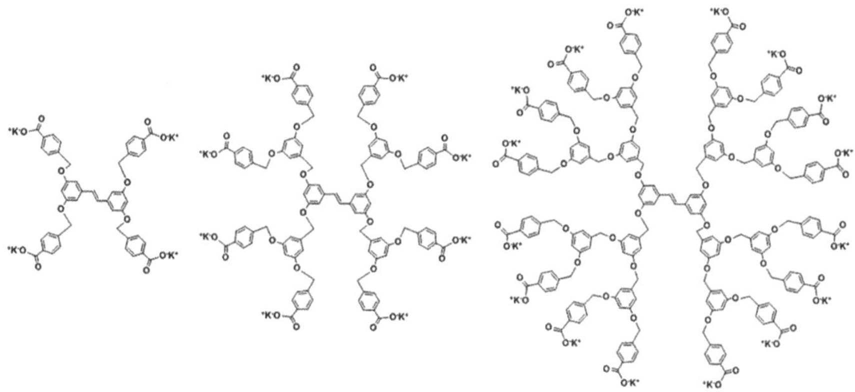

WG1 WG2 WG3

Fig. 13.6 Water-soluble (carboxylate-terminated) dendrimer **4** (WG1–WG3)

Table 13.2 Stokes shifts and *trans* to *cis* ratio at the photostationary state of water-soluble dendrimers

	WG1 (*trans*-**4** G1)	WG2 (*trans*-**4** G2)	WG3 (*trans*-**4** G3)	G3 (*trans*-**2** G3)
$\Delta\nu_{ss}$ (cm)	9,100	7,600	5,900	5,100
$([t]/[c])_{pps}$	6:94	1:99	0:100	8:92

the values of Stokes shift between **4**WG3 and **2**G3 indicates that the stilbene core of **4**WG3 was almost completely isolated from the outside water.

Interestingly, **4**WG3 showed one-way *trans*-to-*cis* photoisomerization [11]. The dendron groups having carboxylate anions at the periphery with counter cations at the surface may induce an electric field. Thus, one can think of the possibility that the effect of change in the electric field may result in change of the electronic state of stilbene core and lead to one-way photochemical isomerization of the stilbene core. Otherwise, the surrounding dendron groups may have a more congested environment for the *cis*-isomer than the *trans*-isomer due to the structural change of hydrophobic inner part and hydrophilic periphery. The *cis*-isomer seems to be unable to isomerize to the *trans*-isomer because of the higher congestion of the dendron groups inhibiting the structural change of the stilbene core within the excited lifetime.

Furthermore, by time-resolved transient grating method, fast and slow molecular structural changes were observed in **4**WG3 (Fig. 13.7). Thus, after the fast process of isomerization around the C=C double bond of stilbene within a lifetime of *ca.* 10 ns, the energy relaxation as well as the conformational change (~1 ms) occurred [12]. This combined system of fast isomerization of the core with a small molecular weight and slow conformational change of whole molecule is similar to the photochemical behavior of Rhodopsin. In Rhodopsin, the isomerization around the double bond takes place as a fast process of less than a picosecond, and the conformational change

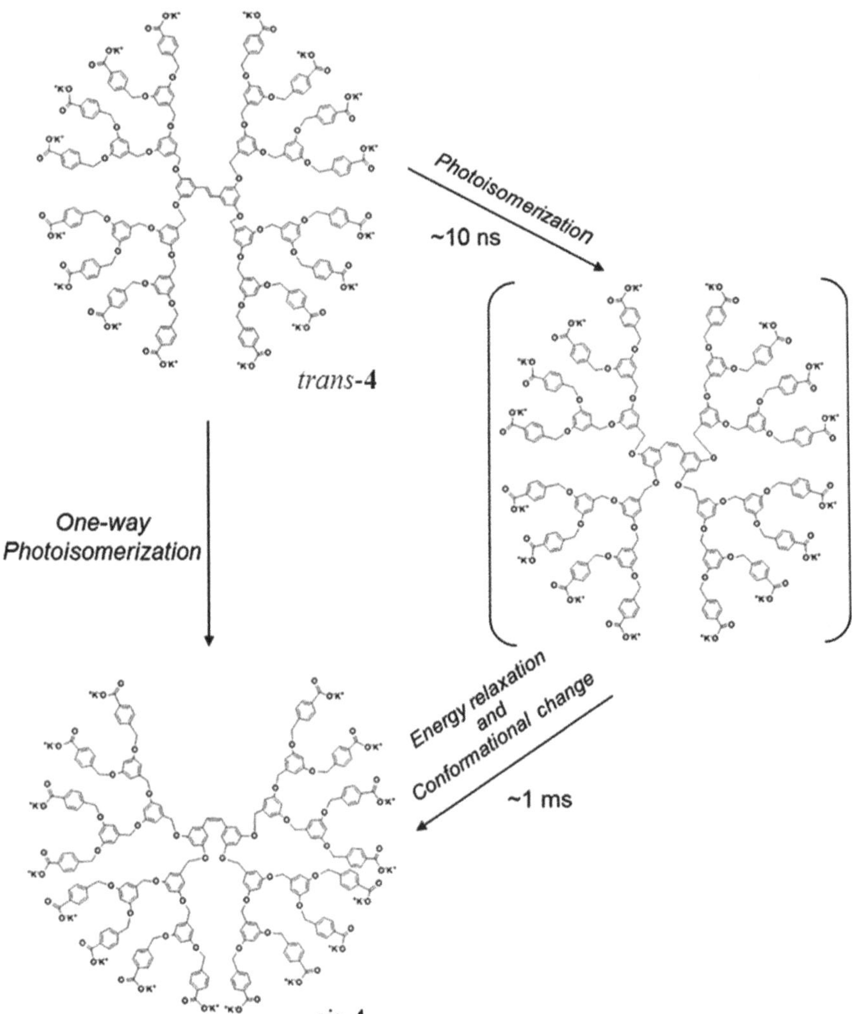

Fig. 13.7 Photoisomerization of water-soluble (carboxylate-terminated) dendrimer **4**WG3

toward the most stable structure of the surrounding opsin goes through several metastable conformations related to the changes in intramolecular hydrogen bonding. One may construct these kinds of molecules by introducing potential hydrogen bond forming dendron groups at the peripheries of the dendrimer molecules.

13.2.1.3 Photosensitizer-Terminated Dendrimers

Stilbene dendrimers **5** having benzophenone as a triplet sensitizer at the terminal of the dendron part were prepared. Both singlet and triplet energy transfers between

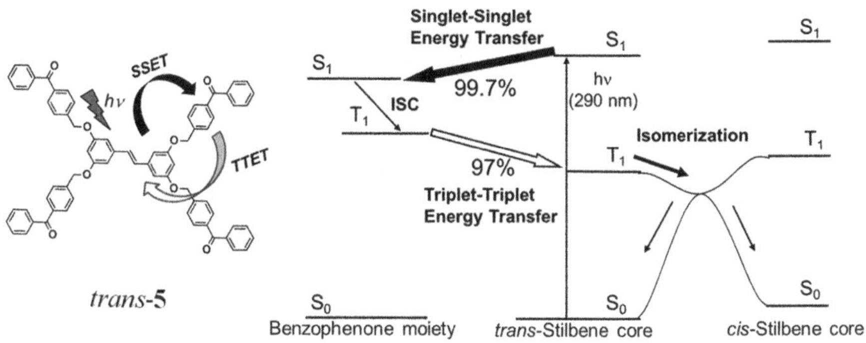

Fig. 13.8 Intramolecular energy transfer and photoisomerization of *trans*-**5**

stilbene core and benzophenone part took place quite efficiently as intramolecular processes (Fig. 13.8). On excitation (290 nm) of stilbene group of *trans*-**5**, intramolecular singlet energy transfer from the excited stilbene core to the benzophenone part took place with considerably high efficiency (99.7 %) [13]. This was confirmed by quenching of fluorescence from the core stilbene. Then, the intramolecular triplet energy transfer from excited benzophenone to stilbene core (97 %) occurred followed by *trans–cis* photoisomerization of the core stilbene occurring from the excited triplet state. Quantum yields for photoisomerization were still high ($\Phi_{t\rightarrow c}=0.41$) and even higher than the value of a similar compound without benzophenone group *trans*-**1**G1 ($\Phi_{t\rightarrow c}=0.32$) despite the involvement of two energy transfer steps.

A naphthalene-terminated stilbene dendrimer *trans*-**6** (Fig. 13.9) underwent *trans–cis* photoisomerization via highly efficient dendron-to-core energy transfer [14]. Because the energy transfer should take place as Förster-type energy transfer process, the spectral overlap between fluorescence spectra of the donor and absorption spectra of the acceptor plays an important role. In this compound the spectral overlap is actually far more efficient for *trans*-isomer than *cis*-isomer resulting in the more efficient energy transfer from naphthalene to the core *trans*-stilbene compared to that of *cis*-stilbene. This difference in energy transfer efficiency caused the change in *cis–trans*-isomer composition at the photostationary state to be *cis*-rich on naphthalene part excitation. Therefore, the direction and efficiency of large structural change could be changed by using the difference in energy transfer efficiencies from the peripheral naphthalene to the core stilbene.

13.2.1.4 Aggregation of Dendrimers

A cone-shaped amphiphilic stilbene dendrimer with hydrophilic triethylene glycol group can be dissolved to form aggregates in pure water even at very low concentration (10^{-6} M). Formation of aggregates induces intermolecular interactions between

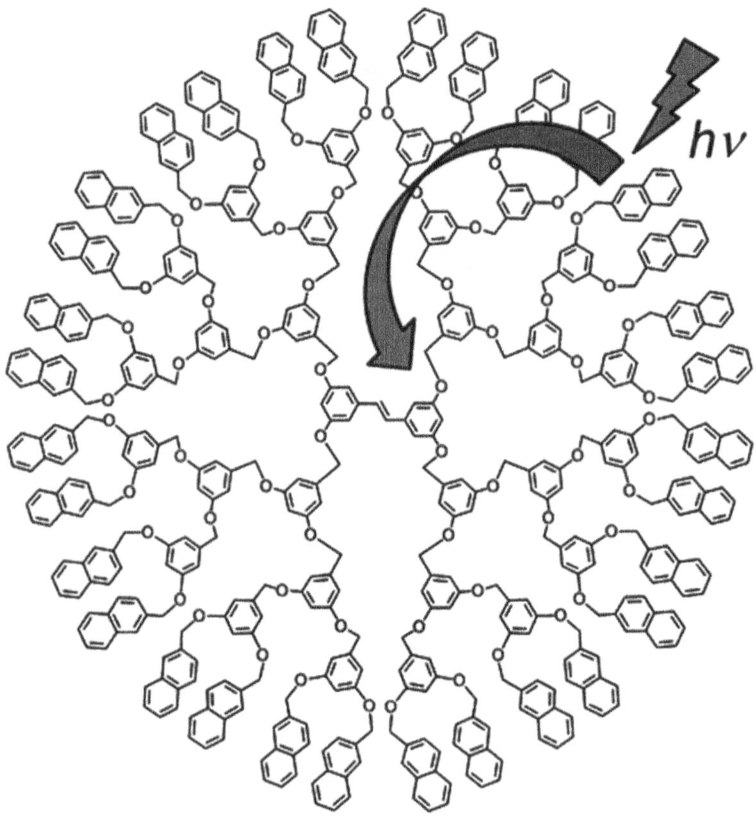

trans-**6**

Fig. 13.9 Intermolecular energy transfer of naphthalene-terminated stilbene dendrimer **6**

stilbene core parts (Fig. 13.10). In fact, excimer fluorescence and [2+2] photocycloaddition reaction of *trans*-stilbene were observed. Generally, intermolecular photochemical reaction of stilbene derivatives occurs only at solid state or relatively high concentration (>10^{-3} M) [15]. Observation of intermolecular photochemical reaction of stilbene in the lowest concentration to date was achieved by using dendritic structure. Furthermore, because of a property of terminated triethylene glycol, the aggregation size and photochemical properties of aggregates highly depended on temperature. Thus, one can make varying size of aggregates and may induce different structural change of aggregates by changing the dendrimer size and temperature.

Formation of hydrogen bonding at the dendron group may also induce the formation of aggregates. Actually, the compounds **8** exhibited CD spectra in benzene solution indicating the presence of chiral aggregates probably due to the formation of intermolecular hydrogen boding among the glutamate groups [16]. If one can

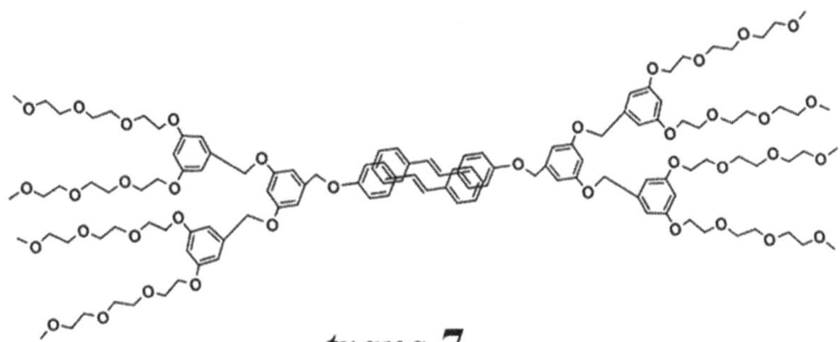

trans-**7**

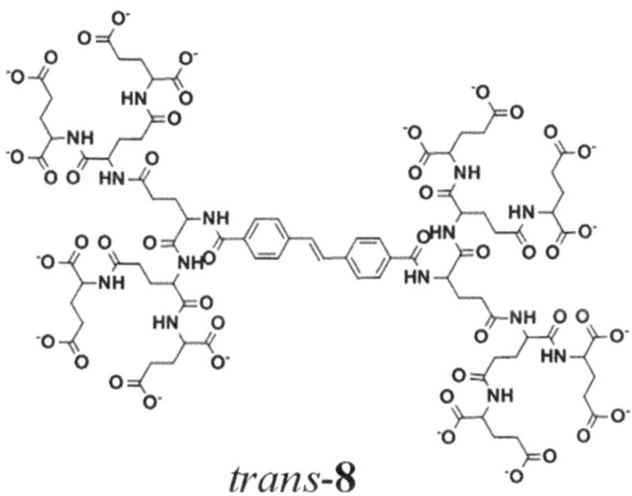

trans-**8**

Fig. 13.10 Aggregation formation of amphiphilic stilbene dendrimer **7** and poly(glutaminate) dendrimer **8**

change the interaction of the dendrimer molecules by light irradiation, one can tune the structure of the aggregates by inducing the structural change.

13.2.1.5 Polyphenylene Dendrimers

The polyphenylene dendron groups are rather rigid compared to the benzyl ether-type dendron. The fluorescence quantum yield (Φ_f) of *trans*-**9** (Fig. 13.11) was greater than 0.6, which is more than 15 times as large as that of parent *trans*-stilbene (0.04) [17]. Despite the rigid structure of the dendrimer, both *trans*-to-*cis* and *cis*-*to*-*trans* photoisomerization reactions occurred. When *cis*-stilbene is used as the core of a polyphenylene dendrimer (**11**), the fluorescence quantum yield increases up to 20 % at room temperature, which is 2000-fold more efficient than that of the parent *cis*-stilbene (Fig. 13.12) [18].

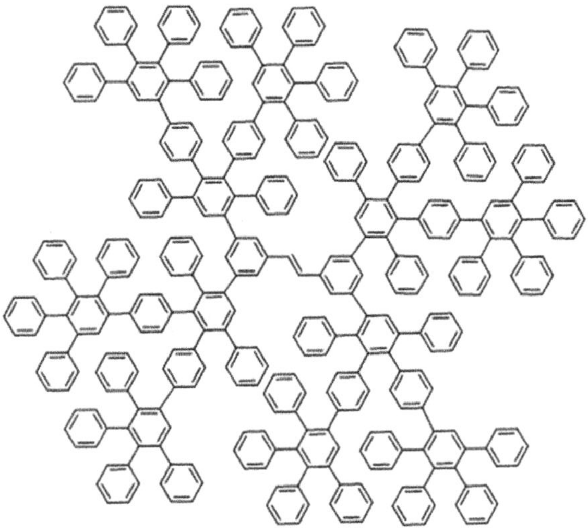
trans-**9**

Fig. 13.11 Structures of polyphenylene dendrimers

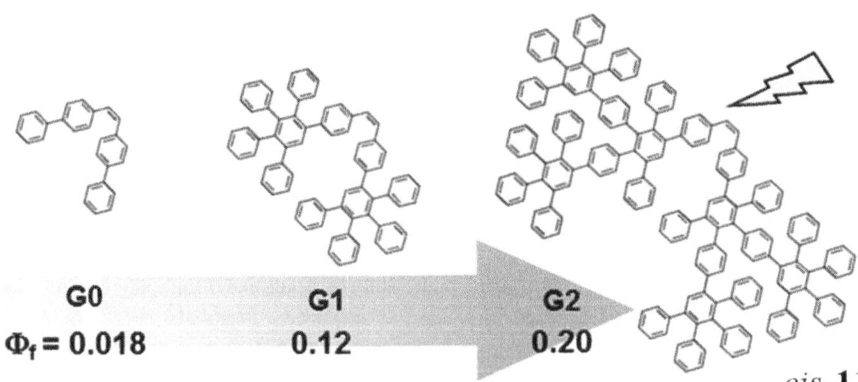

cis-**11**

Fig. 13.12 Efficient room temperature fluorescence of *cis*-stilbene dendrimers with polyphenylene dendrons (*cis*-**11**, G0–G2)

The increase of fluorescence lifetime with increasing generation can be explained as follows. With increasing the generation of polyphenylene dendrimers, activation barrier for isomerization from the excited *cis*-isomer may increase, which may contribute to the longer excited singlet lifetime and efficient fluorescence of *cis*-isomer. Thus, the increase of singlet lifetime and fluorescence quantum yield should be caused by the increase of the barrier for *cis–trans* isomerization from *cis* ($^1c^*$) to twisted state ($^1p^*$), which was brought about by the introduction of rigid polyphenylene group. Moreover, considerably large structural change should take place in these polyphenylene dendron-substituted dendrimers due to the rigidity of the dendron.

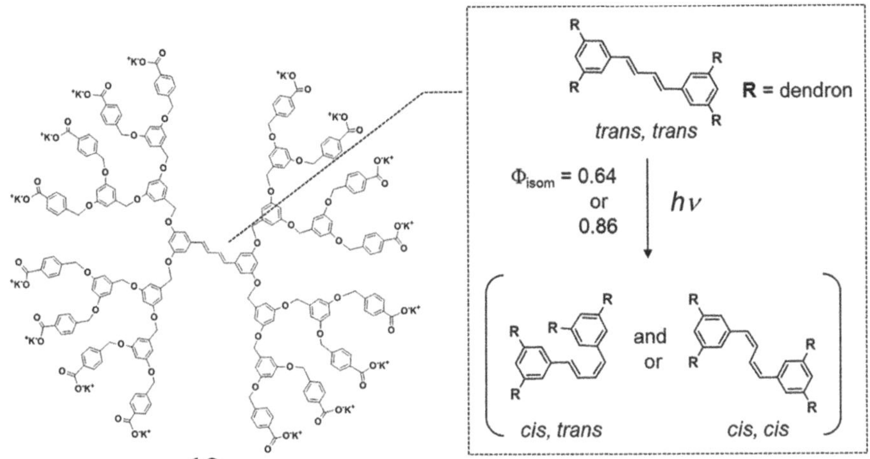

12

Fig. 13.13 Photoisomerization of water-soluble diphenylbutadiene dendrimer **12**

13.2.2 Other Photoresponsive Dendrimers

13.2.2.1 Diphenylbutadiene Dendrimers (12)

Water-soluble diphenylbutadiene-cored dendrimer **12** exhibited a remarkably high quantum yield for *trans*-to-*cis* photoisomerization in aqueous solution (Fig. 13.13). The first-generation water-soluble dendrimer with a diphenylbutadiene core isomerized from *trans*-to-*cis* with a high quantum yield ($\Phi_{isom}=0.64$), which is comparable to that of the value of retinal in bacteriorhodopsin [19]. Furthermore, the second-generation dendrimer **12** showed even greater quantum yield of $\Phi_{isom}=0.86$, which is the highest value known for *trans*-to-*cis* isomerization [20]. From these experiments, it was revealed that water-soluble dendrimers having one carboxylate anion at each one of the benzene ring at the periphery may result in highly efficient *trans–cis* isomerization giving large structural change of the dendrimer molecules.

13.2.2.2 Enediyne Dendrimers (13)

Enediyne dendrimers give efficient fluorescence emission from both the *cis*- and *trans*-isomers and undergo *cis–trans* photoisomerization (Fig. 13.14). The isomerization is expected to produce larger conformational change than that of stilbenes due to the rigid structure of the core enediyne parts having phenyl rings at both ends. By GPC analysis, it was observed that the molecular size of dendrimer decreased upon isomerization from *trans*-isomer to *cis*-isomer [21]. Therefore, the enediyne dendrimer is a potential candidate for future optical molecules.

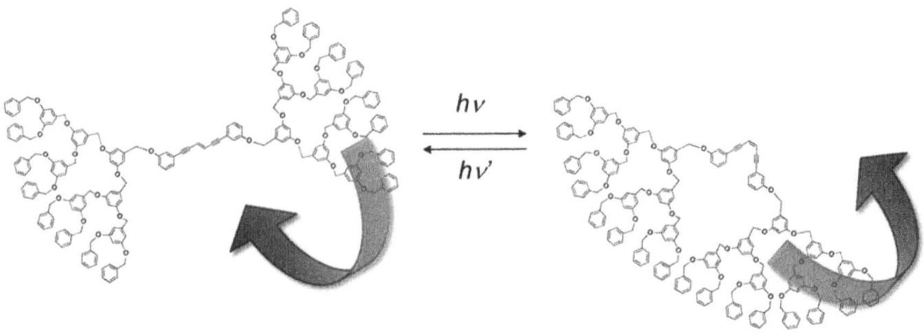

trans-**13** *cis*-**13**

Fig. 13.14 Large conformational change of enediyne dendrimer **13**

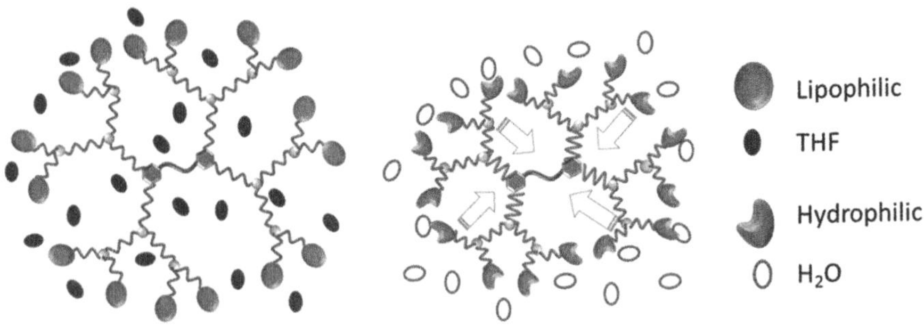

Fig. 13.15 Lipophilic and hydrophilic dendrimers

13.2.3 Summary

In order to achieve large structural change upon photochromic processes, molecular design and synthesis of appropriate compounds is necessary. In this respect, some approaches have been performed including the introduction of hydrophilic substituent at the peripheries of hydrophobic interior having photochromic core. These compounds are soluble in water like a unimolecular micelle, where water may interact with $-CO_2^-$ at the peripheries by hydrogen bonding (Fig. 13.15). The surrounding water molecules (or water clusters) may move with the photochemical change of the core part, accomplishing large structural change of the dendrimer molecules associated with water (Fig. 13.16).

Another possibility of observing large structural change is to use dendrimers having photochromic core surrounded by highly rigid dendrons such as polyphenylene (Fig. 13.17). In this type of compounds, the suppression of the isomerization rate constant from the *cis*-isomer to the perpendicular conformation in the excited singlet state was observed, which indicates the increase of barrier of *cis–trans*

Fig. 13.16 Isomerization of water-soluble dendrimers

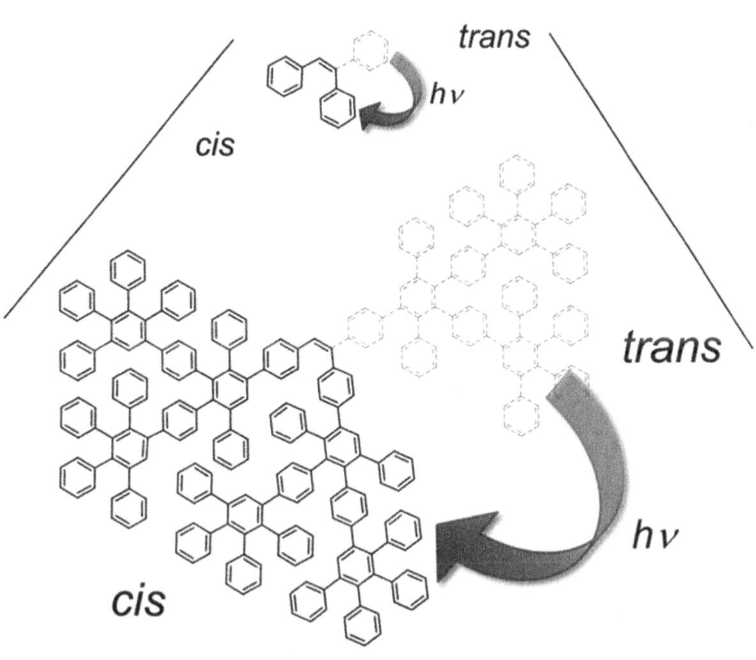

Fig. 13.17 Large structural change of rigid dendrimers

isomerization in the excited state. This increase of barrier of rotation in the excited state should be the consequence of increase of the molecular size related to the increase of the size change and/or conformation by isomerization.

13.3 Hydrogen-Bonded Compounds

As mentioned in Sect. 13.2, dendritic molecules have a virtue of undergoing dynamic change in molecular structure with photoirradiation. However, the shifts in absorption spectra accompanied with photoisomerization around the C=C double

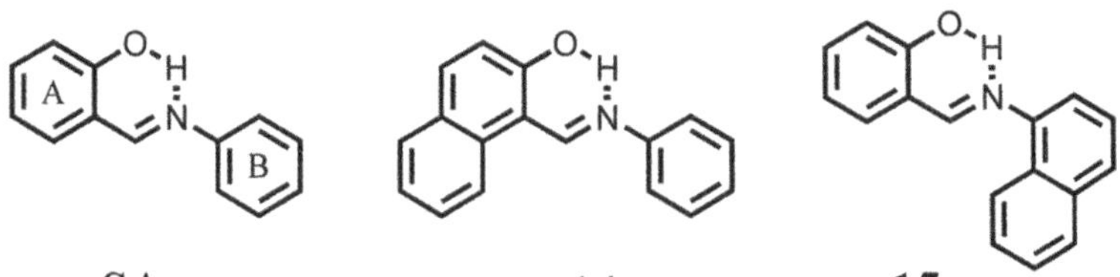

SA **14** **15**

Fig. 13.18 Chemical structures of SA, **14** and **15**

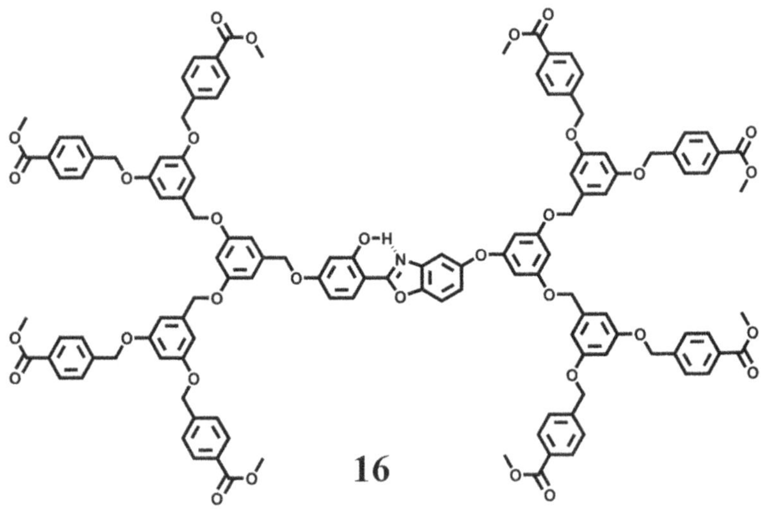

16

Fig. 13.19 Chemical structure of **16**

bond are within only a few nanometers, which is rather small for their dynamic structural change. This indicates that the dynamic structural change does not always mean the dynamic change of colors.

On the other hand, many hydrogen-bonded compounds, such as 2-(2′-hydroxyphenyl)benzoxazole (HBO) and salicylideneaniline (SA), undergo transient photochromism to give the tautomer (Figs. 13.18 and 13.19) [22, 23].

While the fluorescence quantum yield of SA and **14** is usually very low at room temperature, **15** exhibits largely Stokes-shifted fluorescence, indicating the occurrence of hydrogen atom transfer in the excited state. In addition, a dendritic molecule (**16**) with 2-(2′-hydroxyphenyl)benzoxazole as a core [24] undergoes intramolecular hydrogen atom transfer exhibiting the tautomer fluorescence. The fluorescence quantum yield increased with increasing generation.

Although the reaction of the above-mentioned compounds is interpreted as a transient phenomenon within subpicoseconds for forward reaction, the tautomer species responsible for photochromism would be too unstable to isolate. One can easily prepare and separate each geometrical isomer in the *trans–cis* photoisomerization

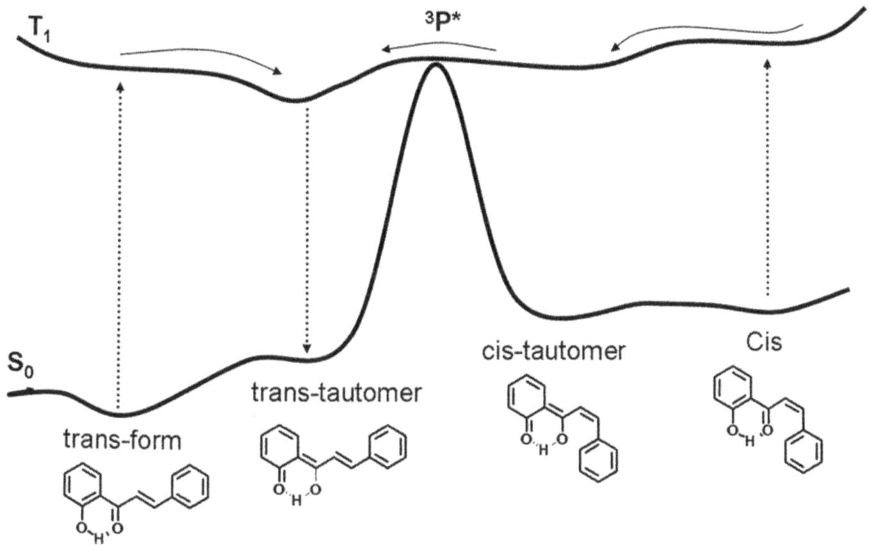

Fig. 13.20 Potential energy surface of 2′-hydroxychalcone

around C=C double bond and related photochromism. Our main purpose to study the compounds having C=C double bond and intramolecular hydrogen bond is to create new types of photochromic compounds due to the isomer specific formation of intramolecular hydrogen bonding and photoinduced hydrogen atom transfer. These two photoresponsive parts may result in both stable color change due to the presence or absence of intramolecular hydrogen bonding depending on the conformation around C=C double bond, *cis*- or *trans*-isomer, and transient photochromism by way of intramolecular hydrogen atom transfer in the excited singlet state which will result in observation of fluorescence emission and transient absorption of the tautomer. We have studied three types of photochromic compounds having both C=C double bond and intramolecular hydrogen bonding: [1] 2′-hydroxychalcone type, [2] pyrrolylethene type, and [3] hemiindigo type. Herein, we wish to introduce these photochromic compounds in terms of their photochemical behavior.

13.3.1 Photochromism of 2′-Hydroxychalcone-Type Compounds

2′-Hydroxychalcone (2′-HC) and its derivatives form strong intramolecular hydrogen bonding between –OH and –C=O and undergo hydrogen atom transfer in the excited state. 2′-Hydroxychalcone underwent one-way *cis*-to-*trans* photoisomerization induced by intramolecular hydrogen atom transfer in the excited state (Fig. 13.20) [25]. Actually, T–T absorption spectrum of 2′-hydroxychalcone in its tautomer form with a lifetime of 1 μs was observed. Since the tautomer has a polyene-like electronic configuration where the adiabatic *cis*-to *trans* isomerization

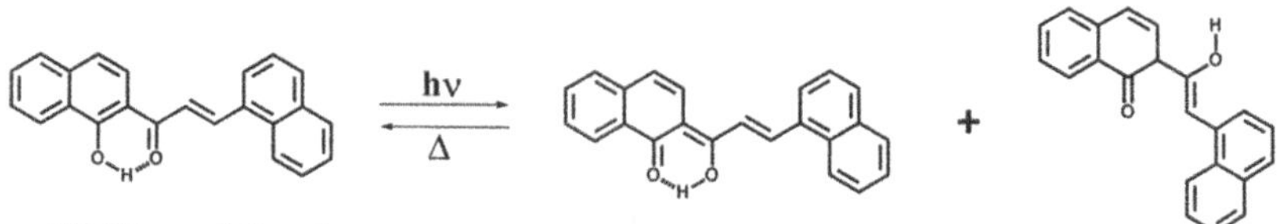

17 (Normal form)
$\lambda_{abs} = 400$ nm

17 (Tautomer closed form)
$\lambda_{FL} = 620$ nm

17 (Tautomer open form)
$\lambda_{abs} = 450$ nm

Fig. 13.21 Chemical structure of **17**

in the triplet state takes place, the potential energy surface of the excited state tautomer of 2′-hydroxychalcone may be similar to that of one-way *cis*-to-*trans* photoisomerization; in the excited triplet state, the perpendicular excited state is located not at the energy minimum but at the energy barrier [25, 26]. Although the observation of any fluorescence evidence of the occurrence of adiabatic intramolecular hydrogen atom transfer is difficult.

One can observe fluorescence from 2′-HC derivatives as described later.

13.3.1.1 1-(1′-Hydroxy-2-Naphtyl)-3-(1-Naphthyl)-2-Propen-1-One (17)

In contrast to 2′-HC, 1-(1′-hydroxy-2-naphtyl)-3-(1-naphthyl)2-propen-1-one (**17**) gave Stokes-shifted fluorescence emission peaking at 620 nm from the tautomer produced through hydrogen atom transfer (Fig. 13.21). Furthermore, the transient absorption of the tautomer was observed at 450 nm [27]. In this case, the absorption spectrum of the normal form appeared at 400 nm and that of the tautomer appeared at 450 nm. Although the color change is a transient phenomenon, we may propose this reaction as a transient photochromism. In addition, we have studied the cooperative effect of photoinduced intramolecular hydrogen atom transfer and the *cis–trans* isomerization around the C═C double bond on the photochromic properties.

13.3.1.2 The Dendritic Molecule with 2′-Hydroxychalcone Core (18)

By introducing the dendritic structure, tautomer fluorescence in the longer wavelength region at 600 nm produced by intramolecular hydrogen atom transfer was observed in **18**, and fluorescence quantum yield increased slightly with increasing generation (Fig. 13.22) [28].

13.3.2 Photochromism of Pyrrolylethene-Type Compounds

The absorption spectra of *cis*-isomer of pyrrolylethenes examined shifted to the longer wavelengths region compared to those of *trans*-isomers in 20–50 nm,

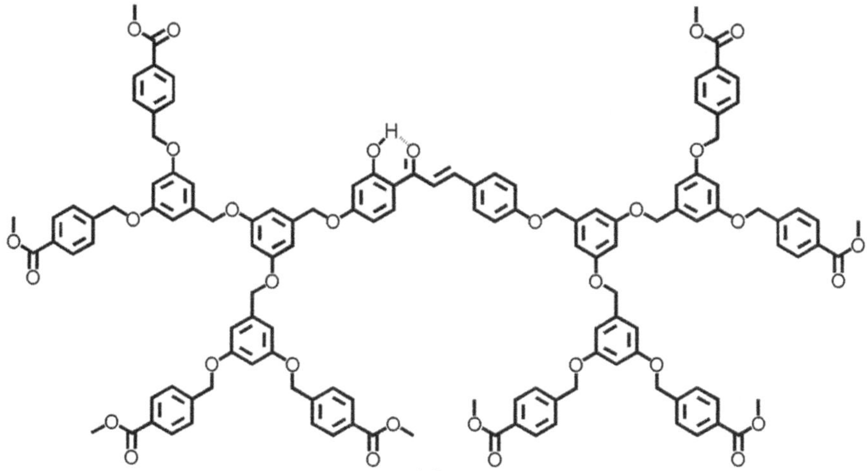

18

Fig. 13.22 Chemical structure of **18**

probably due to the extended conjugation through intramolecular hydrogen bonding [29–31].

As shown in Fig. 13.23, 2-[2-(2-pyrrolyl)ethenyl]pyridine (**19**) underwent one-way *trans*-to-*cis* isomerization not only in aprotic but also in protic solvent. Since T–T absorption spectrum was not observed on excitation of **19**, the one-way *trans-to-cis* isomerization should occur in the excited singlet state. Moreover, *cis*-isomers emitted a tautomer fluorescence produced by an adiabatic intramolecular hydrogen atom transfer in the excited singlet state. Thus, **19** exhibited two kinds of photochromism on direct irradiation: thermally stable one-way photoisomerization and transient process due to hydrogen atom transfer to give the tautomer.

One of the problems in photochemical behavior of **19** is that this compound undergoes one-way *trans*-to-*cis* isomerization. To solve this problem and find a similar molecular system undergoing mutual isomerization, which is essential for photochromism, we have synthesized a series of pyrrolylethenes having a variety of nitrogen-containing heteroaromatic rings and studied their photochemical behavior. When we replace a pyrrole ring with an indole ring in **19**, 2-[2-(2-pyridyl)ethenyl] indole (**20**), we observed change of absorption spectra between *cis*- and *trans-isomers* similar to that of **19** (Fig. 13.24) [29, 30]. However, this compound underwent one-way *trans*-to-*cis* photoisomerization and is not a photoreversible compound. In addition, *cis*-isomer did not show transient photochromism because we could not observe any fluorescence spectra from neither *cis*-isomer nor its tautomer. When we replace a pyridine ring with a quinoline ring, we have succeeded in observing stable photochromism and transient photochromism as described in the next section.

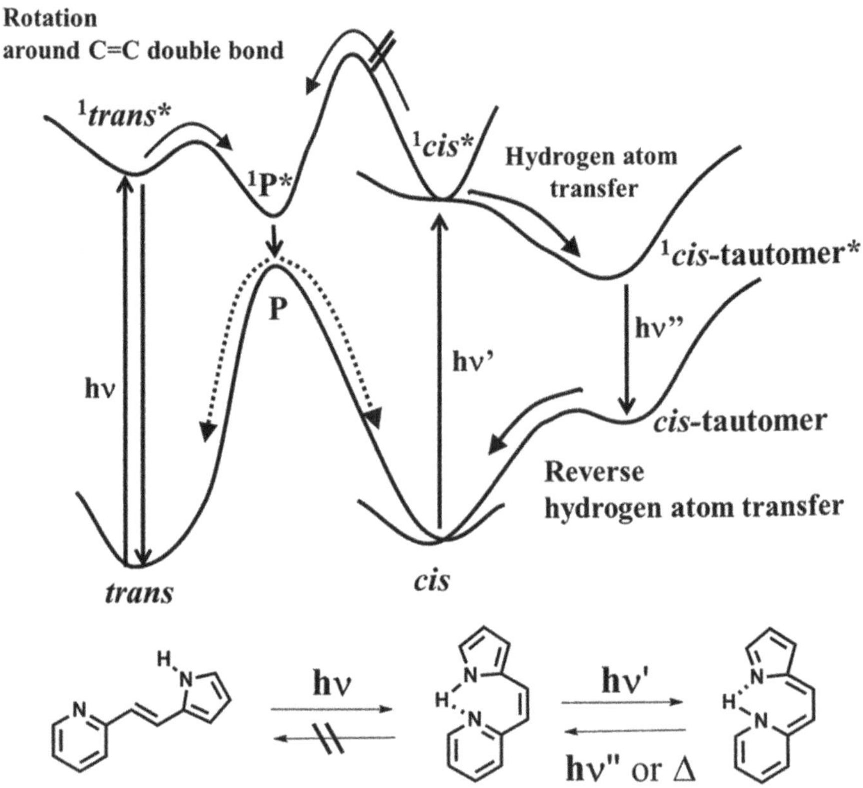

trans-**19**
$\lambda_{abs\ (Hexane)} = 341$ nm
$\lambda_{FL\ (Hexane)} = 430$ nm

cis-**19**
$\lambda_{abs} = 360$ nm

cis-**19**-tautomer
$\lambda_{abs} = 530$ nm
$\lambda_{FL} = 570$ nm

Fig. 13.23 Potential energy surface of isomerization of **19**

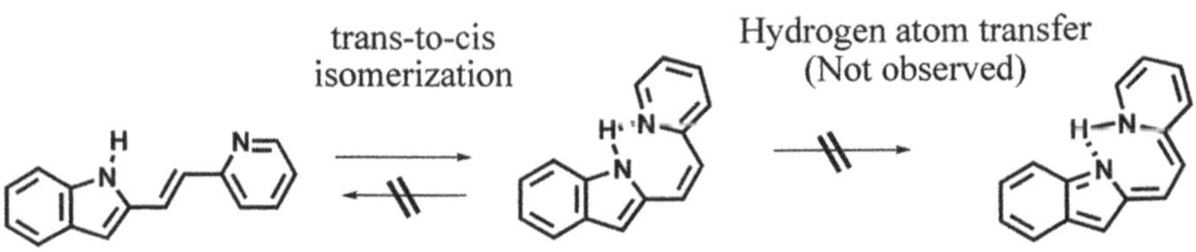

trans-**20**
$\lambda_{abs(Benzene)} = 350$ nm
$\lambda_{FL(Benzene)} = 450$ nm

cis-**20**
$\lambda_{abs} = 379$ nm

cis-**20**-tautomer

Fig. 13.24 Photoisomerization of **20**

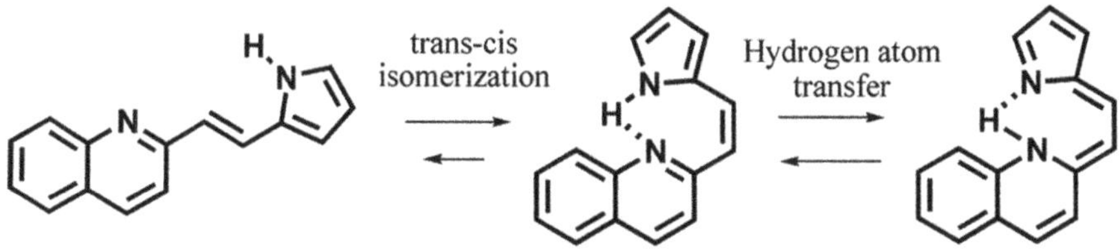

trans-**21**	*cis*-**21**	*cis*-**21**-tautomer
$\lambda_{abs\ (Benzene)} = 377$ nm	$\lambda_{abs} = 414$ nm	$\lambda_{abs} = 580$ nm
$\lambda_{FL\ (Benzene)} = 440$ nm		$\lambda_{FL} = 600$ nm

Fig. 13.25 Photoisomerization of **21**

Fig. 13.26 Chemical structure of indigo as a parent compound

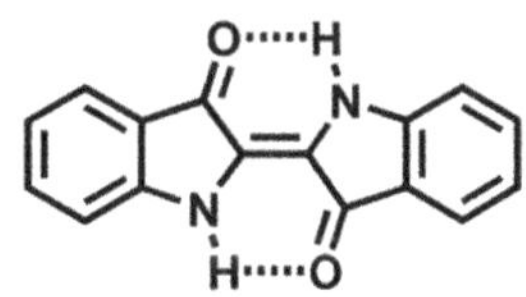

13.3.2.1 2-[2-(2-Pyrrolyl)Ethenyl]Quinoline (21)

Though **19** underwent one-way *trans*-to-*cis* isomerization, 2-[2-(2-pyrrolyl)ethenyl]quinoline (**21**) underwent mutual *trans–cis* isomerization to afford a photostationary mixture of *cis*- and *trans*-isomers in a ratio of 93.4:6.6 in benzene under N_2 (Fig. 13.25) [31]. Moreover, *cis*-isomer exhibited fluorescence with a very large Stokes shift, indicating that a remarkable change of electronic configuration took place to produce a *cis*-tautomer by way of intramolecular hydrogen atom transfer.

13.3.3 Photochromism of Hemiindigo-Type Compounds

Indigo is one of the useful dyes in chemical industry, exhibiting deep blue color and high stability due to the intramolecular hydrogen bonding. Considering the chemical structure of indigo consisting of two indolinone rings having both C=C double bond and intramolecular hydrogen bonding between NH and C=O, we have developed two hemiindigo-type compounds as promising candidates for a new photochromic system. Herein, we wish to introduce their photochemical behavior (Fig. 13.26).

13.3.3.1 2-(2-Pyrrolylidene)Indolin-3-One (22)

As shown in Fig. 13.27, 2-(2-pyrrolylidene)indolin-3-one (**22**) underwent photoisomerization mutually between the *Z*- and *E*-isomers and exhibited dynamic color

Fig. 13.27 Photo-isomerization of **22**

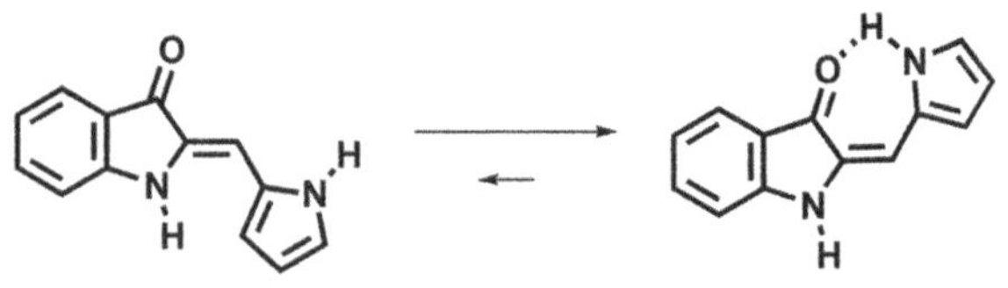

Z-**22**	*E*-**22**
$\lambda_{abs\ (Benzene)} = 470$ nm	$\lambda_{abs} = 524$ nm
$\lambda_{FL\ (Benzene)} = 500$ nm	$\lambda_{FL} = 554$ nm

Fig. 13.28 Chemical structure of **23**

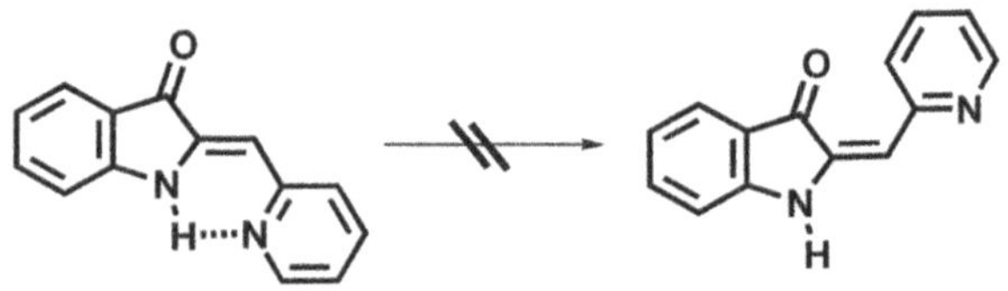

Z-**23**	*E*-**23**
$\lambda_{abs(Benzene)} = 477$ nm	

change between greenish yellow and reddish orange in benzene due to the intramolecular hydrogen bonding in *E*-isomers [32, 33]. The quantum yield of *E*→*Z* isomerization in benzene was lower than 0.01, while that of *Z*→*E* isomerization was 0.3 because the intramolecular hydrogen bonding suppressed the *E*→*Z* isomerization in benzene.

13.3.3.2 2-(2-Pyridylmethylidene)Indoline-3-One (23)

The absorption maximum of *Z*-**23** appeared at 477 and 482 nm in benzene and methanol, respectively. (*Z*)-2-(2-Pyridylmethylidene)indoline-3-one (**23**) did not undergo isomerization to *E*-isomer in neither methanol nor benzene (Fig. 13.28) [33]. Moreover, *Z*-**23** did not emit fluorescence at room temperature. Therefore, *Z-23* is highly photostable.

13.3.4 Summary

Many photochromic compounds have been investigated because of their application in chemical industry. We have also studied and developed new photochromic systems on the basis of dynamic color change induced by small structural change of olefinic C═C double bond and intramolecular hydrogen bonding. In these photochromic compounds, both stable and transient photochromisms are expected and

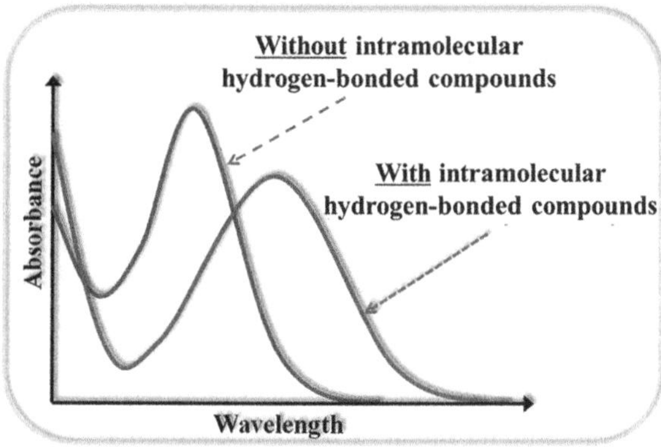

Fig. 13.29 Comparison of absorption spectra in photochromic compounds with/without intramolecular hydrogen bonding

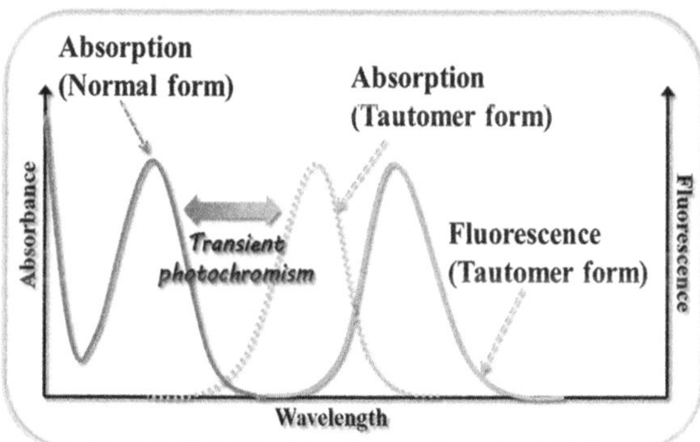

Fig. 13.30 Comparison of absorption and fluorescence spectra of normal and tautomer form

observed. Stable photochromism corresponds to the shift in absorption spectra accompanied by photoisomerization around C=C double bond to produce thermally stable geometrical isomer. The color change by photoisomerization and its efficiency are greatly affected by the presence of intramolecular hydrogen bonding capable of expanding the electronic conjugation, as shown in Fig. 13.29. On the other hand, transient photochromism is the consequence of the occurrence of intramolecular hydrogen atom transfer to give a tautomer in the excited state with very short lifetime. The tautomer may not be isolated, but its properties could be estimated from its spectroscopic observation such as fluorescence measurement and other time-resolved spectroscopy, as shown in Fig. 13.30.

13.4 Summary

If one uses the dendritic structure, large structural change can be expected by appropriately choosing dendrimers having stilbene-type photochromic core and relatively flexible benzyl ether dendrons terminated with hydrophilic substituent at the peripheries. In addition, the introduction of rigid dendrons at the peripheries of stilbene should cause large structural change as well. Although some spectral change in absorption is usually observed for simple olefinic compounds, it is not possible to attain color change by isomerization upon photoirradiation. One of the approaches to induce color change of the core part is to introduce another factor such as intramolecular hydrogen bonding at the core and/or at the peripheries. In this case, color change upon photoirradiation requires considerable change of electronic structure by the effect of intramolecular hydrogen bonding, and both thermally stable photochemical *cis–trans* isomerization around the C=C double bond and thermally transient structural change upon photoinduced hydrogen atom transfer processes may take place. The cooperative effect of isomerization around the C=C double bond and photoinduced hydrogen atom transfer may construct a molecular photochromic system having large structural change accompanied by color change.

References

1. Dürr H, Bouas-Laurent H (1990) Photochromism: molecules and systems. Elsevier, Amsterdam
2. Irie M (2000) Diarylethenes for memories and switches. Chem Rev 100:1685–1716
3. Yokoyama Y (2000) Fulgides for memories and switches. Chem Rev 100:1717–1739
4. Tomalia DA, Naylor AM, III Goddard WA (1990) Starburst dendrimers: molecular-level control of size, shape, surface chemistry, topology, and flexibility from atoms to macroscopic matter. Angew Chem Int Ed Engl 29:138–175
5. Saltiel J, Waller AS, Sears DF Jr (1992) Dynamics of *cis*-stilbene photoisomerization: the adiabatic pathway to excited *trans*-stilbene. J Photochem Phobiol A: Chem 65:29–40
6. Momotake A, Uda M, Arai T (2003) Photochemical cyclization reaction in *cis-tetramethoxystilbene*. J Photochem Photobiol A: Chem 158:7–12
7. Mizutani T, Ikegami M, Nagahata R, Arai T (2001) The first synthesis of stilbene dendrimers and their photochemical *trans–cis* isomerization. Chem Lett 1014–1015
8. Uda M, Mizutani T, Hayakawa J, Momotake A, Ikegami M, Nagahara R, Arai T (2002) Photoisomerization of stilbene dendrimers: the need for a volume-conserving isomerization mechanism. Photochem Photobiol 76:596–605
9. Momotake A, Hayakawa J, Nagahata R, Arai T (2004) The photochemical specific isomerization of dendritic stilbenes. Bull Chem Soc Jpn 77:1195–1200
10. Takizawa T, Arai T (2011) Remote controlled intramolecular exciplex formation and enhanced photoisomerization in stilbene-cored poly(benzyl ether) dendrimers with alkoxycarbonyl surface functional groups. Chem Lett 40:1124–1126
11. Hayakawa J, Momotake A, Arai T (2003) Water-soluble stilbene dendrimers. Chem Comm 94–95
12. Tatewaki H, Baden N, Momotake A, Arai T, Terazima M (2004) Dynamics of water-soluble stilbene dendrimers upon photoisomerization. J Phys Chem B 108:12783–12789
13. Miura Y, Momotake A, Shinohara Y, MD W, Nishimura Y, Arai T (2007) The first observation of the effect of dendritic structure to produce the triplet excited state of the core stilbene by dendron excitation. Tetrahedron Lett 48:639–641
14. Nakazato S, Takizawa T, Arai T (2012) Photoisomerization and energy transfer in naphthalene-terminated stilbene dendrimers. Photochem Photobiol Sci 11:885–888
15. Lewis FD, Johnson DE (1977) Self-quenching and dimerization of singlet *trans*-stilbene. J Photochem 7:421–423
16. Ikeda M, Momotake A, Kanna Y, Shimotori Y, Arai T (2012) Self-assembly of poly(glutamate) dendrimer: solvent-dependent expression of molecular chirality and [2+2] photocrosslinking reaction. Photochem Photobiol Sci 11:1524–1527

17. Imai M, Arai T (2002) Synthesis and photochemical properties of polyphenylene dendrimers with photoreactive stilbene core. Tetrahedron Lett 43:5265–5268
18. Tabuchi M, Momotake A, Kanna Y, Nishimura Y, Arai T (2011) Extremely efficient and long lifetime fluorescence of *cis*-stilbene contained in a rigid dendrimer. Photochem Photobiol Sci 10:1521–1523
19. Miura Y, Momotake A, Kanna Y, Nishimura Y, Arai T (2012) Extremely efficient photoisomerization of water-soluble diphenylbutadiene dendrimers. Dyes and Pigments 92:802–806
20. Miura Y, Momotake A, Kanna Y, Nishimura Y, Arai T (2011) Unusually efficient *trans*-to-*cis* photoisomerization of diphenylbutadiene dendrimers in water. Photochem Photobiol Sci 10:1524–1526
21. Yoshimura N, Momotake A, Shinohara Y, Nishimura Y, Arai T (2007) Synthesis and photoisomerization of enediyne-type dendrimers with benzyl ether dendrons. Bull Chem Soc Jpn 80:1995–2000
22. Ohshima A, Momotake A, Arai T (2005) Photochemistry of salicylideneaniline analogue at low temperature. Chem Lett 34:1288–1289
23. Ohshima A, Momotake A, Arai T (2004) Photochromism, thermochromism, and solvatochromism of naphthalene-based analogues of salicylideneaniline in solution. J Photochem Photobiol A 162:473–479
24. Ohshima A, Momotake A, Nagahata R, Arai T (2005) Enhancement of the large stokes-shifted fluorescence emission from the 2-(2′-Hydroxyphenyl)benzoxazole core in a dendrimer. J Phys Chem A 109:9731–9736
25. Norikane Y, Itoh H, Arai T (2002) Photochemistry of 2′-hydroxychalcone. One-way *cis–trans* photoisomerization induced by adiabatic intramolecular hydrogen atom transfer. J Phys Chem A 106:2766–2776
26. Norikane Y, Nakayama N, Tamaoki N, Arai T, Nagashima U (2003) Quantum chemical studies on photoinduced *cis–trans* isomerization and intramolecular hydrogen atom transfer of 2′-hydroxychalcone. J Phys Chem A 107:8659–8664
27. Kaneda K, Sato S, Hamaguchi H, Arai T (2004) The photoinduced hydrogen atom transfer and *trans–cis* Isomerization of the C=C double bond in 1-(1-hydroxy-2-naphthyl)-3-(1-naphtyl)-2-propen-1-one and related compounds studied using nanosecond time-resolved infrared spectroscopy. Bull Chem Soc Jpn 77:1529–1535
28. Teshima T, Takeishi M, Arai T (2009) Red fluorescence from tautomers of 2′-hydroxychalcones induced by intramolecular hydrogen atom transfer. New J Chem 33:1393–1401
29. Obi M, Sakuragi H, and Arai T (1998) Adiabatic intramolecular hydrogen atom transfer in 2-[2-(2-Pyrrolyl)ethenyl]pyridine. Chem Lett 169
30. Lewis FD, Yoon BA, Arai T, Iwasaki T, Tokumaru K (1995) Molecular structure and photochemistry of (E)- and (Z)-2-(2-(2-Pyridyl)ethenyl)indole. A case of hydrogen bond dependent one-way photoisomerization. J Am Chem Soc 117:3029–3036
31. Arai T, Moriyama M, Tokumaru K (1994) Novel photoinduced hydrogen atom transfer through intermolecular hydrogen bonding coupled with *cis–trans* isomerization in *cis*-1-(2-Pyrrolyl)-2-(2-quinolyl)ethane. J Am Chem Soc 116:3171–3172
32. Ikegami M, Arai T (2003) Photoisomerization and fluorescence properties and hemiindigo compounds having intramolecular hydrogen bonding. Bull Chem Soc Jpn 76(9):1783–1792
33. Ikegami M, Arai T (2005) Effect of the intramolecular hydrogen bonding in the photochromic properties of the hemiindigo dye having a pyrrole ring. Mol Cryst Liq Cryst 431:145–149
34. Arai T, Tokumaru K (1995) Present status of the photoisomerization about ethylenic bonds. In: Neckers DC, Volman DH, Bunau GV (eds) Advances in photochemistry, vol 20. Wiley, New York, pp 1–57
35. Hayakawa J, Ikegami M, Mizutani T, Wahadoszamen MD, Momotake A, Nishimura Y, Arai T (2006) Solvent effect on the photochemical properties of symmetrically substituted trans-3,3′,5,5′-tetramethoxystilbene. J Phys Chem A 110:12566–12571

Chapter 14
Photocontrol of New Molecular Functions by the Isomerization of Azobenzene

Nobuyuki Tamaoki

Abstract Azobenzene shows the reversible *E–Z* photoisomerization upon near ultraviolet (UV) and blue light irradiations. By utilizing changes in the molecular length and polarity of azobenzene in this isomerization, azobenzene has been applied as the molecular switching unit of various molecular functional materials. In this chapter some recent challenges on the photoswitching of new molecular functions using azobenzene derivatives were described. They include the molecular chirality and molecular machines. The feature of these molecular functions is that the control is not accomplished until fine tuning of inter- or intramolecular motions. The racemization of the planar chiral macrocycles **2** and **5** is photochemically and dynamically controlled for the first time by controlling intramolecular free rotation of an asymmetric rotor unit by photoisomerization of an azobenzene moiety in macrocycles. Under circular polarized lights it was possible to enrich one of the enantiomers of **5** as the results of different efficiency of the photoisomerization between racemizing *Z* isomer and the nonracemizing *E* enantiomers. It was also demonstrated that point chirality was dynamically introduced by the *E–Z* photoisomerization of one of the azobenzenes in the methane derivatives substituted with two identical azobenzenes. In the study of molecular machine the sliding motion of a motor protein, kinesin-microtubule, can be controlled by the photoisomerization of azobenzene unit introduced to a monomolecular layer underneath the kinesin or in the adenosine triphosphate (ATP) served as an energy source.

Keywords Adenosine triphosphate • Azobenzene • Chirality • Kinesin • Liquid crystal Macrocyclic compound • Molecular machine

---

N. Tamaoki (✉)
Research Institute for Electronic Science,
Hokkaido University, N20, W10, Kita-ku,
Sapporo 001-0020, Japan
e-mail: tamaoki@es.hokudai.ac.jp

M. Irie et al. (eds.), *New Frontiers in Photochromism*,
DOI 10.1007/978-4-431-54291-9_14,

14.1 Introduction

Azobenzene is a typical photochromic compound that shows *trans–cis* or *E–Z* photoisomerization under UV and visible lights. Due to its distinguished features as a photochromic compound such as easiness of synthesis, possibility of functionalization or structural modification, relatively high quantum yield and high photochemical conversions between photostationary states in both of the photoisomerization routes, relatively high extinction coefficient of both of the isomers, and large change in the molecular shape upon isomerization, it has been used as a photoswitching unit for various molecular functions or properties including ion-binding [1], recognition of small organic molecules [2], phase transition temperature [3] or molecular alignment [4, 5] of liquid crystals, sol–gel transition temperature [6], conformation of synthetic polymers [7] and proteins [8], hybridization of DNA [9], and so on.

In this chapter some recent challenges on the photoswitching of new molecular functions using azobenzene derivatives were described. They include molecular chirality and molecular machines. The feature of these molecular functions is that the control is not accomplished until fine tuning of inter- or intramolecular motions. The racemization of the planar chiral molecules are first photochemically and dynamically controlled by controlling intramolecular free rotation of an asymmetric rotor unit by an azobenzene moiety in macrocycles. The sliding motion of a motor protein, kinesin-microtubule, can be controlled by the photoisomerization of azobenzene unit introduced to a monomolecular layer underneath the kinesin or in the adenosine triphosphate (ATP) served as an energy source.

14.2 Photoswitching of Molecular Chirality

Reversible enantio-differentiating photoisomerization is known for some compounds, where photoresolution and photoracemization occur with circular polarized and normal light, respectively [10–14]. This process has attracted much attention because the application of such compounds to rewritable optical recording media makes it possible to read out memories—without their destruction—by monitoring the difference in the optical rotation of the two states at a wavelength at which the absorbance is weak. Another important reason for studying such photoreactions is that they provide one possible mechanism for the origin of homochirality in bioorganic compounds; this theory proposes that a state displaying a partial enantiomeric excess can be obtained under the influence of circular polarized electromagnetic radiation, and then it can undergo a chiral amplification reaction [15–19]. A very few compounds known to display reversible enantio-differentiating photoisomerization directly between two enantiomers are some chromium complexes and ethylene derivatives. An important point is that circular polarized light (CPL) is used to excite one of the enantiomers more efficiently due to its higher extinction coefficient to the common excited state for both the enantiomers where the racemization

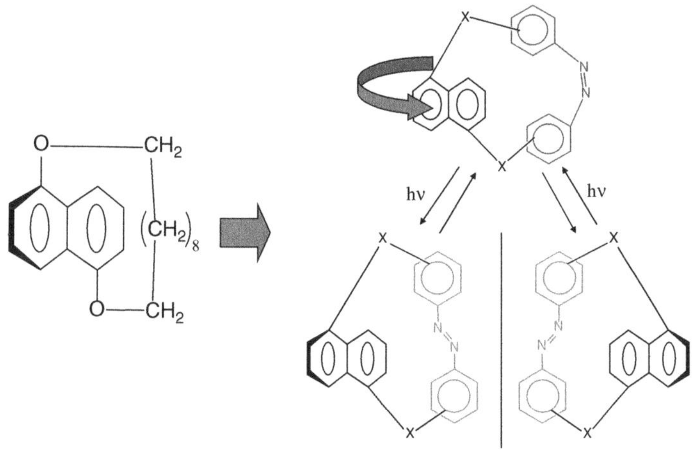

Fig. 14.1 The concept of switching of planar chirality by *E–Z* photo isomerization of azobenzene

is possible. Since the deactivation occurs randomly to either enantiomer, the imbalance in the efficiency of the excitation induces the enrichment of the enantiomer having a lower extinction coefficient under the irradiated CPL. In the ever known molecules that showed the reversible enrichment of one of the enantiomers by CPL irradiation, the excited state was used for racemization.

Recently it was tried to partially enrich one of the enantiomers by CPL irradiation of azobenzene derivatives showing the racemization in one of the ground state of the geometrical isomers. In one trial the intramolecular rotational motion of one asymmetric part of the macrocyclic molecule, which induces a racemization within planar chiral cyclic compounds, was reversibly controlled by *E–Z* photoisomerization of the azobenzene units in the main frame of the macrocycles. In the other trial racemization was attained by the direct symmetrization of the azobenzene dimer containing two identical azobenzene substituents from *E,Z* (one of the azobenzene substituents is in *trans* and the other is in *cis*) to *E,E* (both of the azobenzene substituents are in *trans*) or *Z,Z* (both of the azobenzene substituents are in *cis*). The detail on this molecule will be explained later.

Figure 14.1 shows the concept of chirality switching by *E–Z* photoisomerization of cyclic azobenzene. It is well known that planar chirality innates for ansa compounds having an aromatic ring cyclically substituted with a short enough alkyl chain in the manner where neither symmetric plane perpendicular to the aromatic plane nor free rotation of the aromatic ring against the alkyl chain is allowed to exist (like the compound shown in Fig. 14.1, left). If a part of the alkyl chain is substituted with an azobenzene in the ansa compound, the space for free rotation of the aromatic ring is changed depending on the isomer state of the azobenzene unit. It is expected that a larger space between azobenzene and the aromatic ring is formed in the *cis* state containing a bent azobenzene unit than in the *trans* state containing a rigid straight one. After tuning the length of the rest part of the alkyl chain it may be

Type 1 **Type 2**

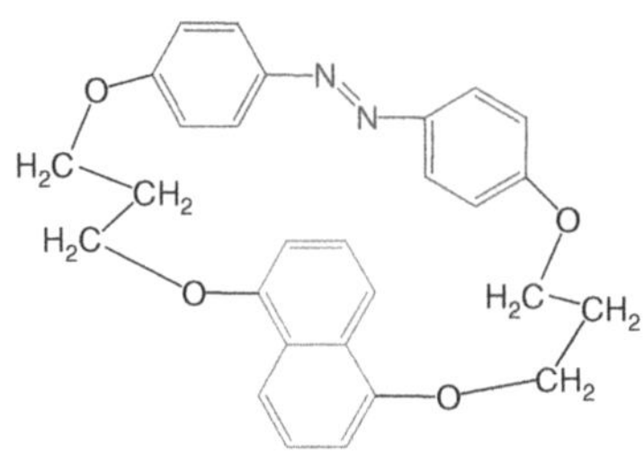

1

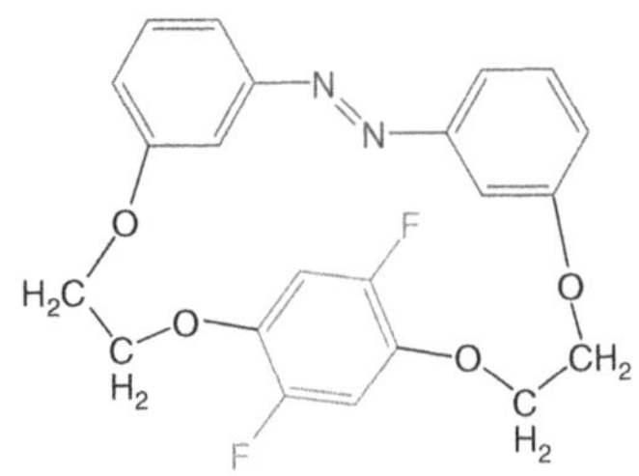

4

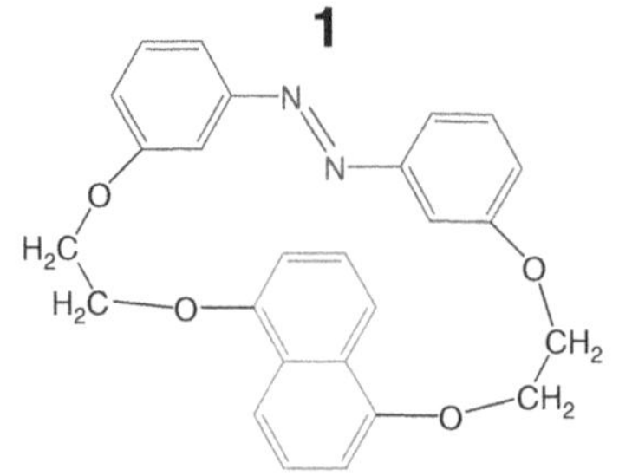

2

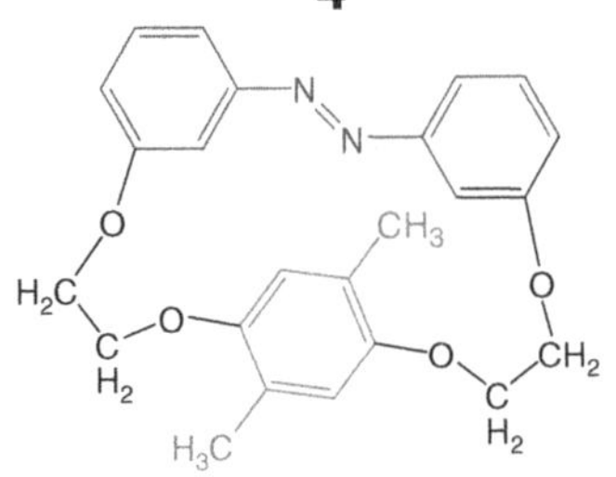

5

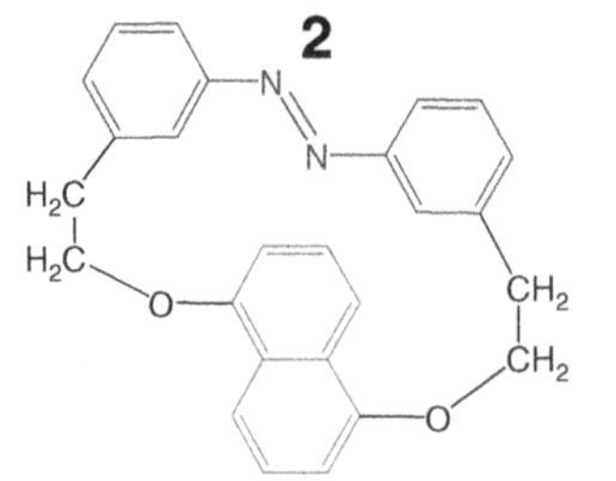

3

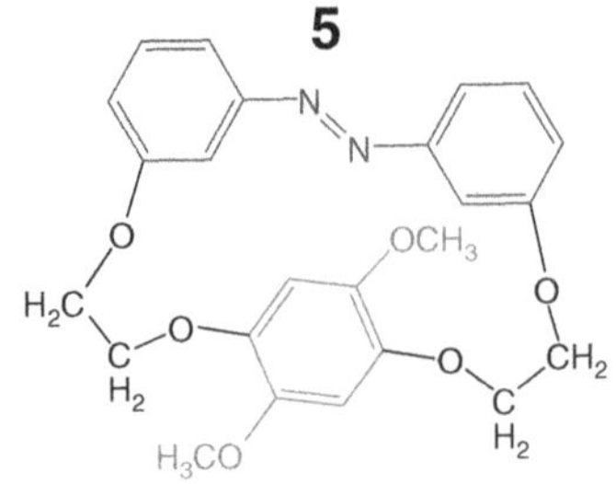

6

Fig. 14.2 Chemical structure of monocyclic azobenzenes **1–6**

possible to attain the complete on/off switching of the free rotation, namely, racemization, by *E–Z* isomerization of the azobenzene unit (Fig. 14.1, right).

Figure 14.2 shows the molecular structures of monocyclic azobenzenes synthesized based on the concept above [20–22]. Two types of the compounds are synthesized. In the type 1 the aromatic ring to rotate is naphthalene and the alkyl chain length was changed. In the type 2 the size of the rotating aromatic ring was changed by controlling the size of the substituents on it while the alkyl chain length was maintained.

Depending upon the length of the alkyl spacer in type 1 or the substituents on the benzene moiety in type 2 and *E* or *Z* conformation of the azobenzene unit, these molecules exhibit various extent of restriction on free rotation of the benzene rotor. In the cyclic azobenzene **1** with a naphthalene rotor and longest spacers of –O–CH_2–CH_2–CH_2–O– or **3** with fluorine substitution on benzene rotor, the rotation of

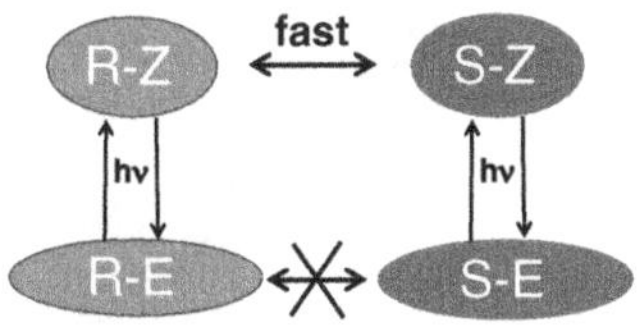

Fig. 14.3 Schematic drawing of the reaction of molecules that show racemization in the *cis* state

the rotor was allowed regardless of *E–Z* photoisomerization of azobenzene moiety. The rotation speed in **1** can be controlled by photoinduced reversible *E–Z* isomerization of their azobenzene at low temperature as evident from the NMR spectra.

The rendered rotation of naphthalene or the substituted benzene rotor in cyclic azobenzene cavity imparts planar chirality to the molecules. Cyclic azobenzene **3** with the shortest spacers of $–CH_2–CH_2–O–$ in type 1 or **6** with methoxy group on 2,5 positions of benzene rotor in type 2 showed enough conformational restriction against free rotation of the rotor in both *E* and *Z* isomers, and the respective planar chiral enantiomers were resolved.

A light-controlled molecular brake was demonstrated for the first time with compounds **2** and **5**. The rotation of naphthalene or dimethylbenzene, respectively, through cyclophane is completely OFF in the *trans* isomer due to smaller cavity size. Such restricted rotation imposes planar chirality to the molecule, and the corresponding enantiomers were resolved on chiral HPLC. The rotation of naphthalene or dimethylbenzene in **2** or **5**, respectively, renders ON in the *cis* isomer due to increased cavity size, and it is experimentally observed by the racemization of separated enantiomers by photoinduced *E–Z* isomerization, which was confirmed by HPLC using a chiral column.

More interestingly, for the first time, the induction of molecular chirality was demonstrated in the simple monocyclic azobenzene by CPL [22]. The photoreactions among the geometrical and stereoisomers, *R–E, S–E, R–Z,* and *S–Z,* are expressed schematically as in Fig. 14.3. Under CPL the efficiency of the photoreaction of $E \rightarrow Z$ or $Z \rightarrow E$ is different between the stereoisomers *S* and *R*. This difference in the efficiency of photoreaction would explain the observed induction of the molecular chirality by CPL. The key characteristics of cyclic azobenzene **5** for the successful enantio-differentiating photoisomerization directly between *E* enantiomers must be the stability of the chiral structure in *E* isomer, fast racemization in *Z* isomer, and the circular dichroism in enantiomers of *E* and *Z* isomers. The maximum enantiomeric excess (ee) under CPL assuming the reaction scheme shown in Fig. 14.3 is calculated based on the difference in the molar extinction coefficient between *S* and *R* stereoisomers. The calculated value for **5** was 0.34 % with the CPL irradiation at 488 nm. The measured value 0.30 % was in reasonably good agreement with the theoretically estimated enantiomeric excess. In conclusion CPL-induced enrichment of one of the enantiomers was demonstrated for **5** showing the fast racemization in *Z* state (Fig. 14.4).

In order to obtain a larger enantiomeric excess a bicyclic azobenzene dimer having relatively large difference in molar extinction coefficient between enantiomers was synthesized (Fig. 14.5) [23]. This bicyclic system with two photoresponsive

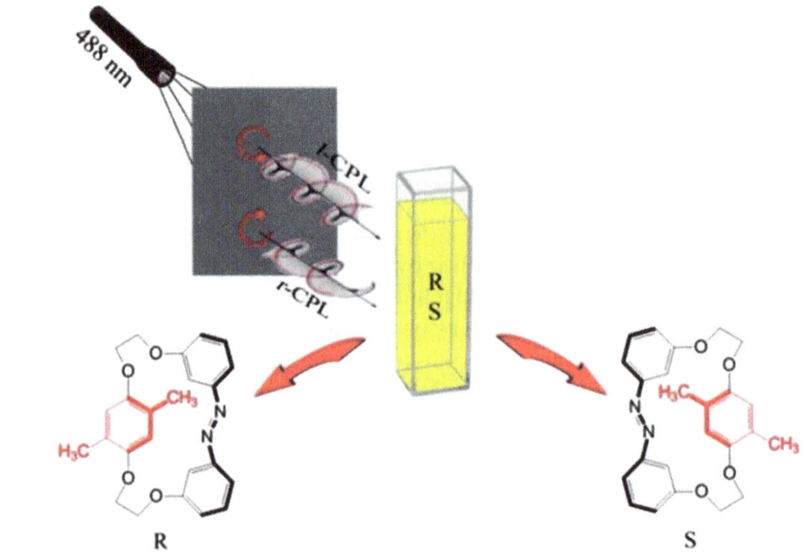

Fig. 14.4 Enrichment of one of the enantiomers of **5** by the action of circular polarized lights

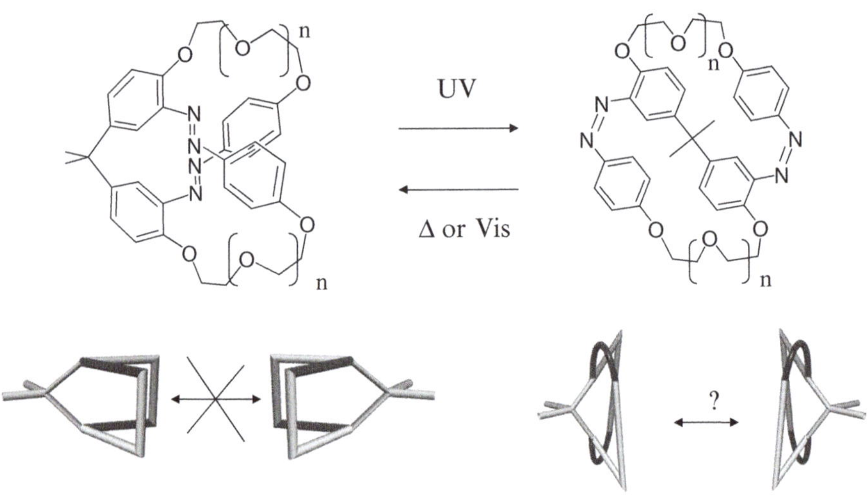

Fig. 14.5 The concept of chiral bicyclic azobenzene and its photoisomerization between *E,E* and *Z,Z*-isemers (*top*). The schematic drawings of stereoisomers for *E,E* (*left*) and *Z,Z* (*right*) and possible difference in the racemization property of them (*down*)

component exhibits three geometrical isomers, *E–E, E–Z*, and *Z–Z* (Fig. 14.6). The bicyclic compound **7** with shorter spacers gave stable enantiomers for all of the geometrical isomers. Contrary to that, the *Z–Z* isomer of compound **8** shows the racemization, which was confirmed by HPLC experiments with a chiral column for a separated pure enantiomer of *E–E*-**8** before and after irradiation. It was confirmed

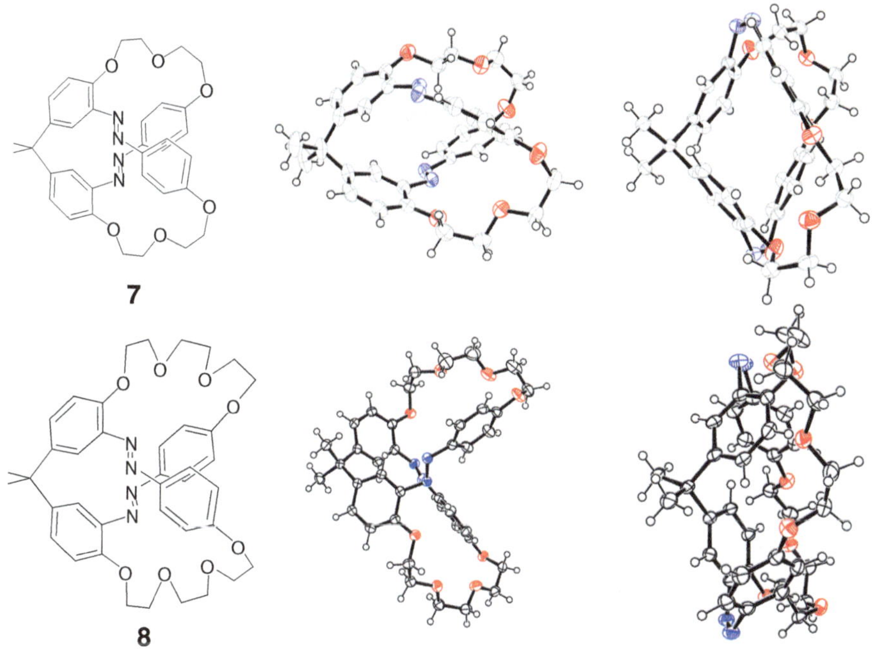

Fig. 14.6 Chemical structure of synthesized bicyclic azobenzenes **7** and **8** (*left*) and their crystal structures in *E,E* (*center*) and *Z,Z* (*right*) states

that the separated pure enantiomer of *E*–*E*-**8** racemized to the mixture of *R*– and *S*–*E*–*E*-**8** via a photogenerated *Z*–*Z* isomer.

8 was examined for the possibility of enrichment of one of the enantiomers through the photoisomerization of racemic **8** under the influence of *l*- or *r*-CPL, using CD spectroscopy as the detection technique. Irradiation of racemic *E,E*-**8** with *r*- or *l*-CPL at 436 nm resulted in an active CD spectrum. Because the peak position and the sign of the CD spectrum for the solution irradiated with CPL matched to that of one of the pure enantiomers of *E,E*-**8**, it is reasonably concluded that one of the enantiomers was partially enriched upon photoirradiation with CPL. Alternating irradiation using *r*- and *l*-CPLs induced an alternation in the sign of the peaks in the CD spectra. Further irradiation with nonpolarized light returned the solution to the inactive CD state, i.e., the compound returned to a racemic state. From the value of $\Delta\varepsilon_{436}$ and the induced CD value of 0.3 mdeg at 361 nm for the 6.3×10^{-5} M solution of **8** in its PSS at 436 nm, which contained the *E,E* isomer as 26 % of the total isomers, we calculated the photoinduced *ee* of the *E,E* isomer to be 1.1 %.

In the paragraphs above molecules showing photoswitchable racemization were introduced, where interconversion between chiral stable stereoisomers' state (*E* or *E,E*) and the common achiral or racemizing state (*Z* or *Z,Z*) was performed by photoirradiation.

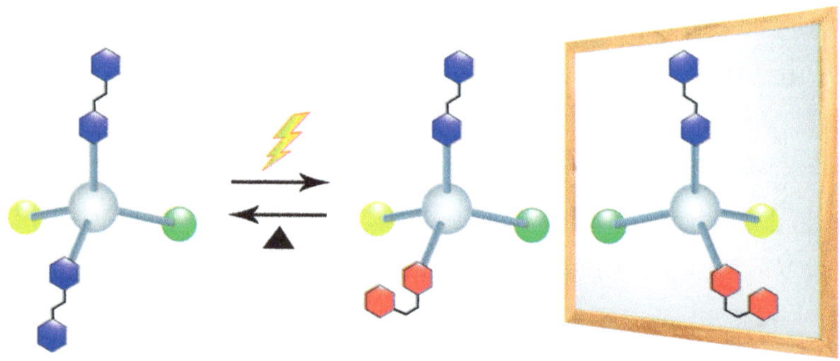

Fig. 14.7 Schematical drawing of a molecular structure and the reaction with which a new photoswitching of chirality can be demonstrated

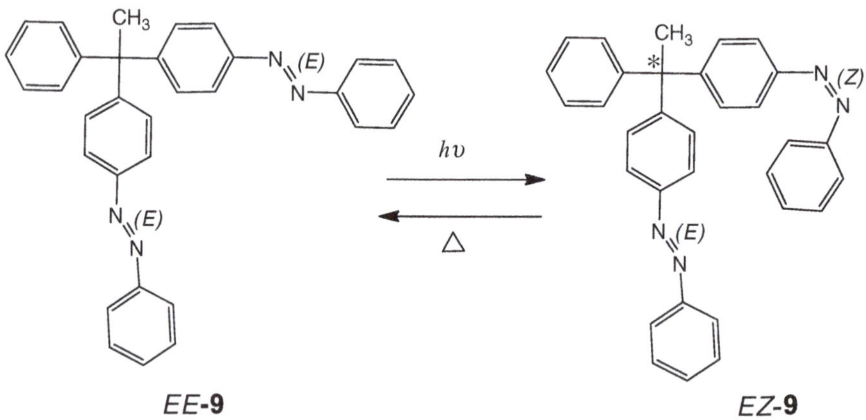

Fig. 14.8 Synthesized azobenzene dimer based on the new concept and its *E–Z* isomerization of one of the azobenzene units inducing reversible molecular chirality

A new idea to establish photoswitchable racemization was demonstrated recently with a compound schematically shown in Fig. 14.7. Newly designed was a molecule consisting of a carbon atom having two photoisomerizable azobenzene moieties and a methyl and a phenyl groups (Fig. 14.8) [24]. The conformational difference caused by the *E*→*Z* photoisomerization of one of the azobenzene moieties was successfully utilized for the generation of point chirality in the molecule. To the best of our knowledge this is the first example of the induction of point chirality in which two of the substituents around an sp^3 carbon atom are geometric isomers. The generation of chirality by the isomerization of one of the azobenzene units was confirmed by the separable peaks for the *E,Z* isomer assignable to the S and R enantiomers in the HPLC using a chiral column and the CD spectra with mirror images each other of the separated peaks. By thermal back reaction chiral *E,Z* isomers returned back to nonchiral *E,E* isomer.

It is theoretically possible to enrich one of the enantiomers of *E,Z* geometrical isomer of **9** by utilizing partially selective photoisomerization by CPL from racemic mixture of chiral *E,Z* isomer to nonchiral *E,E* or *Z,Z* isomers assuming the reaction

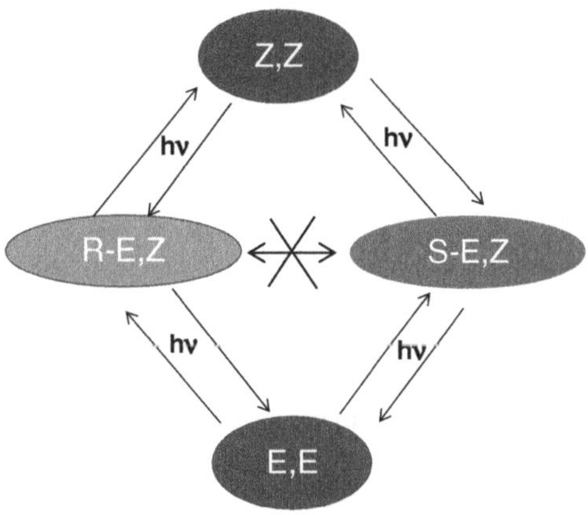

Fig. 14.9 Schematic drawing of the reaction of molecules that show racemization via achiral *Z,Z* or *E,E* isomer states

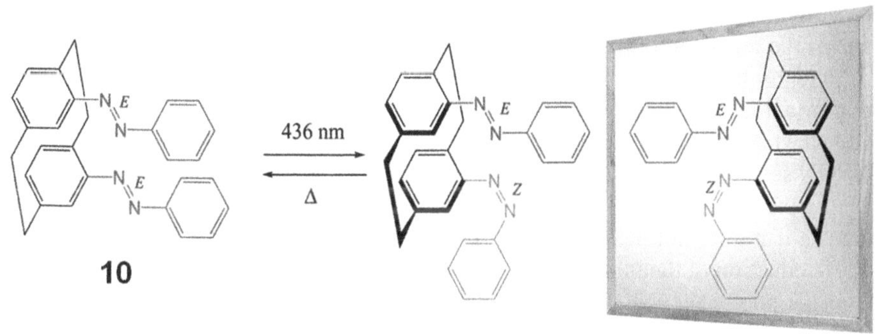

10

EE EZ_A EZ_B

Fig. 14.10 Chemical structure of prochiral compound **10**. In the *E,Z* state stereoisomers exist

paths shown in Fig. 14.9. Until now such enrichment was not experimentally demonstrated most probably due to the too small difference in the extinction coefficient between the enantiomers of *E,Z*-**9**.

A similar photoswitchable chirality was also demonstrated with potentially planar chiral molecule **10** showing much larger difference in the extinction coefficient between the enantiomers of *E,Z* (Fig. 14.10) [25].

14.3 Driving and Controlling Molecular Machines

14.3.1 *Photoswitching the Motility of a Motor Protein, Kinesin-Microtubule*

As a next challenge to regulate the molecular motion by *E–Z* photoisomerization of azobenzene, the target was set to molecular machines. The molecular machine is defined as molecules or molecular systems that convert some energy to work via a mechanism of the molecular level. It can be categorized into two, namely, natural and artificial. In nature several motor proteins are working. The motor proteins in our bodies perform several mechanical functions initiated by chemical reactions; for example, the transformation of chemical energy to mechanical energy as seen in muscle tissue, the transportation of vesicles or organelles in cells, and the synthesis of energy molecules [e.g., adenosine triphosphate (ATP)]. Kinesin is one of the most important linear motor proteins; it transports nanoscale objects along microtubules, rails constructed of tubulins, within cells [26, 27]. If we could apply a function to an artificial molecular system, it might be possible to use it to transport nanoscale objects precisely between desired positions. The realization of such artificial regulation of a motor protein might open up a new field in nanotechnology [28–31].

Several earlier studies have examined the artificial control of kinesin [32, 33]. The ideal control would be one that provides ON/OFF switching of the motility function of kinesin at any desired time and any desired position in space. Light is one of the most appropriate signals for switching because it allows regulation with sufficiently high temporal and spatial resolution. The switching of kinesin's function from the OFF state to the ON state through the action of light was demonstrated quite a while ago using a caged ATP [34, 35]. In contrast, the reverse switching from the ON state to the OFF state was only recently demonstrated using a caged peptide having the same structure as kinesin's tail, which is known to inhibit the motility of kinesin if it does not bear a cargo [36]. In that study, the authors demonstrated an 80 % reduction of the initial gliding velocity of microtubules on the kinesin surface after the photochemical deprotection of the *o*-nitrobenzyl protecting group on the caged peptide. Nevertheless, using light to reversibly control the motility of kinesin—as desired for the complete regulation of the linear motor protein both temporally and spatially—has never previously been demonstrated until quite recently.

The reversible control of motile property of kinesin by light was quite recently demonstrated by two methods using azobenzene derivatives. In the first method the azobenzene derivative was introduced in the monomolecular layer under kinesin in the in vitro motility assay where the positional change of fluorescently labeled microtubules driven by the fixed kinesin on the substrate upon the hydrolysis of ATP is investigated on a fluorescent optical microscope [37, 38]. Various azobenzene monolayers are tested, and some containing amino acid residue with free amino groups at the terminal as shown in Fig. 14.11 showed the higher motility *velocity* of the microtubule in the *Z* state of the azobenzene monolayer than in the *E* state, making it possible to control the velocity repeatedly. The controllable change in the velocity was 13–15 % of the initial velocity in the *E* state (Figs. 14.12 and 14.13). It was further confirmed that the changed motility is originated from the difference in the hydrolysis rate of ATP depending on the isomer state of the azobenzene derivatives in the monolayer under kinesin [38].

The other method for the reversible control of the motility of kinesin was the introduction of azobenzene into ATP [39]. The chemical structures of the synthesized compounds are shown in Fig. 14.14. Unsubstituted or 4-*tert*-butyl-substituted azobenzene was introduced at the 2′-position of the ribose ring of ATP.

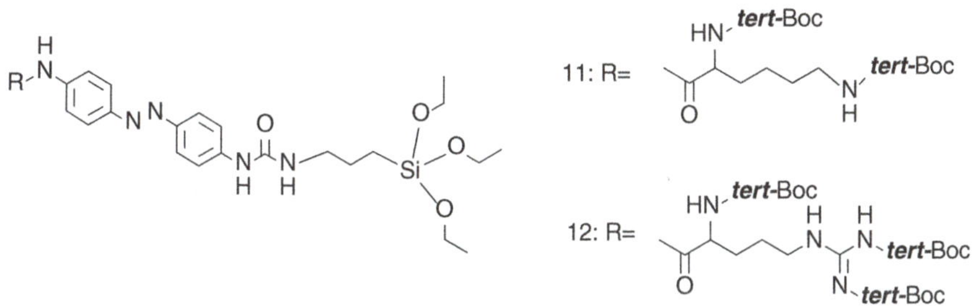

Fig. 14.11 Chemical structure of the compounds used for the monolayer formation. The compounds are applied with the protecting groups on the amino groups and then deprotected on the glass substrate by using trifluoroacetic acid

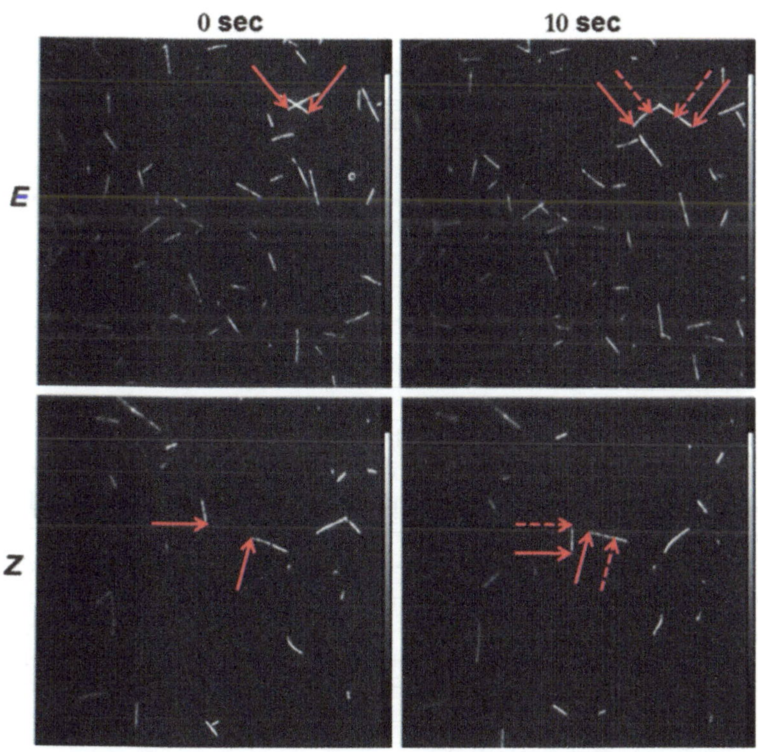

Fig. 14.12 Photographs of rhodamine-labeled microtubules driven by the kinesin-mediated ATP hydrolysis on the deprotected lysine-azobenzene monolayer. *Top*: in the *trans* state, at 0 s (*left*) and 10 s (*right*) in the motility experiment. *Bottom*: at the *cis* rich state after 366 nm irradiation, at 0 s (*left*) and 10 s (*right*) in the motility experiment. *Red arrows* point the front position of a microtubule at the time. The *dotted arrow* on the right photographs points the front position of the same microtubule at 0 s. Field of view: 82 μm × 82 μm

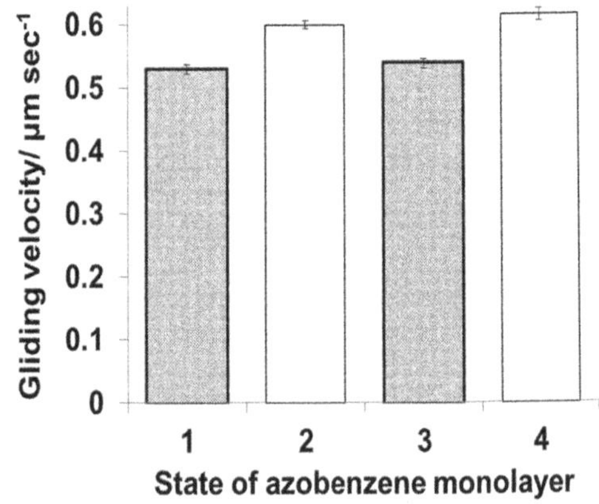

Fig. 14.13 Changes in microtubule motility upon photoirradiation. State 1: prior to irradiation. State 2: after irradiation at 366 nm for 30 s. State 3: after subsequent irradiation at >500 nm for 3 min. State 4: after subsequent irradiation at 366 nm for 30 s

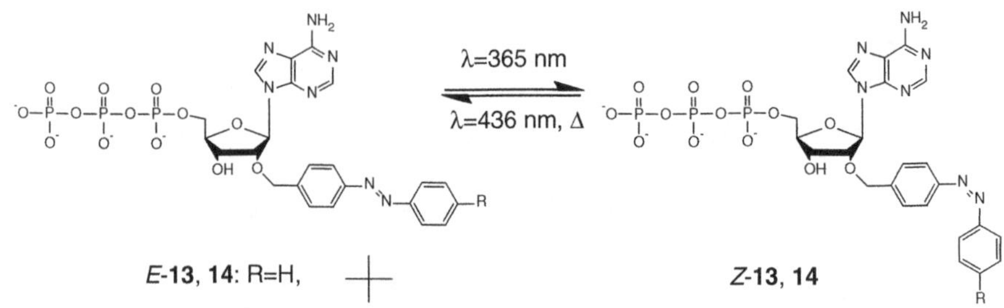

Fig. 14.14 Chemical structure of azobenzene-introduced ATP analogues and their photoisomerization between *trans* and *cis* isomers

Both of the compounds **13** and **14** showed the activity as the energy source for the motility of kinesin. The sliding motion of microtubules was observed on the kinesin in the in vitro motility assay mentioned above using **13** or **14** instead of ATP. When **14** was used as the energy source, the reversible changes of the motility velocity was observed by UV and blue light irradiations which induce *trans*→*cis* and *cis*→*trans* photoisomerization of azobenzene, respectively. The motility of microtubules was faster with the state of *cis* of **14**, and the controllable change in the velocity was about 30 % of the initial velocity in the *E* state (Figs. 14.15 and 14.16).

14.3.2 Trial of the Synthesis of Azobenzene Molecular Machines Continuously Converting Light Energy to Work

The design and construction of motors of ever smaller sizes has fascinated the scientists of worldwide owing to the growing interest in miniaturization of components for the construction of useful devices, which is an essential feature of modern technology [40–42]. In this perspective, there are several attempts towards the

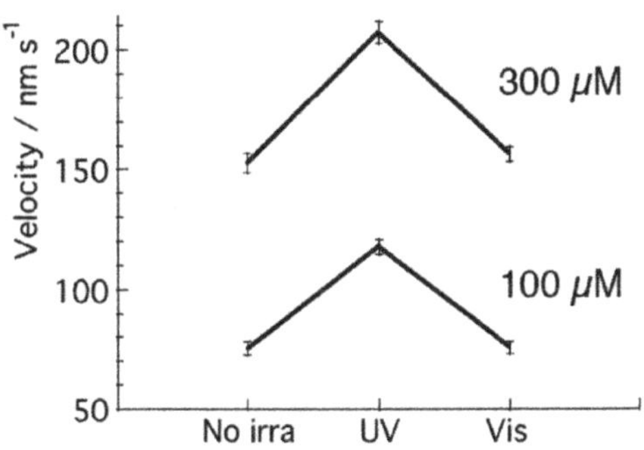

Fig. 14.15 Change in the velocity of microtubules in the in vitro motility assay using the ATP analogue **14** upon UV (at 366 nm to the photostationary state) and further vis (at 436 nm to the photostationary state) irradiations

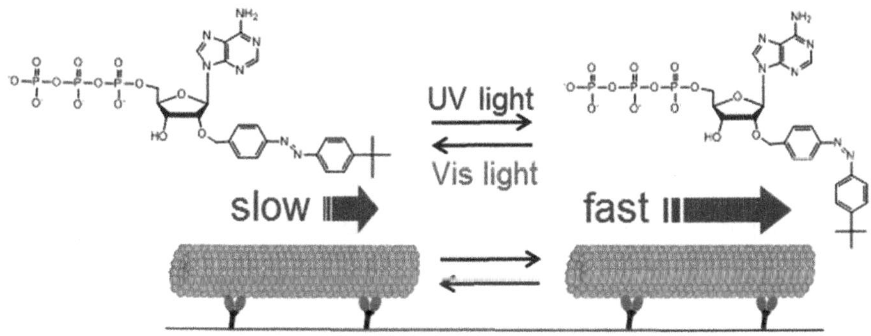

Fig. 14.16 Schematic drawing showing the photocontrolled gliding velocity of microtubules using the photoisomerizable ATP analogue

development of synthetic molecular systems that function as shuttles [43, 44], rotors [45–47], muscles [48–50], elevators [51], and motors [52–57], where these systems have given a new outlook to the field of synthetic molecular-level machines. Among various reports on molecular machine constructs, those generating the linear and rotary motion of the micro/nano objects received greater attention attributed to the expected applications as delivery systems [35, 58, 59].

Ichimura et al. [60] reported one of the first examples of the translational motion of objects on surfaces, where the surface energy gradient formed by photoisomerization of azobenzene units resulted in the movement of droplets on azobenzene modified surfaces. Recently, the photochromic liquid crystalline polymers [61–65] or crystals [66, 67] found applications as artificial muscles and actuators exploiting the structural changes induced upon photoisomerizations.

Feringa et al. [68–70] reported the rotational motion of the micro glass rods on a switchable chiral nematic liquid crystalline film, induced by the isomerization of the molecular motors in the nematic liquid crystalline host. In this system, the forward rotation was induced by the photoisomerization of the sterically overcrowded ethylene motor molecule, while the reverse rotation was achieved by its thermally induced reverse isomerization. This study demonstrated how the collective action of molecular scale motors embedded in a liquid crystalline host is translated as the rotational motion of the macroscale object. In a similar study, Kurihara et al. [71] reported a fully photocontrolled rotary motion of the micro glass rods which is induced by the photoisomerization of the central chiral azobenzene molecule. In both these systems, the rotational motion of the micro objects is achieved by the rotational reorganization of the cholesteric helix, which is reversibly switched by isomerization of the chiral dopants.

It is important to know the relationship between the molecular scale events such as isomerization of the chiral dopant and the macroscopic event of glass rod rotation. The literature of the last decade contains several studies dedicated to the design and synthesis of phototunable chiral dopants functionalized with various photochromic units for the purpose of reversible switching of cholesteric pitch and thereby the color reflections [20, 72–86]. When a chiral solute is dissolved in nematic liquid crystal at the limit of low concentration, the induced helical pitch P is correlated with the weight concentration Cw of the dopant (weight of dopant/weight of solution) and its enantiomeric purity r according to the equation

$$P^{-1} = \beta \cdot C_{\mathrm{w}} \cdot r \tag{14.1}$$

The proportionality constant β is referred to the helical twisting power (HTP) of the chiral dopant. From the accumulated experimental [87–97] and theoretical studies [98–100] towards the photocontrolled reversible tuning of the cholesteric pitch, it is clear that the initial HTP and its difference between the bistable states are the two important parameters which decide the reversible control of the reflection colors in a wide spectral region. The change in reflection wavelength band ($\Delta\lambda$) is related to the change in helical twisting by the following equation:

$$\Delta\lambda = n\left(\frac{1}{C_{\mathrm{w}} \cdot r \cdot \beta_{\mathrm{fin}}} - \frac{1}{C_{\mathrm{w}} \cdot r \cdot \beta_{\mathrm{ini}}}\right) \tag{14.2}$$

where "n" is the refractive index of the doped liquid crystal mixture and the "β_{ini}" and "β_{fin}" are defined as the observed HTP of the chiral dopant before and after photoirradiation, respectively. Although there are few reports on the molecular motor application of the reversibly switchable doped chiral nematic liquid crystalline phases [68–71], there are no studies which directly link the nature of the chiral dopant to rotation degree of the micro objects. The experimental and theoretical studies by Feringa et al. examined the mechanism of the rotational reorganization of the cholesteric helix based on the nature of the liquid crystal host and the photoisomerization of the chiral dopant [69, 70]. Although it is apparent that the change in HTP must be the reason for the rotation of the glass rods, there are no detailed studies in literature which probe the exact parameter of the chiral dopant determining the degree of rotation of the micro objects on these films.

In order to study the role of chiral dopants in fully photocontrolled rotational reorganization of cholesteric texture, five new planar chiral azobenzenes were synthesized (Fig. 14.17) [101]. In this study, the modified designs of planar chiral cyclic

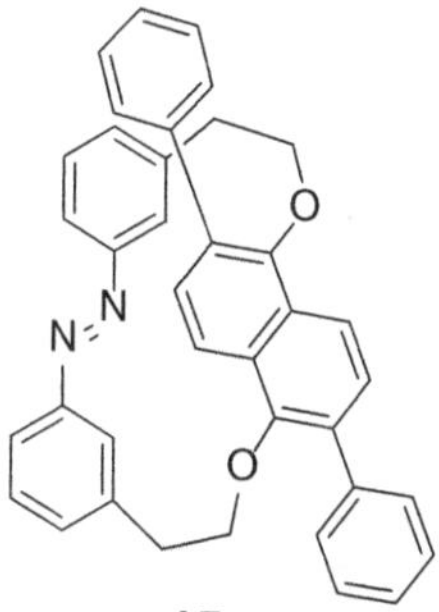

15

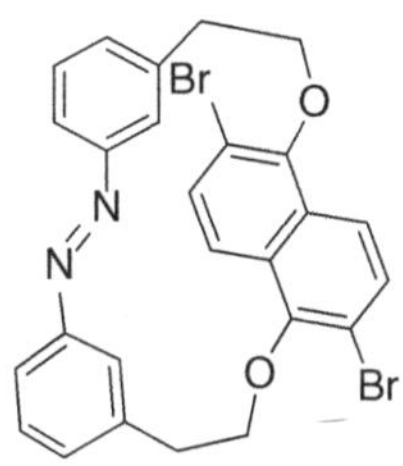

16

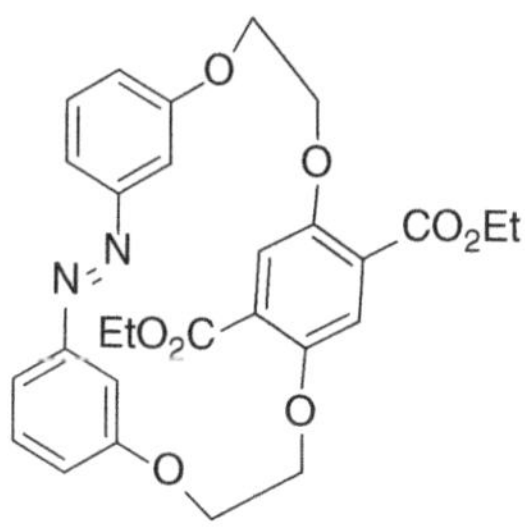

17

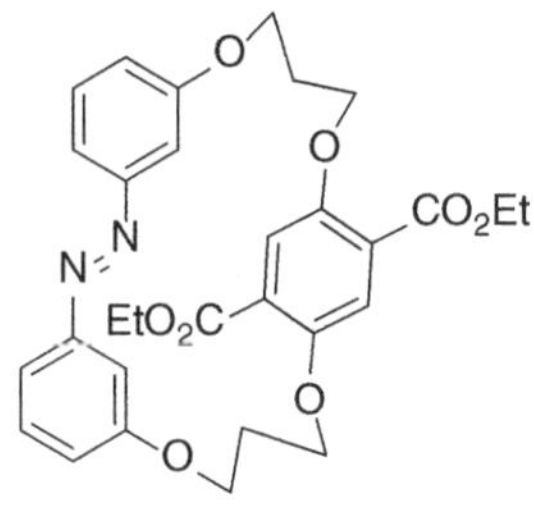

18

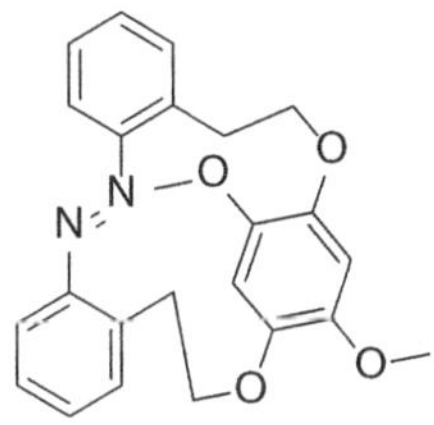

19

Fig. 14.17 Chemical structures of planar chiral azobenzene derivatives for the engines of the LC-based molecular machine

azobenzenes were used to understand the effect of molecular geometry and mesogenic properties on HTP of the molecules and to explore their application as chiroptical switches in four structurally different nematic liquid crystalline phases.

The initial value of HTP, the change in the HTP upon photoisomerization, and also the photoisomerization rate were changed depending on the chemical structure. The compound **15** showed the largest HTP in its *E* state while **16** showed the largest change in HTP upon photoisomerization which was faster than other derivatives. All of the chiral azobenzene derivatives induced upon photoisomerization of the rotation of glass rods on the LC surface doped with them (Fig. 14.18). Most importantly, detailed experimental results clearly bring out the linear dependence of the degree of rotation of the micro glass rod to the ratio of HTP of the chiral dopant before and after irradiation, not just the absolute value of change in HTP between the initial and final states of the chiral dopant against its initial HTP.

Another interesting phenomenon was the non-corresponding texture change upon photoisomerization of the dopant between forward (from *trans* to *cis*) and backward (from *cis* to *trans*) reactions if the LC was doped with **16** when high intensity of light and a thicker sample were used [102]. In the forward photoreaction the texture with the line shape was maintained with the broadening and showed the rotation of it. In the backward reaction, however, the line texture suddenly disappeared and a new round one was formed. This phenomenon resulted in

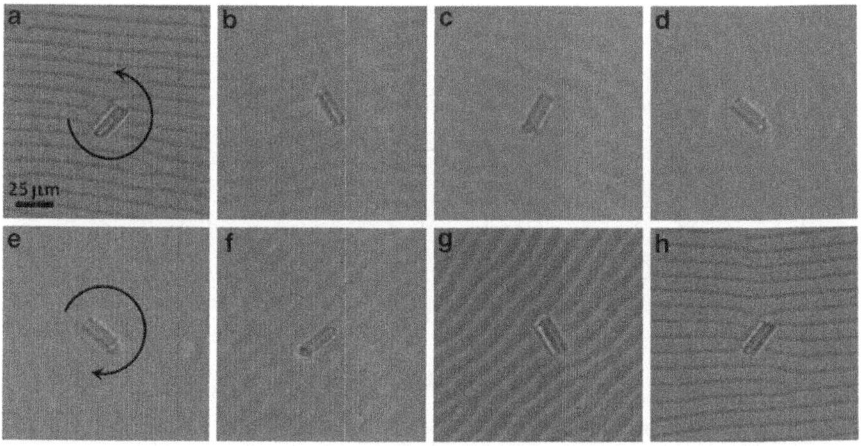

Fig. 14.18 Optical micrographs showing the rotation of a glass rod on the surface of chiral nematic liquid crystalline film formed by doping with *E*-$\mathbf{16}_A$ ("A" means the first eluted enantiomer) (1 wt.%) in 5CB. Rotation of glass rod on 366 nm (**a**–**d**) and 436 nm (**e**–**h**) irradiations recorded at an interval of 15 s

non-corresponding motion of a small glass object when it was put on the surface of the LC film. In the forward photoisomerization under UV the small glass object rotated while the object stayed without rotation during blue light irradiation. By repeating alternated irradiation with UV and blue lights, the glass object continuously rotated unidirectionally. This is the real demonstration of the conversion of the light energy to work.

14.4 Conclusion

It was demonstrated that motion of molecules could be controlled dynamically by the *E*–*Z* photoisomerization of azobenzenes. In the case of the intramolecularly rotating motion of a rotor, the fine tuning of the bulkiness of the rotor or the macrocyclic ring size was the key to the success for the complete ON–OFF photoswitching of the rotational motion. It was possible to utilize this photoswichable motion for chirality control of planar chiral molecules at the stopped rotation state. The *E*–*Z* photoisomerization could be applied to control the motion of molecular machines. It was successfully demonstrated that the velocity of the motility of a motor protein, kinesin-microtubule, was reversibly photocontrolled by the isomerization of azobenzene embedded into ATP or the mono-molecular layer contacting with kinesin. Azobenzene derivatives were also used for controlling or driving the artificial molecular machines. Chiral azobenzene derivatives could be applied to the Feringa's molecular machine composed of a liquid crystal and a photoresponsive chiral dopant. Interesting relationship between HTP and the rotation angle of the glass rod upon the photoisomerizations between photo stationary states was obtained. It was also possible to drive the machine continuously in the unidirectional manner, converting light energy to the mechanical work as the real molecular machine not just the mechanical switch.

References

1. Shinkai S, Ogawa T, Nakaji T (1979) Photocontrolled extraction ability of azobenzene-bridged azacrown ether. Tetrahedron Lett 20(47):4569–4572
2. Ueno A, Yoshimura H, Saka R, Osa T (1979) Photocontrol of binding ability of capped cyclodextrin. J Am Chem Soc 101(10):2779–2780
3. Tazuke S, Kurihara S, Ikeda T (1987) Amplified image recording in liquid crystal media by means of photochemically triggered phase transition. Chem Lett 16(5):911–914
4. Sackmann E (1971) Photochemically induced reversible color changes in cholesteric liquid crystals. J Am Chem Soc 93(25):7088–7090
5. Tamaoki N (2001) Cholesteric liquid crystals for color information technology. Adv Mater 13(15):1135–1147
6. Murata K, Aoki M, Suzuki T, Harada T, Kawabata H, Komori T, Ohseto F, Ueda K, Shinkai S (1994) Thermal and light control of the sol–gel phase transition in cholesterol-based organic gels. novel helical aggregation modes as detected by circular dichroism and electron microscopic observation. J Am Chem Soc 116(15):6664–6676
7. Irie M, Hirano Y, Hashimoto S, Hayashi K (1981) Photoresponsive polymers. 2. Reversible solution viscosity change of polyamides having azobenzene residues in the main chain. Macromolecules 14(2):262–267
8. Willner I, Rubin S, Riklin A (1991) Photoregulation of papain activity through anchoring photochromic azo groups to the enzyme backbone. J Am Chem Soc 113(9):3321–3325
9. Asanuma H, Ito T, Yoshida T, Liang X, Komiyama M (1999) Photoregulation of the formation and dissociation of a DNA duplex by using the *cis–trans* isomerization of azobenzene. Angew Chem Int Ed 38(16):2393–2395
10. Suarez M, Schuster GB (1995) Photoresolution of an axially chiral bicyclo[3.3.0]octan-3-one: phototriggers for a liquid-crystal-based optical switch. J Am Chem Soc 117(25):6732–6738
11. Huck NPM, Jager WF, deLange B, Feringa BL (1996) Dynamic control and amplification of molecular chirality by circular polarized light. Science 273(5282):1686–1688
12. Feringa BL, van Delden RA, Koumura N, Geertsema EM (2000) Chiroptical molecular switches. Chem Rev 100(5):1789–1816
13. Feringa BL (ed) (2001) Molecular switches. Wiley, Weinheim
14. Everitt SRL, Inoue Y (1999) Asymmetric photochemical reactions in solution. In: Ramamurthy V, Schanze KS (eds) Organic molecular photochemistry. Marcel Dekker, New York
15. Kuhn W, Braun E (1929) Photochemische Erzeugung optisch aktiver Stoffe. Naturwissenschaften 17(14):227–228
16. Bonner WA, Rubenstein E (1987) Supernovae, neutron stars and biomolecular chirality. Biosystems 20(1):99–111
17. Feringa BL, van Delden RA (1999) Absolute asymmetric synthesis: the origin, control, and amplification of chirality. Angew Chem Int Ed 38(23):3418–3438
18. Avalos M, Babiano R, Cintas P, Jimenez JL, Palacios JC (2000) From parity to chirality: chemical implications revisited. Tetrahedron Asymmetry 11(14):2845–2874
19. Kawasaki T, Sato M, Ishiguro S, Saito T, Morishita Y, Sato I, Nishino H, Inoue Y, Soai K (2005) Enantioselective synthesis of near enantiopure compound by asymmetric autocatalysis triggered by asymmetric photolysis with circularly polarized light. J Am Chem Soc 127(10):3274–3275

20. Mathews M, Tamaoki N (2008) Planar chiral azobenzenophanes as chiroptic switches for photon mode reversible reflection color control in induced chiral nematic liquid crystals. J Am Chem Soc 130(34):11409–11416
21. Basheer MC, Oka Y, Mathews M, Tamaoki N (2010) A light-controlled molecular brake with complete ON–OFF rotation. Chem Eur J 16(11):3489–3496
22. Hashim PK, Thomas R, Tamaoki N (2011) Induction of molecular chirality by circularly polarized light in cyclic azobenzene with a photoswitchable benzene rotor. Chem Eur J 17:7304–7312
23. Tamaoki N, Wada M (2006) Dynamic control of racemization rate through *E–Z* photoisomerization of azobenzene and subsequent partial photoresolution under circular polarized light. J Am Chem Soc 128(19):6284–6285
24. Hashim PK, Tamaoki N (2011) Induction of point chirality by *E/Z* photoisomerization. Angew Chem Int Ed 50(49):11729–11730
25. Hashim PK, Basheer MC, Tamaoki N (2013) Chirality induction by *E–Z* photoisomerization in [2.2] paracyclophane-bridged azobenzene dimer. Tetrahedron Lett 54(2):176–178
26. Brady ST (1985) A novel brain ATPase with properties expected for the fast axonal transport motor. Nature 317:73–75
27. Vale RD, Reese ST, Sheetz MP (1985) Identification of a novel force-generating protein, kinesin, involved in microtubule-based motility. Cell 42(1):39–50
28. Agarwal A, Hess H (2010) Biomolecular motors at the intersection of nanotechnology and polymer science. Prog Polym Sci 35(1–2):252–277
29. Goel A, Vogel V (2008) Harnessing biological motors to engineer systems for nanoscale transport and assembly. Nature Nanotech 3:465–475
30. van den Heuvel MGL, Dekker C (2007) Motor proteins at work for nanotechnology. Science 317:333–336
31. Fischer T, Hess H (2007) Materials chemistry challenges in the design of hybrid bionanodevices: supporting protein function within artificial environments. J Mater Chem 17:943–951
32. Stracke R, Böhm KJ, Wollweber L, Tuszynski JA, Unger E (2002) Analysis of the migration behaviour of single microtubules in electric fields. Biochem Biophys Res Commun 293(1):602–609
33. Tucker R, Katira P, Hess H (2008) Herding nanotransporters: localized activation via release and sequestration of control molecules. Nano Lett 8(1):221–226
34. Higuchi H, Muto E, Inoue Y, Yanagida T (1997) Kinetics of force generation by single kinesin molecules activated by laser photolysis of caged ATP. Proc Natl Acad Sci USA 94(9):4395–4400
35. Hess H, Clemmens J, Qin D, Howard J, Vogel V (2001) Light-controlled molecular shuttles made from motor proteins carrying cargo on engineered surfaces. Nano Lett 1(5):235–239
36. Nomura A, Uyeda TQP, Yumoto N, Tatsu Y (2006) Photo-control of kinesin-microtubule motility using caged peptides derived from the kinesin C-terminus domain. Chem Commun 2006:3588–3590
37. Rahim MKA, Fukaminato T, Kamei T, Tamaoki N (2011) Dynamic photocontrol of the gliding motility of a microtubule driven by kinesin on a photoisomerizable monolayer surface. Langmuir 27(17):10347–10350
38. Rahim MKA, Kamei T, Tamaoki N (2012) Dynamic photo-control of kinesin on a photoisomerizable monolayer—hydrolysis rate of ATP and motility of microtubules depending on the terminal group. Org Biol Chem 10:3321–3331
39. Kamei T, Fukaminato T, Tamaoki N (2012) A photochromic ATP analogue driving a motor protein with reversible light-controlled motility: controlling velocity and binding manner of a kinesin–microtubule system in an in vitro motility assay. Chem Commun 48:7625–7627
40. Ballardini R, Balzani V, Credi A, Gandolfi MT, Venturi M (2001) Artificial molecular-level machines: which energy to make them work? Acc Chem Res 34(6):445–455
41. Balzani V, Gomez-Lopez M, Stoddart JF (1998) Molecular machines. Acc Chem Res 31(7):405–414

42. Balzani V, Credi A, Venturi M (2007) Molecular devices and machines. Nano Today 2(2):18–25
43. Pease AR, Jeppesen JO, Stoddart JF, Luo Y, Collier CP, Heath JR (2001) Switching devices based on interlocked molecules. Acc Chem Res 34(6):433–444
44. Keaveney CM, Leigh DA (2004) Shuttling through anion recognition. Angew Chem Int Ed 43(10):1222–1224
45. Kottas GS, Clarke LI, Horinek D, Michl J (2005) Artificial molecular rotors. Chem Rev 105(4):1281–1376
46. Frantz DK, Linden A, Baldridge KK, Siegel JS (2012) Molecular spur gears comprising triptycene rotators and bibenzimidazole-based stators. J Am Chem Soc 134(3):1528–1535
47. Lubbe AS, Ruangsupapichat N, Caroli G, Feringa BL (2011) Control of rotor function in light-driven molecular motors. J Org Chem 76(21):8599–8610
48. Dawson RE, Lincoln SF, Easton CJ (2008) The foundation of a light driven molecular muscle based on stilbene and α-cyclodextrin. Chem Commun 2008:3980–3982
49. Juluri BK, Kumar AS, Liu Y, Ye T, Yang Y-W, Flood AH, Fang L, Stoddart JF, Weiss PS, Huang TJ (2009) A mechanical actuator driven electrochemically by artificial molecular muscles. ACS Nano 3(2):291–300
50. Chuang C-J, Li W-S, Lai C-C, Liu YH, Peng S-M, Chao I, Chiu S-H (2009) A molecular cage-based [2]rotaxane that behaves as a molecular muscle. Org Lett 11(2):385–388
51. Badjic JD, Balzani V, Credi A, Silvi S, Stoddart JF (2004) A molecular elevator. Science 303:1845–1849
52. Kelly TR, De Silva H, Silva RA (1999) Unidirectional rotary motion in a molecular system. Nature 401:150–152
53. Kelly TR, Cai X, Damkaci F, Panicker S, Tu B, Bushell SM, Cornella I, Piggott MJ, Salives R, Cavero M, Zhao Y, Jasmin S (2007) Progress toward a rationally designed, chemically powered rotary molecular motor. J Am Chem Soc 129(2):376–386
54. Vicario J, Walko M, Meetsma A, Feringa BL (2006) Fine tuning of the rotary motion by structural modification in light-driven unidirectional molecular motors. J Am Chem Soc 128(15):5127–5135
55. Vicario J, Walko M, Meetsma A, Feringa BL (2005) Controlling the speed of rotation in molecular motors. Dramatic acceleration of the rotary motion by structural modification. Chem Commun 2005:5910–5912
56. Koumura N, Zijlstra RWJ, van Delden RA, Harada N, Bl F (1999) Light-driven monodirectional molecular rotor. Nature 401:152–155
57. Tierney HL, Murphy CJ, Jewell AD, Baber AE, Iski EV, Khodaverdian HY, McGuire AF, Klebanov N, Sykes ECH (2011) Experimental demonstration of a single-molecule electric motor. Nature Nanotech 6:625–629
58. Hirokawa N, Takemura R (2005) Molecular motors and mechanisms of directional transport in neurons. Nature Rev 6:201–214
59. Dennis JR, Howard J, Vogel V (1999) Molecular shuttles: directed motion of microtubules along nanoscale kinesin tracks. Nanotechnology 10:232–236
60. Ichimura K, Oh S-K, Nakagawa M (2000) Light-driven motion of liquids on a photoresponsive surface. Science 288(5471):1624–1626
61. Finkelmann H, Nishikawa E, Pereira GG, Warner M (2001) A new opto-mechanical effect in solids. Phys Rev Lett 87(1):015501-1–015501-4
62. Warner M, Terentjev EM (2003) Liquid crystal elastomers. Oxford University Press, Oxford
63. Yamada M, Kondo M, Mamiya J, Yu Y, Kinoshita M, Barrett CJ, Ikeda T (2008) Photomobile polymer materials: towards light-driven plastic motors. Angew Chem 120(27):5064–5066; Angew Chem Int Ed 47(27):4986–4988
64. Yoshino T, Kondo M, Mamiya J, Kinoshita M, Yu Y, Ikeda T (2010) Three-dimensional photomobility of crosslinked azobenzene liquid-crystalline polymer fibers. Adv Mater 22(12):1361–1363
65. Seki T, Nagano S, Kawashima Y, Zettsu N, Ubukata T (2005) Light-driven organized layer materials. Mol Cryst Liq Cryst 430(1):107–114

66. Irie M, Kobatake S, Horichi M (2001) Reversible surface morphology changes of a photochromic diarylethene single crystal by photoirradiation. Science 291(5509):1769–1772
67. Morimoto M, Irie M (2010) A diarylethene cocrystal that converts light into mechanical work. J Am Chem Soc 132(40):14172–14178
68. Eelkema R, Pollard MM, Vicario J, Katsonis N, Ramon BS, Bastiaansen CWM, Broer DJ, Feringa BL (2006) Molecular machines: nanomotor rotates microscale objects. Nature 440:163
69. Eelkema R, Pollard MM, Katsonis N, Vicario J, Broer DJ, Feringa BL (2006) Rotational reorganization of doped cholesteric liquid crystalline films. J Am Chem Soc 128(44): 14397–14407
70. Bosco A, Jongejan MGM, Eelkema R, Katsonis N, Lacaze E, Ferrarini A, Feringa BL (2008) Photoinduced reorganization of motor-doped chiral liquid crystals: bridging molecular isomerization and texture rotation. J Am Chem Soc 130(44):14615–14624
71. Kausar A, Nagano H, Kuwahara Y, Ogata T, Kurihara S (2011) Photocontrolled manipulation of a microscale object: a rotational or translational mechanism. Chem Eur J 17(2):508–515
72. Lin T-H, Chen Y-J, Wu C-H, Fuh AY-G, Liu J-H, Yang P-C (2005) Cholesteric liquid crystal laser with wide tuning capability. Appl Phys Lett 86(12):161120
73. Shibaev P, Chiappetta SD, Milner V, Genack A, Bobrovsky A (2005) Light controllable tuning and switching of lasing in chiral liquid crystals. Opt Exp 13(7):2358–2363
74. van Delden RA, van Gelder MB, Huck NPM, Feriga BL (2003) Controlling the color of cholesteric liquid-crystalline films by photoirradiation of a chiroptical molecular switch used as dopant. Adv Funct Mater 13(4):319–324
75. Ma J, Li Y, White T, Urbas A, Li Q (2010) Light-driven nanoscale chiral molecular switch: reversible dynamic full range color phototuning. Chem Commun 46:3463–3465
76. Li Q, Li Y, Ma J, Yang D-K, White TJ, Bunning TJ (2011) Directing dynamic control of red, green, and blue reflection enabled by a light-driven self-organized helical superstructure. Adv Mater 23(43):5069–5073
77. Li Q, Green L, Venkataraman N, Shiyanovskaya I, Khan A, Urbas A, Doane JW (2007) Reversible photoswitchable axially chiral dopants with high helical twisting power. J Am Chem Soc 129(43):12908–12909
78. Lee H-K, Doi K, Harada H, Tsutsumi O, Kanazawa A, Shiono T, Ikeda T (2000) Photochemical modulation of color and transmittance in chiral nematic liquid crystal containing an azobenzene as a photosensitive chromophore. J Phys Chem B 104(30):7023–7028
79. Green L, Li Y, White T, Urbas A, Bunning T, Li Q (2009) Light-driven molecular switches with tetrahedral and axial chirality. Org Biomol Chem 7:3930–3933
80. Mathews M, Zola RS, Hurley S, Yang D-K, White TJ, Bunning TJ, Li Q (2010) Light-driven reversible handedness inversion in self-organized helical superstructures. J Am Chem Soc 132(51):18361–18366
81. Ruslim C, Ichimura K (2001) Conformation-assisted amplification of chirality transfer of chiral Z-azobenzenes. Adv Mater 13(1):37–40
82. Pieraccini S, Masiero S, Spada GP, Gottarelli G (2003) A new axially-chiral photochemical switch. Chem Commun 2003:598–599
83. Pieraccini S, Gottarelli G, Labruto R, Masiero S, Pandolini OS, Spada GP (2004) The control of the cholesteric pitch by some azo photochemical chiral switches. Chem Eur J 10(22):5632–5639
84. Yoshioka T, Ogata T, Nonaka T, Moritsugu M, Kim SN, Kurihara S (2005) Reversible-photon-mode full-color display by means of photochemical modulation of a helically cholesteric structure. Adv Mater 17(10):1226–1229
85. Kawamoto M, Aoki T, Wada T (2007) Light-driven twisting behaviour of chiral cyclic compounds. Chem Commun 2007:930–932
86. Wang Y, Urbas A, Li Q (2012) Reversible visible-light tuning of self-organized helical superstructures enabled by unprecedented light-driven axially chiral molecular switches. J Am Chem Soc 134(7):3342–3345

87. Janicki SZ, Schuster GB (1995) A liquid crystal opto-optical switch: nondestructive information retrieval based on a photochromic fulgide as trigger. J Am Chem Soc 117(33): 8524–8527
88. Yokoyama Y, Sagisaka T (1997) Reversible control of pitch of induced cholesteric liquid crystal by optically active photochromic fulgide derivatives. Chem Lett 26(8):687–688
89. Denekamp C, Feringa BL (1998) Optically active diarylethenes for multimode photoswitching between liquid-crystalline phases. Adv Mater 10(14):1080–1082
90. Yamaguchi T, Inagawa T, Nakazumi H, Irie S, Irie M (2000) Photoswitching of helical twisting power of a chiral diarylethene dopant: pitch change in a chiral nematic liquid crystal. Chem Mater 12(4):869–871
91. Yamaguchi T, Inagawa T, Nakazumi H, Irie S, Irie M (2001) Photoinduced pitch changes in chiral nematic liquid crystals formed by doping with chiral diarylethene. J Mater Chem 11:2453–2458
92. van Leeuwen T, Pijper TC, Areephong J, Feringa BL, Browne WR, Katsonis N (2011) Reversible photochemical control of cholesteric liquid crystals with a diamine-based diarylethene chiroptical switch. J Mater Chem 21:3142–3146
93. Rameshbabu K, Urbas A, Li Q (2011) Synthesis and characterization of thermally irreversible photochromic cholesteric liquid crystals. J Phys Chem B 115(13):3409–3415
94. Bossi ML, Murgida DH, Aramendia PF (2006) Photoisomerization of azobenzenes and spirocompounds in nematic and in twisted nematic liquid crystals. J Phys Chem B 110(28): 13804–13811
95. Jin L-M, Li Y, Ma J, Li Q (2010) Synthesis of novel thermally reversible photochromic axially chiral spirooxazines. Org Lett 12(15):3552–3555
96. Pieraccini S, Masiero S, Ferrarini A, Spada GP (2011) Chirality transfer across length-scales in nematic liquid crystals: fundamentals and applications. Chem Soc Rev 40:258–271
97. Davis R, Mallia VA, Das S, Tamaok N (2004) Butadienes as novel photochromes for color tuning of cholesteric glasses: influence of microscopic molecular reorganization within the helical superstructure. Adv Funct Mater 14(8):743–748
98. Earl EJ, Wilson MR (2003) Predictions of molecular chirality and helical twisting powers: a theoretical study. J Chem Phys 119(19):10280–10288
99. di Matteo A, Todd SM, Gottarelli G, Solladie G, Williams VE, Lemieux RP, Ferrarini A, Spada GP (2001) Correlation between molecular structure and helicity of induced chiral nematics in terms of short-range and electrostatic–induction interactions. The case of chiral biphenyls. J Am Chem Soc 123(32):7842–7851
100. Gottarelli G, Hibert M, Samori B, Solladie G, Spada GP, Zimmermann R (1983) Induction of the cholesteric mesophase in nematic liquid crystals: mechanism and application to the determination of bridged biaryl configurations. J Am Chem Soc 105(25):7318–7321
101. Thomas R, Yoshida Y, Akasaka T, Tamaoki N (2012) Influence of a change in helical twisting power of photoresponsive chiral dopants on rotational manipulation of micro-objects on the surface of chiral nematic liquid crystalline films. Chem Eur J 18(39):12337–12348
102. Tamaoki N, Akasaka T, Thomas R (to be published)

Index

M. Irie et al. (eds.), *New Frontiers in Photochromism*,
DOI 10.1007/978-4-431-54291-9, © Springer Japan 2013

Lightning Source UK Ltd.
Milton Keynes UK
UKOW06n1320010616

275397UK00001B/29/P

9 784431 542902